Engineering Ethics

Concepts, Viewpoints, Cases and Codes

Published by the

National Institute for Engineering Ethics
Murdough Center for Engineering Professionalism
College of Engineering, Texas Tech University

Edited by

Jimmy H. Smith
and
Patricia M. Harper

Articles by National Institute for Engineering Ethics Staff

Jimmy H. Smith, Professor of Civil Engineering and Director

William D. Lawson, Lecturer, Senior Research Associate and Deputy Director

Patricia M. Harper, Program Coordinator and Assistant to the Director

Bill W. Baker, Former Murdough Center Editor and Program Development Executive

Jay M. Newhard, Former Research Associate

Articles by Distinguished Guest Writers

Wm. A. Wulf, President, National Academy of Engineering

Steven P. Nichols and W. F. Weldon, Mechanical Engineering, UT/ Austin

Taft Broome, Jr., Professor of Civil Engineering, Howard University

Thomas Smith, III, Managing Director, American Society of Civil Engineers

Joe Paul Jones, Past President, National Society of Professional Engineers

E. D. "Dave" Dorchester, Past President, National Institute for Engineering Ethics

August 2004

Table of Contents

ISBN 0-9760663-0-0

Table of Contents
(Continued)

Table of Contents
(Continued)

NSPE Board of Ethical Review Cases (Continued) — Page

Note: To find NSPE-BER cases by Subject, See Appendix A, then go to www.niee.org or http://www.pitt.edu/~bmclaren/ethics/caseframes/index.html

Table of Contents
(Continued)

“…to see what is right and not do it is cowardice.”

Confucius (551 - 479 B.C.)

“Example has more followers than reason.”

Bovee

Preface

Engineering Ethics – Concepts, Viewpoints, Cases and Codes, produced by the National Institute for Engineering Ethics (NIEE), is not designed to be used as a stand-alone text in a formal academic course in engineering ethics. However, it may be used effectively as a reference and for guidance in engineering ethics. It covers a wide variety of ethical issues related to engineering practice and is believed to be especially useful for independent study by individuals in all engineering disciplines.

The publication contains details of relatively few ethical theories; so in that regard, it is not intended to serve as an effective reference on theoretical ethics. Its primary strengths:

- Variety of viewpoints expressed by experienced engineers, many of whom are national leaders, on ethical issues covering concepts of the importance of ethical leadership, of professionalism, and of responsibilities and obligations of individual engineers as well as responsibilities and obligations of the engineering profession; and
- Numerous case studies, most based on actual events, related to fundamental and professional ethics, plus two major cases using videos on engineering ethics, *Incident at Morales* and *Gilbane Gold*, which are described in detail and guidance presented for those wishing to view and discuss the videos.

Major sections include:

- Engineering Ethics: an Overview
- Principles, Concepts and Viewpoints Regarding
  - Basic Ethics
  - Engineering Ethics
- Two Major Case Studies Available on Video and Interactive DVD
  - *Incident at Morales* (2003) – Detailed Information and Study Guide
  - *Gilbane Gold* (1989) - Brief Description and Suggested Assignment
- Over 50 Cases in Basic Ethics, Professional Ethics and Engineering Ethics
  - General Interest Cases on Fundamental Ethics
  - Cases on Critical Thinking, Honesty, and Responsibility
  - Applied Ethics in Professional Practice Cases
  - NSPE Board of Ethical Review Cases
- Appendices
  - List of NSPE BER Cases Available on the Internet
  - References Related to Engineering Ethics
  - Codes of Ethics from Several Engineering Societies and Groups
  - A Variety of Other Appendices Related to Engineering Ethics

The Study of Engineering Ethics has the Following Objectives

The objectives of studying Engineering Ethics are to develop abilities to:

1. Communicate willingly and effectively with others on ethical issues

2. Differentiate among
   a) Personal ethics
   b) Legally required ethics
   c) Ethics based on the engineer's responsibility to serve the public good

3. Recognize and resolve ethical problems by
   a) Learning about ethics resources available for guidance
   b) Considering numerous case studies
   c) Understanding the ethical component of problems by extensive analysis and discussion of case studies
   d) Analyzing specific situations presented by case studies

4. Formulate solutions to ethical problems illustrated in case studies by
   a) Recognizing the consequences of actions taken
   b) Applying different perspectives on ethical problem solving such as duties, intentions, consequences, rules and relationships
   c) Knowing what's expected, knowing what's right, and doing what's right

5. Comprehend, Compare, Evaluate and Act on these solutions

To accomplish these objectives, the reader is encouraged to:

1. Review basic knowledge and fundamental definitions of
   a) Professionalism
   b) Ethics

2. Develop an understanding and comprehension of
   a) What it means to be a professional, and what is expected
   b) Ethics as it relates to the professional
   c) Concepts of codes of ethics and other guidelines for decision-making

3. Apply the concepts of ethics codes and other guidelines to
   a) Simple actions of living and working
   b) Complex actions in the workplace
   c) Numerous case studies of actual and illustrative work situations

4. Relate consequences resulting from both simple and complex actions to
   a) Their immediate supervisor
   b) Employees they supervise
   c) The public
   d) Themsleves

5. Analyze case study examples and situations in order to distinguish between
   a) Choosing between right and wrong
   b) Choosing among competing goods

6. Develop skills to formulate, analyze, and compare solutions to
   a) Ethical dilemmas encountered in the workplace
   b) Ethical dilemmas encountered in relationships with others

7. Learn to evaluate the value and effect of the various solutions by
   a) Obtaining all the facts, listing and testing the options
   b) Making a decision and knowing when and how to take action...and being willing to do so.

Engineering Ethics: An Overview

By Bill W. Baker
Former Murdough Center Editor & Project Executive, Retired
Former Assistant Dean, College of Engineering, Texas Tech University

Because of many examples of contemporary greed, the public is demanding that virtue be taught to those who would be professionals and that ethical conduct and professionalism be demanded of those who are practicing within the professions. This may be an impossible-to-meet demand because controversy has swirled since the days of ancient Greece and Rome as to whether values can be taught after the formative years which some refer to as "sandbox" years. This article will address the purpose of teaching ethics, some basic concepts, the importance of moral reasoning, standards and law, safety and risk, and responsibilities of professionals. It will also include a brief section on international ethics.

Purpose

The feeling by some that current attempts to teach ethics do not seem to do much good may have its roots less in the philosophical arguments as to whether ethics can be taught than in the real-world where many present-day scholars think that such attempts would be less worthy than more traditional courses. Beyond that, most professors in such areas as science, engineering and business have had little exposure to formal instruction in ethics and values and even less experience in teaching them. Furthermore, they are less inclined to teach such subjects than those which have definite right and wrong answers. Teachers tend to teach those things which they are most comfortable teaching.

Current disagreements about the teaching of ethics are not new. If we go back about 2,500 years to Athens, Protagoras was writing and teaching that truth is just a matter of convention and that it is relative to each person's point of view. Socrates' position was that this is exactly wrong, that real truths can be determined, and that they determine human values. Plato's argument was the same, and he said that Protagoras and his pluralistic followers had no respect for the truth. Once values are made relative and subject to scholastic or social whimsy, wisdom is conformity and truth is politics, he claimed. Not a single ancient Greek philosopher ever defended Protagorean relativity, yet modern pluralists explicitly espouse Protagoreanism. Many in U.S. society today hold pluralistic beliefs.

Perhaps truth falls somewhere in the middle, being unable to live with complete dogmatism or relativity. Instead, there must clearly be some room for mutual understanding. Perhaps virtues cannot be taught, but each person can learn to reason. That, then, becomes the goal of those who are attempting to teach courses in ethics. It is not to provide hard and fast "yes" and "no" answers but to generate awareness of ethical concerns and dilemmas, to disarm prejudices, and to bring about more value laden professional behavior.

Basic Concepts

Depending on how one wants to look at it, the history of engineering may be thought to be nearly as old as man. It could be related to the development of crude stone tools, the use of fire and, later, the wheel. If one seeks to define its historical limits by its search for professional status, then its history is much more recent and is basically the history of its societies, or roughly the past century. To get to the roots of its organized concern with engineering ethics, a study of the past two decades will cover the issue.

Those interested in a historical statement are referred to a 1971 book by Edwin T. Layton, Jr., *The Revolt Of The Engineers*, published by the Press of Case Western Reserve University. It traces the evolution of engineering and its search for professional and political status. Others would say that what is professional in technology may go back farther than the 1800's and the establishment of professional societies. They point to the various crafts and guilds and the apprenticeship programs of the medieval period. Rather than a book, they would refer you to a performance of the Wagnerian opera, "Die Meistersinger", for a look at what early engineering was like with its system of masters and apprentices. In any case, it was at this time of the breakdown of the feudal system that crafts found it necessary to join together into guilds for mutual protection and regulation.

Even before 2,000 BC, Egyptian hieroglyphs used the title "chief of works" to describe that era's technologist whose duties were somewhat like those of today's city engineer. The University of Miami's Murray I. Mantell, writing in his book, *Ethics and Professionalism in Engineering*, says that almost every form of engineering was known to the civilizations of antiquity which included Egypt, China, Persia, Chaldea, Babylon, Assyria, Phoenicia, Etruria, Palestine, Moab, and Peru where canals, public water supplies, docks, harbors, lighthouses, bridges, roads, and massive structures were built. Mantell continues by saying that people doing engineering type work tested the strength of materials, planned fortifications, designed engines, devised methods for transporting heavy objects, developed systems of navigation and communication and were skilled metallurgists.

Mantell relates a series of 19th Century inventions to the recent radical change in engineering. Included in these inventions were Carnot's theory of the development of the steam engine in 1824, the introduction of the dynamo and the electric generator in 1831, the telegraph in 1836 and the electric lamp in 1880. He also credits the westward rush to gold by the 49ers as the event which provided the incentive and the money for major advances in mining engineering. Bessemer's 1856 improvements in steel making and the 1868 introduction of reinforced concrete were major steps forward. All of these presented engineers with new needs and new opportunities and gave the profession a body of mathematics and science that could replace or supplement the art and crafts of earlier times. Time-wise, this corresponded with the rise of the American university and with the emergence of a fundamental cultural belief in the value of learning, or "scientific expertise."

In a section called "The Engineer and Society" Mantell postulates that it was the rapid growth of the American West which resulted in a shortage of experienced engineers to train apprentices, thus leading to the initiation of college training for those who would be engineers and who would take advantage of the new science and math, even if it did mean being subjected to the ridicule of old "practical" engineers. Remnants of that attitude are encountered even today. The rapid growth in the formal training of engineers is shown in this statistic: At the time of the Civil War, the United States had four engineering schools. Thirty years later that number had grown to 89.

Internationally, the first engineering professional societies began in France. French army engineers organized as the Corps du Genie in 1672, and the French national highway department's engineers formed the Corps des Ponts et Chaussees in 1716. More than a century later, in England, the Institution of Civil Engineers was founded in 1818. This was followed in 1847 by the Institution of Mechanical Engineers. Early engineering societies in America developed in the following order:

American Society of Civil Engineers, 1852;

American Institute of Mining, Metallurgical and Petroleum Engineers, 1871;

American Society of Mechanical Engineers, 1880;

Institute of Electrical and Electronic Engineers, 1884, and the

American Institute of Chemical Engineers, 1908.

Called the Founder Societies, these groups were subsequently joined by the National Council of State Boards of Engineering Examiners, the American Society for Engineering Education, the American Institute of Aeronautics and Astronautics, the Accreditation Board for Engineering and Technology, the National Society of Professional Engineers, Canada's Engineering Institute and a number of other important professional societies. Though each group has a different set of goals, a thread of commonalty running through them all is the goal of insuring professional performance while seeking professional recognition.

Moral Reasoning and Dilemmas

The Hastings Center was established in Washington, DC slightly more than a score of years ago to study ethical problems. A number of the ideas expressed in this section of the workbook are based on Hastings Center studies.

There is a strong relationship between the teaching of ethics and the development of moral behavior even though virtues may not be something which can be taught once the formative years are passed. Our approach in this course is to introduce disciplinary and methodological perspectives of ethics that relate to professionals. We seek to sensitize students to ethical concerns by providing them with concepts and skills that will help them recognize and deal with moral issues which affect them personally, their profession and society in general.

As the students progress through this course, the expectation is that, using Hastings' words, their "moral imagination" will be whetted and that they will come to quickly recognize ethical issues. Moral imagination refers to emotions and feelings, but for the course to be worthwhile, it needs to carry beyond that. It should carry to the point where emotional response gives way to conscious rationality when appraising and judging ethical dilemmas.

Because technical problems may be fundamentally right or wrong, a course in ethics should aim to provide skills and knowledge to understand the moral implications of decisions concerning the applications of technology. When thinking of skills, it is customary to think in terms of tools whose use requires special skills. Tools for evaluating moral consequences are concepts, rules and principles. With them, students may work their way to ethical conclusions. What can't be left out of the final conclusion is that the individual has the freedom to choose and the responsibility for his choice. Answers won't be found in this course, but what may be discovered is the way to arrive at a personal best answer.

Professionals, accustomed to dealing with problems for which there are provable wrong and right answers may not be turned on to the idea of using concepts, rules and principles of philosophy, but classical philosophy has real worth as a problem solving tool. The necessary function of analytical philosophy is to make clear the most appropriate way to approach problems in professional ethics, even though it does not provide definite answers to ethical dilemmas. Without a philosophical structure for guidelines, answers are apt to be directionless and inconsistent. Thus, at least some study and understanding of philosophy must be a part of any course on professional ethics.

Without a patterned way to approach the problem of making choices, freedom to choose may not be worth much.

It follows from the above that ethics is a subject which provides guidance for conduct while ethical theory tries to answer questions about ethics. Some degree of familiarity with both is necessary for an understanding of this course. Some have said that ethical theory is the concern of philosophers while ethics is or should be the concern of ministers and guidance counselors. Fortunately for those who are faced with ethical dilemmas which require them to decide how and why to act, this attitude has largely disappeared. It has been replaced by one which combines ethical theory with a study of moral issues.

Applied ethics has its roots in this development.

Standards and Law

Forces of machines, forces of nature, forces resulting from economic, social and political life — these all come together in technological developments. Because of the unknowns in all of the diverse areas, there are risks that can only partially be forecast or diagnosed. This, in itself, makes engineering an experiment and means that engineers are the experimenters. From this complex milieu rise some aspects of the engineer's concerns about safety and his/her responsibility for the general health and welfare of nature and humans.

As MIT's Langdon Winner commented in his 1977 book, *Autonomous Technology*, technological developments shift the fashions of the moment and open areas both of insight and blindness. Technology is problematic and a source of concern because it changes within itself and it generates other changes in its wake. Winner continues by saying that modern history is characterized by continuous change and that "somehow machines and other manifestations of new technology are at the center of this process. ... To look for crucial questions is to look for inventions, innovations, and the myriad of ramifications that follow from technological change."

Another question is not whether technological changes will occur but whether, with the changes, risks can be controlled and safety reasonably assured. If it is to be controlled, how and by whom become the next questions. Engineers, individually and as a group, have roles to play in the controlling process that are as great as their roles in the development process. The engineer is pivotal because his education and experiences have equipped him to bring about new technology, to anticipate its benefits, to forecast problems or risks, to assess alternatives and to inform the publics which make up society. His/her responsibilities stem from the fact that the work is experimentation and he/she is the experimenter.

Many examples abound of technology gone astray, such as Chernobyl, Challenger Shuttle, Three Mile Island and Bophal. These types of new technological developments offered tremendous social promise but delivered unanticipated or unannounced social disasters, and they are at the root of much of today's public outcry for greater exercise of ethics and responsibility by technologists. Related to this is the fact that within the profession, some claim that codes of ethics adopted by the societies provide neither adequate guidance nor protection to help them meet their expanding responsibilities.

Because the effects of technology reach into every area of modern living, it is not surprising that laws have evolved to protect the public interest. So have regulatory bodies. It would seem to follow from this that laws which protect ethical engineers would have emerged since the ethical practice of engineering has come to be so highly regarded as a means of protecting the public. Unfortunately, that has not occurred. Still, to a greater extent than before, the law has become a major element in the practice of engineering.

With increasing frequency, engineers find themselves in the position of being expert witnesses in courts and other kinds of legal hearings. In this instance, there is a major distinction between ethics as they relate to the lawyers and to the expert witnesses. It is the lawyer's obligation, while avoiding untruths, to present the strongest possible case for his client. There is no obligation to present all pertinent information or to present a balanced case. It is the adversarial courtroom setting that is supposed to bring out the balance and the full truth as a result of the combined presentations of the litigants.

Scientists and engineers, when appearing as expert witnesses, have an obligation to present to the best of their abilities, accurate and complete accounts. As Stephen Unger pointed out in his book, *Controlling Technology: Ethics and the Responsible Engineer*, the conscious suppression of data to create a biased impression is a serious breach of ethics. Expert witnesses are also obligated not to accept assignment as expert witnesses unless they, in fact, are experts in the area. Even then, they are obligated to familiarize themselves as fully as possible about the subject on which they are to express their expertise.

It might be lucrative in the short run for an expert witness to slant testimony in favor of the person who is paying his bill, but the long-term result of that is a loss of credibility for the engineering profession and a lowered personal reputation. Slanted or incomplete testimony is a clear violation of ethics.

It is interesting that opinion is normally excluded from trials, yet opinion is what the expert witness gives. It is one of the things which makes that person special and which imposes a keen level of ethical responsibility. An expert is a person who has the experience and training to provide him with special knowledge or skills in a particular field that most people would not have. In the case of an expert witness, qualifying before the court is a part of the testimony. In this qualifying, the expert witness is apt to be asked about his background, education, registration, society membership, projects and other related items while the court determines whether he truly is qualified to give expert testimony.

The ethical obligations that are so apparent in expert witness testimony are equally demanding in other areas where an expert may be involved in assisting an attorney with the interpretation of technical matters in case preparation and assisting in examining and cross-examining technical witnesses. It is not only necessary that the expert make himself understood by the judge and the attorneys but also by the jurors, most of whom may have no technological sophistication. Obviously, it follows that an effective expert witness must not only have technical skills but also communication skills. Because it is the job of the opposing counsel to pick apart and tear down the expert's testimony, it is necessary that the witness be professional in appearance, conduct and attitude and that his answers be reasoned and calm. Breakdowns in courtesy and self-control simply must not occur.

Being an expert witness is demanding and may be exasperating. The consequences can be tremendous. Therefore, considerable thought should be given to whether the task ought to be accepted. The truth must be told as the expert sees it, so he should not accept as a client someone in whose position he does not have confidence. Once accepted, the professional approach is to charge fees that would be charged for other important work requiring the same level of time and participation. Setting a fee on a contingency basis should not even be considered and would be regarded as highly unprofessional.

ASFE, professional firms practicing in the Geosciences, (http://www.ASFE.org) has excellent materials for professional practice issues such as expert witness testimony.

Concern for Safety and Risk

Cost, benefit, and risk analysis are critical areas in engineering concerns for safety. Since total safety is more a goal than an achievement, trade-offs are involved in most such decisions. The area is covered to some extent in the text by Martin and Schinzinger. Still, risk assessment and cost/benefit analysis, especially when they move from dollar areas to the human values and moral values arena become highly nebulous and getting a handle on them is difficult. The fact that we can neither fully explain them nor avoid them makes it appropriate that additional information on the subject be introduced.

Two books which might be read by those with a particular interest in cost/benefit/ risk analysis are *Risk In The Technological Society* by Christopher Hohenemser and Jeanne X. Kasperson, and *Societal Risk Assessment* by Richard C. Schwing and Walter A. Albers, Jr. A third good source is *Project Evaluation* by Arnold Hardberger. The March 1982 issue of IEEE *Technology And Society* magazine deals with an interesting concept of cost/benefit analysis with the introduction of what author, Carl Barus, calls "malefits" as an additional item in the cost/benefit equation. He says that both the desired effects (benefits) and the undesired effects (malefits) are outputs of technology. Rather than trying to separate costs and benefits, he says the costs are the money paid for the mixture of benefits and malefits or undesired effects actually produced. He goes on to claim that dollar quantification of benefits and external costs obscures, not clarifies, the actual physical and social effects of technology. Cost/benefit analysis, to him, is a better tool for advocacy in technological decisions than for informing the public.

Given the complex nature of cost/benefit analysis and the ethical concern that it substitutes money values for social values, one should not be surprised that it is a technique which is frequently criticized. The problem is that until other criteria are put forth, public interest will continue to be measured chiefly in terms of dollar flow. It is very likely that an objective measure of social benefit simply cannot be developed. If that is the case, then the criticism must be answered by better communication between policy makers and those who will be affected. Only then will the public feel that it has some leverage for controlling decisions.

According to Barus' theory, if policy makers use only economic costs that can be measured to arrive at benefits, society stands to lose priceless assets. If analysis shows dollar benefits to exceed dollar costs, projects are presented as desirable, when "real" costs, even if unquantifiable, may exceed "real" benefits. This amounts to misleading the public, intentionally or otherwise. It is in this area that ethics enters what is otherwise a "bean counting" exercise.

Given that technology is one of society's major instruments of change, it follows that its design and assessment ought to be based on broader considerations than amoral economics. The resulting quandary is that when unquantifiables dominate, the quantitative portion of the analysis is meaningless. The obligation, at this point, becomes one of explaining assumptions and premises as an unbiased professional rather than as an advocate. The politicians will become the advocates and that is unavoidable, because some technological decisions ultimately become political decisions. In such cases, the engineer does not get the final say. From a professional perspective, this is called "market subordination." The public, therefore, needs honest words from engineers if it is to reasonably evaluate technological activity.

Utilitarians think of cost benefit analysis as being a study of what will be the greatest good for the greatest number for the longest period of time at the least cost. Those seem, on the surface, to be exquisite measures until one considers that what yields the most good to the majority sometimes comes at a horrible cost to the minority. In such a case, decision makers, where they decide to do the greatest good for the greatest number, may be both unethical and immoral because of what their decisions do to a few. When measuring costs and benefits, the tendency is to assign more worth to certain individuals or groups than to others.

From a purely economic standpoint, net benefits will be maximized when the scale of development is extended to the point where the benefits added by the last increment are equal to the cost of adding that increment. This is called marginal theory and means that the low cost point is the one where marginal benefits or marginal revenue equal marginal cost.

When we begin to consider human benefit and human worth and good and evil and whether all are of equal value and the other myriad ethical concerns which surround every major project, the final decision hinges on making the best choice from the available alternatives. Considerations have to not only include what is the most economically efficient — and this must be included because economic efficiency does contribute to social welfare — but also that which achieves ethical goals such as justice, health and amity. Welfare maximization is as important in cost/benefit analysis as is the achievement of economic objectives. The following is extracted from *Societal Risk Assessment* and is presented as an example of the complexity of cost/benefit/risk analysis which attempts to apply a holistic approach:

"The most obvious characteristic of this class of problem is what Daniel Callahan in his 1979 book, *Morality and Risk Benefit Analysis*, calls inherent uncertainty. The uncertainty built into the very nature of the problem is dominated by the difficulty in obtaining good, solid, scientific data and, possibly more important, by the even greater difficulty of getting a consensus on what is important. With differences in age, experience, goals, demographics and psychology, various groups are bound to approach a problem with a different set of perspectives and the alleged 'facts' brought to the analysis — that is what are chosen to be called facts — may differ radically from those chosen by a different group having a different value system.

'Values' and 'facts' are inseparable and we must avoid the hazard of assuming that better science could perfectly answer the question: 'How safe is safe enough?' We cannot be distracted from the more difficult issue of making decisions in the face of substantial uncertainty in either a technological or non-technological society, each carrying a burden of evident and hidden risks. These risks await the individual and even society itself, but to avoid all risk would lead to the ultimate risk of stagnation and 'rotting in a trap of safety'."

Maybe the best summation of value differences within nature was shown in a drawing of two seabirds, one a herring gull and the other an oystercatcher. One sketch showed a contented gull, sitting on an empty nest, oblivious to the fact that an ethnologist had removed the eggs to see which the creature would prefer, and it settled for the nest. The other sketch showed an oystercatcher trying to sit on a monstrous egg, larger than itself, ignoring its own egg which was a manageable size. Again the ethnologist had provided the dummy to test the discrimination of the bird. Gulls prefer the nest to the eggs; with oystercatchers, bigger is better. Human values range through a similar scale, and these values have to be considered when safety and risk are analyzed.

There are times when, regardless of the benefits associated with a project, risks are at such a threshold that the decision is an automatic "no go." The estimation of risk is as complex as is the estimation of benefits or total costs. A complicating factor is that risk analysis is imprecise partly because people assign different levels of acceptability based on who subjects them to it and who it is that is being put at risk. People who rock climb, sky dive or fly hang gliders subject themselves to risks that they would not accept from an employer. Still, they will accept risks from an employer that they will not tend to accept from society in general. As Aaron Wildansky said of risk reduction: "Each of us would do less for ourselves than we would insist that the government do for us."

Risk is also complex to assess because risks from technologically related projects fall into two separate categories. One is the routine, expected risks associated with the design and found to be acceptable at the time the project is studied, while the other is from abnormal conditions that are not part of the normal operation of the basic design concept. There is also the case of "acceptable risk." Acceptable to whom? Who makes that decision? Are the considerations physical risk or financial risk or both? Are these risks distributed across time as with the oil industry where a number of people are killed annually or simply subject to disastrous accidents where many might be killed in a rare accident, such as a nuclear accident. In another case, is the air and acid rain pollution from coal preferable to the risk of a nuclear accident?

As if they were needed, there are still more complications such as the clash of analytical and intuitive risk assessment. Analytical assessment says that airplane travel is safer than automobile travel, yet more people have an intuitive disabling fear of flying than of riding in an automobile.

As with the case of cost/benefit analysis, with risk/benefit issues, one of the engineer's ethical responsibilities is to explain scientific and technical options to the public so the layman can make an informed judgment, because consensus is unlikely to develop and political decisions will probably be called upon to settle the issue. Said differently, some say the special responsibility of the engineer is to determine risks; it is not to judge their acceptability. Others say, by virtue of his or her expertise, the engineer is uniquely positioned to do a superior job of "judging acceptability."

Responsibilities and Employer Authority

Late in the World War II period W. J. King wrote a three-part article which in 1944 was published in the May, June and July issues of *Mechanical Engineering*. It has been republished a number of times with his permission in books by other authors. It would take much searching to find a better prescription of "ought to" than the King article in regard to the engineer's responsibilities in his work relationship, his relationship with his boss, and his responsibility to associates and outsiders.

Part II of the King article deals with duties and responsibilities of engineers who are in executive or supervisory positions. Here he covers such areas as individual behavior and techniques, his responsibilities to his organization and his obligations to his personnel.

Part III covers purely personal considerations for engineers such as character and personality, both of which are inextricably tied to professionalism. A good source for this article, perhaps better than nearly 50 year old issues of the publishing periodical, is the earlier referenced Mantell book, *Ethics and Professionalism in Engineering*. In that book, the King article, along with its bibliography, is Appendix A and is recommended reading. It reduces ethics to its most practical and applied sense.

Describing responsibility from a different perspective, H.L.A. Hart, in his book *Punishment And Responsibility*, (Oxford University Press, 1968) wrote of four different kinds of responsibility:

1. role responsibility,
2. causal responsibility,
3. liability responsibility, and
4. capacity responsibility.

Reviewers have said that he might as well have been writing about corporate responsibility as professional responsibility, because a corporation can have all of those responsibilities. Role responsibility goes with roles, tasks and jobs; causal responsibility is what you get when you cause something to happen; liability responsibility deals with who pays; and capacity relates to competence.

What sets the professional apart from the corporation is moral responsibility. Whereas corporate responsibility tends to look backwards to responsibility for what happened, much of engineering responsibility must be forward looking and not only deal with "can we do it," but also "ought we do it and, if we do, can we manage it?"

Corporations need not be immoral, they are just amoral or non-moral, whereas engineers must be moral to be responsible, ethical and professional.

Another area in which there is sometimes conflict is the matter of unionism. It matters not that employee engineers have frequently benefited from union activity, the aversion that professionals have for unions is long and deeply rooted. Membership in unions is regarded as a clear erosion of professionalism, but in some professions, this is occurring. See *"Proletarianization Of The Professional"* by Martin Oppenheimer. Professional ethics require complete honesty while union politics rewards, even necessitates, diplomatic double-talk. This conflict makes it nearly impossible for engineers who want to be seen as "hound's tooth" clean in their professional work to engage in the less savory aspects of union politics.

Further complicating the issue of engineer unions is the fact that some engineers are employees, while other engineers are employers or part of the management team. Unionizing one group and exempting the other has the potential for a variety of conflicts and the likelihood of dividing the profession against itself.

A primary difficulty in the area of responsibility relates to sometimes conflicting values confronted by engineers who work in industry. Robert Jackall, writing in his book *Moral Mazes*, points out that part of the ethical conflict or dilemma that faces engineers in industry relates to industrial management. Managers constitute less than 10% of the work force, but they are the quintessential bureaucratic work group in our society. They are by, of and for the organization. Unlike members of professions or public servants, they avow no allegiance to civil service codes or to a public service ethic. Their allegiances are only to the principles of the organization, to the markets which are themselves bureaucratically organized, to the groups and individuals in their world who can demand and command their loyalties, and to themselves and their careers.

Continuing his listing of accusations, Jackall says managers' bureaucratic work poses a series of dilemmas that often demand compromises with traditional moral beliefs. Management wants, even demands, a team that works well together and does not tolerate those who say: "I don't want to do that." What this really means sometime is going with the flow and not making waves. This is especially true during times of crisis. This is also the time when ethical protest or whistle blowing is most likely to occur and it is not the time for it. Clearly Jackall is not an admirer of contemporary managers.

Management's ethics are notable for their lack of fixedness. In the corporate world, morals do not emerge from a set of internally held convictions or principles, but rather from ongoing and changing relationships. Because bureaucracies are always situational and relative, so are ethics. In this environment, ethics are more pragmatic than idealistic.

Some managers say that in business, when the planning is over and the operation begins, there is no room for abstract ethical and moral principles. They further say that truth is socially defined and is not absolute, making them little more than pluralistic Sophists. Therefore, compromise about anything and everything is not a moral defeat but an inevitable fact of organizational life. Right is irrelevant when authorities declare it to be wrong. They add that this is necessary because "what makes a corporation work at all is the support we give to each other no matter what happens...We have to support each other and we have to support the hierarchy. Otherwise, you have no management system." The logical result of altering goals to fit expediency is the elimination of any ethical lines at all, concludes Jackall.

Whether one agrees with Jackall's strong attack on managers, one must agree that corporate values are thrown into frequent and sometimes violent conflict with engineering ethics and standards of professionalism. Even where corporate responsibility is high, it tends to be historical and backward looking in its perspective, whereas, as Hunt wrote, engineering ethics and responsibilities have to be forward looking. What does this say for the ethical dilemmas faced by engineers who are also managers? If there is a difference between corporate or managerial ethics and engineering ethics, which should prevail when the manager or corporate officer is an engineer?

Moral Obligations and Rights of Professionals

As is regularly pointed out in the literature, being a member of a recognized profession carries with it certain responsibilities, responsibilities which comprise a large part of what this course is all about. But professionalism is not a single edged sword which cuts only one way. Being a member of a profession also carries with it rights. Writing on this subject in American Society of Civil Engineers' Journal of Professional Issues in Engineering, October 1985, Murray A. Muspratt of Chisholm Institute of Technology, Victoria, Australia, suggests a bill of rights which apart from constitutional rights and legal rights would include such ethical rights as:

1. The right to act according to ethical conscience and to decline assignments where a variance of moral opinion exists.
2. The right to express professional judgment, and to make public pronouncements that are consistent with corporate constraints on proprietary information.

3. The right to corporate loyalty and freedom from being made a scapegoat for natural catastrophes, management ineptitude or other forces beyond the engineer's control.
4. The right to seek self-improvement by further education and involvement in professional associations.
5. The right to participate in political party activities outside of working hours.
6. The right to apply for superior positions with other companies without being blacklisted.
7. The right to due process and freedom from arbitrary penalties or dismissal.
8. The right to appeal for ethical review by a professional association, ombudsman or independent arbitrator.
9. The right to personal privacy.

All people may not agree with each of these proposed rights, and it is clear that some have been violated in well publicized cases. While not in the same format, they are similar to the Employee Bill of Rights that authors Schinzinger and Martin reference in their text.

A major difference between employee rights and professional rights is the right of professional conscience which is of utmost importance to the professional. Though the concept of professional conscience is nebulous, it is the shelter for a variety of moral responses, among them being whistle-blowing. While the right of professional conscience is the umbrella for whistle-blowing, often it is a leaky umbrella and is not an effective protector.

Not new, but an excellent article for review by those who want a thorough treatment of whistle blowing, from knowing how to knowing when to knowing where some of the most prominent of the whistle blowers are now, is "Knowing How to Blow the Whistle" by Tekla S. Perry. This article was published in the September 1981 issue of *IEEE SPECTRUM* and should be generally available in the periodical rooms of libraries.

Another good article on the subject of whistle blowing is an essay "In Defense of Whistle Blowing" by Gene G. James, Philosophy Department, Memphis State University. Copyrighted in 1984, it is published on pages 249-269 of a book *Profits And Professions* which was edited by W.L. Robinson, M.S. Pritchard and Joseph Ellin and published by Humana Press.

The very societies and institutions which stress ethical values that are grounded in personal responsibility and public accountability have been weak in protecting whistle-blowers from harassment, dismissals, and the expense of law suits. In making this point, Bertrand G. Berube, an engineer, a former GSA regional administrator, a whistle blower and now an owner of his own business, told American Society for Engineering Education members at their 1987 meeting: "If you blow the whistle on a boss, you are likely to be without a job for three to four months and legal fees will be in the range of $30-$40 thousand; for blowing the whistle on a government agency, you may expect to be out of work for one to two years and your legal fees may run from $125-$150 thousand. If you blow the whistle on the political administration in power, you may be off the job for four to seven years and legal fees may be in the $400-$550 thousand range."

That is a high price to pay for subsequent recognition by your professional society for your dedication to professionalism, but it, unfortunately, has been the experience of many who chose to exercise their right to blow a whistle when they felt that engineering ethics demanded such drastic action.

International Ethics

Since the early 1970's, approximately one half of all graduate engineering degrees and 20% of the undergraduate degrees granted by US schools of engineering have gone to foreign students. During this same period, very few US students have earned engineering degrees in foreign institutions. This means two things. The multinational nature of much of US industry is such that large numbers of US engineers are practicing in foreign countries though working for US firms. The flip side is that great numbers of foreign engineers are working in the US, many having US degrees and most of them working for US firms. The US engineers working overseas for US firms are not allowed, if they subscribe to the codes, to follow the "When in Rome" philosophy. Conversely, many of the foreign engineers practicing in the US have achieved or seek US permanent residence or citizenship and by virtue of their employment by US firms, "When in Rome" or in this case "When in the US" does apply.

Ethics and moral values from which many of these come are quite different from those in this country, and concerns are arising that the potential exists for a serious and growing problem. Some examples are:

When in 1992 the European Economic Community became a fact rather than a plan, the engineering profession was faced with new challenges, opportunities and problems because of what it implies for the practice of foreign engineers in the US and for the US engineers in practice in foreign countries, especially within the European community. We had to get our registration act together. If we do not even have consistent reciprocity between the states, how can we expect free and open opportunities for our engineers within the EEC? If we are given that, how can we accord their engineers the same access to US practice?

Professional Ethics Report, newsletter of the American Association for the Advancement of Science, summer 1989 issue, comments that the professional recognition programs that are developed among the nations comprising the European Economic Community will make for tremendous opportunities or tremendous problems for US engineers. It raises hopes that for the first time an international credentialing procedure may develop whereby professional engineers from the US and other nations could practice freely throughout the world. Critical to this will be whether the US gets its act together, because US engineers can become familiar with barriers without going abroad; they can simply cross a state line. It seems that the EEC is ahead of the US on this point of creating a single market for recognizing professional credentials.

NSPE, NCEES and ABET have established the Council for International Engineering Practice, a group which has been meeting with the Canadian Council of Professional Engineers to solve recognition problems caused by the U.S.-Canada Free Trade Agreement. A copy of the Principles of Conduct and Ethics Under NAFTA is included at the end of this book. It is likely that CIEP will soon begin similar discussions with other countries.

Environmental Issues: Recent events in Eastern Europe also hold ethical challenges for US professionals. As the formerly communist bloc nations seek to develop capitalistic economies and establish their own hard currencies and financial independence, there is almost certain to be an influx of US personnel and firms into the newly free nations. Most of them have industrial bases which are a shambles while they have atmospheres that make smog-shrouded Los Angeles or Denver seem almost pristine in comparison. Professionals and their firms will have to make choices between impacts on the economic environment and the atmospheric environment that will be difficult, long reaching and, maybe, long term.

Though a persuasive argument may be made that engineers as a group had strong environmental concerns long before most laymen heard of the word "ecology," it still is within the past two decades that environmental concerns have come to be regarded as an ethical responsibility of professional engineering. If we accept as correct the charge that engineers do have social as well as technical responsibilities, then those who engineer, design and operate municipal treatment plants, industrial facilities and other elements of technology that have the potential for pollution also have responsibilities for preventing the pollution.

Ian Barbour, in his book *Technology, Environment and Human Values*, comments that it has usually been assumed that science and technology are inherently good and have contributed to progress. In many cases, though, it has been the success of a technology in achieving limited goals that has created unanticipated problems. While this does not take away from the fact that new technologies have contributed to health, material welfare and higher standards of living, it does highlight the fact that these advances have sometimes had high environmental costs, costs which were unanticipated and, hence, not factored into the cost-benefit analysis. It has caused engineers and others to question the future environmental costs of new technology. Both asking and answering this question is the emerging ethical responsibility referred to in the opening paragraph. Barbour, continuing, suggests that there are three primary grounds for environmental ethics: short run human benefits from the environment, duties to future generations and duties towards non-human creatures.

Because of the breadth of concern for environmental issues and results of technology, there is emerging a concept called participatory technology in which engineers have responsibilities of informing the public as well as clients. Since many go/no go decisions wind up being political ones, then the public, in a way, is invited to participate in the use and regulation of technology. Such considerations as computers and privacy, cars and pollution, nuclear power and energy shortages, etc., require broad public input which, in turn, requires honest engineering appraisal. If it is true that prudence is foresight, caution and utility, then it is up to the technologists to provide that prudence while thinking about safety and remote results.

Technology has the potential to be a liberator, the power to be a threat and may be an instrument of power. This is why decisions about technology should include environmental and human values as well as decisions about technical feasibility and economic efficiency.

Computer Abuse: Just as the use of computers has become world wide, so too has the opportunity for their abuse and ethical misuse become global. Neither national security information, industrial secrets nor personal privacy are safe from computer hackers or thieves whose mission may be a lark or serious business and whose geographic area of mischief may range from a neighborhood or a single company's or school's files to international intrigue.

Areas in which computer users need to pay particular attention, lest ethical violations be committed, are simulation and design. In many cases, computer simulation has begun to replace what have been regarded as essential tests or experiments. Users must be cautious that safety is not sacrificed for expediency. Equal caution must be evidenced where Computer Aided Design, Computer Aided Manufacture and other engineering computer programs are used in such a way that non-engineers are performing what have been traditional engineering tasks, sometimes with inadequate knowledge of the engineering elements involved to check and understand the output. In effect, many companies are allowing non-engineers to make engineering decisions and engineering designs without engineers' input.

Though it hardly seems necessary to mention it, many need to be reminded that computers can handle large amounts of data and make information from that data available quickly. Computers provide the opportunity to make decisions rapidly, but they do not assure the quality of those decisions.

Ethics has particular pertinence with respect to computer usage because much of a computer's effectiveness depends on its networking, and computer networks are built on reciprocal trust. With computers, as elsewhere in engineering, ethics, trust and professionalism are terms which are interdependent and cannot be separated.

Responsibilities to Society, Clients, Employers and the Profession

The professional has a wide range of responsibilities in four primary areas:

1. to society,
2. to clients,
3. to employers, and
4. to his/her profession and fellow professionals.

It follows that the profession also has responsibilities to the professional.

In today's technological society, the engineer's responsibility for the overall well-being of our society is as great as that of any citizen or professional person. Engineers who do not think about the philosophical problem of what is best for mankind may be the means of destroying humanity rather than of bettering it. Since perhaps 90-95 percent of all engineers are employed rather than in private practice, it follows that many of them work in profit making enterprises. While private profit does not necessarily conflict with public interest, public and private objectives do not completely coincide. Business, partly because of engineer input, is beginning to realize that it has a social responsibility in addition to its goals of making profits and perpetuating itself. In that vein of discussion, Adolph Ackerman, quoted in *Ethical Problems In Engineering*, writes that ethics become the dominating influence on the truly professional engineer and that knowing more than the general public does about the effect his work will have, his first duty is to serve the public interest above all others no matter what his employer may want or some government regulation may permit.

Carrying along that same reasoning, Victor Paschki writes: "A cause and effect relationship exists between an engineering action and human beings. Every person recognizes his responsibility for his actions in general and there is no good reason to exempt from this responsibility the main content of one's life that is one's professional work." While Paschki values forums where engineers may explore ideas of values and ethics and social consequences, he concludes that the final consequences must be with the individual.

Mantell, in his book *Ethics and Professionalism in Engineering*, says that the truly great contribution of engineering and its methodology to society is its willingness and ability, in spite of much that is not known about materials and forces of nature, to perform needed services within a time limit and with whatever financial resources are available. The right answer for society, according to Mantell, may not always be the best possible answer to the problem because of these constraints.

Though none question the engineer's responsibility to society, codes are somewhat less than clear in this area. The code words, "public health and safety," are less inclusive than "the welfare of humanity at large" which is a much wider concern than the immediate public health and safety. The welfare of humanity involves fundamental moral principles and long-term consequences of present engineering actions. As important as the ability to render technically competent services is the obligation to make for society a better place to live.

Unless they maintain an ethical attitude of unbending integrity which results in placing public interest above all else, there is the risk that engineers will be regarded not as professionals, but as mere technicians. This is because professional status carries with it an implied contract to serve society beyond all other obligations in return for the protection and status that society grants to the profession. Still, those who argue that the engineer should assume even more responsibility have an obligation to push, also, for a means of protecting those professionals who suffer financial or career setbacks as a result of meeting their ethical responsibilities.

As a part of his responsibility to clients and employers, because technology is ever changing, the professional engineer must be ever learning. When they graduate, engineers have been given the best education that can be crammed into their time of attendance at school. Unfortunately, some have not developed an appreciation of the need for continuing education, either formal or informal. They have a clear obligation not to let this scientific training fall into arrears. Other professionals, such as doctors and lawyers, find in their daily work situations which require them to continue to learn and study. The practicing engineer may encounter such situations less frequently, but it is no less a requirement. Increasingly, state registration boards are imposing continuing education requirements. This will call attention to the problem but only partially solve it. Many engineers are not required to register, so board actions do not impact them, yet their ethical obligations are undiminished by the fact that they are exempted from a registration requirement.

Grist for the thought mill is contained in the "Ten Commandments of Engineering Professionalism" penned by R. Harvey in 1988 as part of an article titled "Professionalism and the Civil Engineer," published in *Journal Of Professional Issues In Engineering*, Vol. 115, #4 and listed again here:

1. Thou shall think professionally, act professionally, and demand the same from those around you.
2. Thou shall not be manipulated by those around you.
3. Thou shall not expect perfection from your fellow engineers.
4. Thou shall not become entrenched in your ways.
5. Thou shall be receptive to new ideas and innovative approaches to old problems.
6. Thou shall become active in your community and make yourself known to the press and to your local and regional politicians.

7. Thou shall not engage in intra-professional cutthroat competition.
8. Thou shall select engineers based on expertise and not on price.
9. Thou shall not bid engineering work.
10. Thou shall seek just compensation for engineering services.

Just as engineers have responsibilities to their profession, so too does the profession have responsibilities to its members. Recently in a seminar conducted at a state wide Society of Professional Engineers meeting by the developers of this course, participants were asked to respond "yes" or "no" to a dozen questions about engineering ethics. The imprecise nature of the subject was such that on 11 of the 12 questions, there was nothing approaching complete agreement as to the answer. The one question on which there was almost total agreement was: "Do you feel that professional societies generally do a satisfactory job of protecting members who make great personal sacrifice in the face of ethical dilemmas?" The near unanimous answer was "No." Clearly, for this perception to change, societies will have to do more than grant some award or recognition to an engineer after a job has been lost and savings decimated by legal defense costs.

Membership in professional societies as a sign of professionalism on the part of individual engineers is an accepted concept. Until the societies are able to do more to protect the rights of their most ethical and dedicated members, the road will be perceived as a one-way street, and many may not join who could profit by and contribute through such membership.

Three Moral Theories[1]

By Jay Newhard, Ph.D.[2]

There are two general ethical questions which we might want answers to. One is, ***which actions are the morally right actions?*** A second is, ***why is a morally right action morally right?*** Or in other words, ***what is it that makes an action morally right?***

An answer to the first question will be some sort of list, or set of instructions indicating which actions are morally right. The following short list, though oversimplified, may be considered as an example.

1. Always tell the truth
2. Always keep your promises
3. Always treat others and their property with respect

Two general ethical questions

1. Which actions are the morally right actions?
2. What is it that makes an action morally right?

Even if this isn't a complete list, this is a good start, with three very sound moral principles. The point they are introduced to illustrate is that even with this short set of moral principles, there are fairly common circumstances which lead to moral dilemmas.

Suppose you've promised a friend of yours to help him move this Saturday. Your friend is counting on using your truck to get some large pieces of furniture moved, and to get everything moved in one day. Suppose that you go out to your truck on Saturday morning, and just after you start it up, you hear a loud snap and a bang—your timing belt broke. There's no way for you to drive your truck now—in fact, you'll need to get a ride to work on Monday. As a conscientious friend, you call the local truck rentals, but everything is already reserved for the day. Suppose that you have one last option to keep your promise, as principle 2 directs you to. You are good friends with your next-door neighbor, who has an old truck. Your neighbor is away for the weekend, but you keep a key to their house for them, in case they lock themselves out, and you know where they keep the keys for the truck. They probably wouldn't mind if you used it to drive to work, but because moving is hard on the truck, and the truck is old, they would probably be reluctant to lend it for that purpose. Of course, in any case, you don't have their permission.

These circumstances, together with the second and third principles, present you

[1] This article may be reproduced for others to study; reference to NIEE and this workbook is all that is needed.

[2] During Academic Year 2001-2002, Dr. Jay Newhard served as a Visiting Assistant Professor of Philosophy and a Senior Research Associate, National Institute for Engineering Ethics, Murdough Center for Engineering Professionalism, Texas Tech University. He holds a BS and MS in Chemical Engineering from Penn State University and an AM and PhD in Philosophy from Brown University. Newhard is now at Bloomsburg University, PA.

with a moral dilemma. If you borrow the truck without permission, you keep your promise, but you fail to treat your neighbor and his property with respect. If you don't borrow the truck, you treat your neighbor and his property with respect, but you fail to keep your promise.

One response is to make an exception to one of the moral principles; for example, "always keep your promises, if you can" or "always treat others and their property with respect, if you can." Neither of these amended principles helps with the dilemma, since you *can* either keep your promise or treat your neighbor with respect. The dilemma is to decide which one. If only one principle were modified, that would help us out of our dilemma; but this raises the same question: which principle is it right to make an exception to?

Ultimately, the question we need answered is the second one, namely: What is it that makes an action morally right? If we knew the answer to that question, we would have help out of our dilemma. And in general, if we know what makes an action morally right, we can then determine which actions are the morally right actions. In other words, the second question is more fundamental than the first.

Philosophers have been trying to answer that question for at least 2400 years, and unfortunately, there is no consensus regarding the answer. Nonetheless, the answers given by three philosophers—Immanuel Kant, John Stuart Mill, and Aristotle—are generally regarded as the most significant.

Theory 1- Good Intentions
An action is morally right if it is done with intentions which every rational person would approve
Immanuel Kant

To get at the ideas behind Kant's and Mill's theories, consider a new scenario. Valerie is out on a walk on her day off. As she turns a corner, she notices some smoke coming out from the kitchen window of a house across the street. She stops to see whether the smoke is from a fire or just from some cooking, when she hears the screams of two young children, who she spots in a window upstairs. Valerie decides to go ring the doorbell to make sure the parents or guardian have things under control. On her way across the street, she notices that the garage doors are open, and that the garage is empty—so when no one answers the doorbell, Valerie is pressed to make a decision.

The smoke is getting thicker, and she can hear the children still screaming. She stops to listen for sirens, just in case someone already called the fire department, but there is no sign of a fire truck. Valerie decides to let herself in. She finds that the front door is locked, but is able to let herself in through the garage. With these delays, the house is already thick with smoke, so she decides to rescue the kids before calling the fire department. It is hard for her to find her way around an unfamiliar house filled with smoke. Once upstairs, she has no trouble finding the kids, since they're still screaming, but they've locked themselves in their bedroom.

It takes her a few minutes to get them to unlock the door, but by this time, the fire department has arrived and has extended a ladder to the bedroom window. Valerie hands the children to a fireman on the ladder one at a time. Both children are scared and screaming. The second child is especially scared. As Valerie passes him to the fireman, he squirms out of her hands, and falls onto the sidewalk and dies.

Although her rescue efforts finished tragically, it is not hard to see Valerie as having done the morally right thing in trying to rescue the children. After all, we could hardly ask more of her: she was alert to notice the fire in the first place, she was courageous and brave in deciding to rescue the children, she was responsible enough to listen for fire engines before she entered the burning house, and she was level-headed throughout her attempt to locate and rescue the children. Despite the death of the second child, Valerie's actions were heroic.

These features of Valerie's actions are captured by the notion of good intentions, which is central to Immanuel Kant's moral theory. According to Kant, ***an action is morally right if it is done with intentions which every rational person would approve***. Since any rational person trapped in a burning building would want to be rescued, Kant's theory explains why Valerie's actions are morally right.

One type of action Kant held as a paradigm of a morally right action was telling the truth. Kant thought that it was never morally right to lie, under any circumstances. First of all, no rational person wants to be lied to. Second, if lying were morally permissible, then there would be nothing to prevent someone from lying. Very quickly, the entire practice of communication would halt. Since no rational person could want this result, there are two reasons why lying is immoral for Kant.

There does seem to be something right about this idea. But while Kant's theory is very successful in capturing good intentions as a feature of morally right actions, there are some problems with it. For example, Kant would obviously accept the first two moral principles in the short list above. Turning our attention to the third, it seems that respecting other people and their property is also something every rational person would approve. In fact, this is part of Kant's view. But as we saw in the first example above, accepting all three principles can lead to moral dilemmas. So it looks like Kant's theory doesn't help us out of them.

Recall Valerie's attempt to rescue the two children. Imagine the parents returning home to find that there had been a fire in their house, and a fireman explaining to the parents that one son died while an heroic passerby attempted to rescue him from the blaze. One very likely and understandable reaction for the parents is to be upset with Valerie, since it appears very likely that both children would be alive if she hadn't attempted to rescue them. After all, the firemen arrived in time to save both children, and they are specially trained to rescue people from fires. If Valerie had just minded her own business, if she had just taken her walk along a different route, if she had just taken a different day off, both of their children would still be alive. These considerations are bound to be very vexing for the parents. That Valerie was actually very alert, courageous, responsible, level-headed, and generally heroic does nothing to bring their son back, and so is bound to console the parents very little. If you are able to imagine yourself in the position of the parents, it is actually a quite natural conclusion to say that despite her heroic actions and good intentions, Valerie's actions were morally wrong, because they led to losing an innocent person's life, which is a morally bad outcome.

The moral importance of an action's outcome or consequences is captured by John Stuart Mill's theory, which is called "utilitarianism." Utilitarianism is based on a notion Mill called "utility," which is a composite notion consisting of three morally relevant components of the consequences of an action, namely, happiness, health, and well-being. So, for instance, making someone happier increases utility, and is a morally right action according to Utilitarianism. Likewise, making someone healthier—say, by giving them a vitamin, or by saving their life—increases utility, and is a morally right action. Along the same lines, making two people healthier is morally better than making one person healthy, though both are morally good actions.

Suppose a sick person needs to take a syrup which tastes pretty bad. In this case, the increase in health must be greater than the decrease in happiness caused by the bad taste for it to be a morally right action. In other words, utilitarianism is concerned with net changes in utility. According to Utilitarianism, if an action produces a net increase in utility, it's morally right; if an action produces a net decrease, it's morally wrong.

Theory 2 - Principle of Utility
An action is right in proporion as it tends to maximize overall utility, and wrong as it tends to produce the reverse.
John Stuart Mill

Now it might seem strange to compare the amount of change in happiness with the amount of change in health. Both happiness and health are abstract to begin with; to compare changes in them might seem a little bizarre. The general idea is that all of the effects of an action in terms of happiness, health, and well-being are added up in some way, and if the sum is a positive—if there is a net increase in utility—the action is morally right, and if the sum is negative—if there is a net decrease in utility—then the action is morally wrong.

Although changes in utility most commonly occur through changes in happiness and health, well-being is also an important component of utility, because it allows the notion of utility to include other sorts of actions which seem morally good, but which don't produce changes in happiness or health. A good example is learning. Even if learning makes someone neither more nor less happy, and even if it makes them neither more nor less healthy, learning makes them better off, it increases a person's well-being. Thus utilitarianism can explain why learning is morally good, since it increases utility through the well-being component.

John Stuart Mill formulated the Principle of Utility as follows: ***an action is right in***

proportion as it tends to maximize overall utility, and wrong as it tends to produce the reverse. Although the mathematical aspect of utilitarianism may seem a little strange at first, it explains quite a few of our intuitions about morality. Giving 2 meals to a homeless person is better than giving 1 meal—twice as good, according to utilitarianism. Giving 5 dollars to a starving homeless person is a much better action morally than giving 5 dollars to a rich person, because the 5 dollars does the homeless person much more good—it will provide some necessary items the homeless person wouldn't have otherwise had, but would provide little if anything for the rich person which they didn't already have. In terms of utility, there is a greater increase in utility to give 5 dollars to the homeless person.

Another feature of utilitarianism to glean from these examples is that the morally right action is the one which tends to maximize utility among the available alternatives. If someone has a choice between two actions, both of which will increase utility, then the morally right action is the one producing the greater increase in utility. Suppose the fire department was unavailable to respond to the fire in our story, and that Valerie could only carry one child out of the house at a time. If there was time to save only one child, the action of saving only one child would be considered morally right by utilitarianism, since the alternative is to save neither child. However, since the fire department did respond, Valerie's actions were morally wrong according to Utilitarianism, since the alternative of doing nothing would have resulted in both children being rescued. It doesn't matter to Utilitarianism that Valerie didn't know the fire department would arrive in time; all that matters is the change in utility compared with the alternatives.

As for Valerie's alertness, courage, and level-headedness, the utilitarian can say that perhaps she should not be *blamed* for her morally wrong action, but that does not change the fact that it was morally wrong. She's the kind of person who normally does things which increase utility, with this tragic exception.

Perhaps the most important feature of Utilitarianism is that if everyone followed it, there would be a state of maximal overall happiness, health, and well being. If we picture a utopian state, we probably picture something a lot like maximal utility. According to Utilitarianism, this is the end result if everyone just does what is morally right—it's built right into the principle of utility. And if we think about it, we probably think that if everyone did what was morally right, we would have a utopia. By contrast, it's possible for everyone to act with intentions every rational person would approve, and yet for things to go awry with enough frequency that the result wouldn't resemble utopia.

Nevertheless, there are some important objections to Utilitarianism. Consider a completely innocent person strolling down the sidewalk past a hospital emergency room. A doctor rushes out and grabs her arm, pricks her finger to do a quick blood test. "O positive!" the doctor cries out. "Follow me," he says to the innocent passerby. The doctor explains to the passerby that her O positive blood type makes her a universal donor, which is very fortunate, since the doctor has seven patients each needing a vital organ. One needs a heart, another a kidney, two patients each need a lung, another a liver, another bone marrow, and a comatose patient needs the left hemisphere of her brain. By using the passerby for parts, the doctor can save seven lives, at the cost of one! This is a clear case of a morally right action for Utilitarianism, though plainly it is too extreme and morally wrong to sacrifice an innocent passerby for the sake of the seven patients.

Utilitarianism is quite a good theory to use where there are a limited amount of resources, as long as extreme situations such as these can be avoided. For example, despite the emergency room example, much of medical ethics is based on utilitarianism, since the medical community has a limited amount of resources, and their declared aim is to maximize their patients' health and happiness. Utilitarianism suits these aims of medicine remarkably well—although we also want caring, dedicated doctors with good intentions, even more important is that they bring about good results—we want them to cure us, and keep us healthy.

Similarly, the engineering community has a limited amount of resources. While we want engineers who are caring and dedicated, and who act with good intentions, still more important is that they bring about good results. This is reflected in various engineering codes of ethics by the precept to hold paramount public health, safety, and welfare. Just as for the medical community, utilitarianism suits the aims of the engineering community remarkably well, though it is important to beware of extreme situations, where Utilitarianism gives results which are plainly incorrect.

Ideally, a moral theory would capture both good intentions and good consequences. While this may sound straightforward, it turns out to be very difficult for a theory to incorporate both: notice that Kant and Mill simply disagree over whether Valerie's actions were morally right. In general, wherever an action is controversial, the two theories will disagree over whether it is morally right. Put another way, if good intentions ***and*** maximal utility are required for an action to be morally right, then wherever there is a moral dilemma, there is a good chance that none of the alternatives will be morally right. We could neither fire nor spare the CEO, since firing him does not involve good intentions every rational person would approve—especially the CEO—and sparing him does not maximize utility.

Notice, though, that there is some common ground between these two theories. What both theories *do* grant about Valerie's efforts to rescue the children from the fire is that she acted alertly, courageously, responsibly, and with a level head. These are virtues Valerie has. Aristotle's view is that morally right actions are virtuous actions. This theory, based on virtues, rather than intentions or consequences, is our third moral theory. For Aristotle, a virtue is a mean between extremes, where one extreme is the vice lacking the virtue, and the other extreme is

Theory 3 - Virtue Ethics
Morally right actions are virtuous actions. This theory is based on virtues rather than intentions or consequences
Aristotle

the vice of having the virtue in excess. As an example, consider the virtue of bravery. A person lacking bravery has the vice of cowardice. But too much bravery is actually a vice, also. Being so brave that one does not flee from machine-gun fire is foolhardy. This is a vice of excess. Friendliness is a virtue, too. A person who lacks friendliness has the vice of sternness. But too much friendliness is not a good thing, either, for one becomes a social butterfly.

Aristotle points out that what lies in the mean and what lies at the extreme cannot be said without relevant details. The proportion of the virtues it is good for a person to have depends on who they are, what gifts they have, and what they do. Firemen and policemen need to have more bravery than accountants or attorneys. While surgeons who perform long surgeries need more concentration and endurance than a pipe fitter, a pipe fitter needs more strength.

What is fundamental to understanding and having virtues is ***judgment***. Using good judgment doesn't guarantee maximal utility, but if we do have virtues such as being alert, courageous, responsible, and level-headed, tragedies will be rare. Also, that a person uses good judgment does not guarantee that they act with intentions every rational person would approve, but since good judgment takes all relevant things into account, there is little room for criticism from Kant. To act virtuously is to do the right thing for the right reason.

Aristotle argues that while we are all given the capacity to be virtuous, a person needs the proper education and training in order to be virtuous. It takes a good deal of experience and practice to develop virtues, and to refine our judgment so that we judge well, in a way which suits our gifts and circumstances. Circumstances can be critical to deciding which action is virtuous. For example, knowing the temperament of the neighbor and the condition of the truck are critical in deciding whether borrowing the truck is the virtuous action. Only if the neighbor is exceedingly generous and congenial toward you, and there is little risk of the wear and tear breaking down the truck, would it be virtuous to borrow the truck to help your friend move. Otherwise, you must explain to your friend what happened; sometimes things go wrong, and you can't keep your promise.

Despite its possible advantages over Kant's theory and Utilitarianism, virtue ethics is hard to implement in professional settings for three reasons.

First, since judgment is fundamental to having virtues, applications of virtue ethics in particular cases are bound to be controversial.

Second, because the controversies stem from differences in judgment, which is somewhat subjective, controversies will often be difficult to resolve. Obviously, in a professional setting, there must be an effective way to resolve controversies in order to minimize distraction from professional aims and efforts.

The third reason is related to the second. Notice that the fundamental question Aristotle is answering is: What kind of person should I be? rather than What is it that makes an action morally right? An answer to either question will lead to an answer to the other. But because Aristotle's question centers on a person, it is difficult to implement into professional settings.

It is much more practical, and much less susceptible to controversy, to establish professional guidelines which answer the question: What is it that makes an action morally right? so that controversies can be handled on a case-by-case basis.

- - -

Suggested Reference: George Sher, ed. *Moral Philosophy: Selected Readings*, Second Edition. (Fort Worth: Harcourt Brace College Publishers, 1996).

Fundamentals of Ethics

By Patricia M. Harper

National Institute for Engineering Ethics
Program Coordinator and Assistant to the Director

Many times ethical problems encountered in engineering practice are very complex and occasionally involve conflicting ethical principles.[1] To assist professionals in dealing with ethical issues, NIEE has developed seminars, workshops, and courses on professional ethics. These programs focus on:

- Encouraging professionals to analyze complex problems and to resolve them in the most ethical manner
- Sensitizing professionals to important ethical issues before they have to confront them in practice
- Acquainting professionals with situations other engineers have faced, and providing the tools for determination of what to do when faced with similar situations

The goal of engineering ethics courses is generally to foster moral autonomy, which is the ability to think critically and independently about moral issues and to apply this moral thinking to situations encountered in engineering practice.

Engineers find it more difficult to consider ethics issues than technical issues because:

- Ethics issues are open-ended and not susceptible to formulaic answers
- Ethics issues rarely have a single correct answer

Ethics Problems are Similar to Design Problems[1]

Specifications of design are determined .... Ethical details and facts are determined

A variety of designs are considered .... A variety of ethical actions/options are considered

Design possibilities are evaluated .... Ethical options are evaluated

There may be many workable designs .... There may be a range of possible solutions

Most appropriate design is selected .... Most appropriate action is selected

Both apply a large body of knowledge to the solution

Both involve the use of analytical skills

Rarely is there a unique correct answer

What is Engineering Ethics?

Engineering ethics amounts to the set of justified moral principles of obligations, rights, and ideals that ought to be endorsed by those engaged in engineering. Engineering ethics is also the currently accepted codes and standards of conduct endorsed by various groups of engineers and engineering societies.[2]

Ethical Theories

Ethical theories play a role in understanding moral dilemmas according to Mike Martin and Roland Schinzinger[2] by placing them within wider contexts of basic ideals and principles. They are not moral algorithms that can be mechanically applied to remove perplexity; they help by providing frameworks for understanding and reflecting upon dilemmas, and often they provide at least some systematic guidance.

There are a relatively large number of ethical theories. Charles Fledderman[1] contends this does not indicate a weakness in theoretical understanding; rather it reflects the complexity of moral problems and the diversity of approaches to ethical problem solving that have been developed over the centuries.

Three types of ethical theories attempt to formulate the fundamental principles of obligation applicable to both professional and personal conduct in everyday life. These theories differ according to what they treat as the most fundamental moral concept: good consequences for all, duties, or human rights.[2]

Utilitarianism - (John Stuart Mill)

The emphasis of Utilitarianism is not on maximizing the well-being of the individual, but rather on maximizing the well-being of society as a whole, and as such it is somewhat of a collectivist approach. Utilitarianism tries to balance the needs of society with the needs of the individual, with an emphasis on what will provide the most benefit to the most people.

There are several drawbacks to this theory:

- Sometimes what is best for everyone may be bad for a particular individual or group of individuals.
- Implementation depends greatly on knowing what will lead to the most good.
- Frequently, it is impossible to know exactly what all the consequences of an action are.

Depending on whether the production of goodness is to be assessed with respect to action or to the consequences of general rules about actions, utilitarianism can be developed in two different directions: Rule- and Act-utilitarianism.

Fledderman[1] portrays <u>Rule-utilitarianism</u> where moral rules are most important. Although adhering to the rules might not always maximize good in a particular situation, it will ultimately lead to the most good.

Martin and Schinzinger[2] state that Act-utilitarianism focuses on individual actions rather than general rules. Using this theory, everyday maxims like "keep your promises" and "don't deceive" can be broken whenever doing so will produce the most good in a specific situation.

Newhard[3] notes that another feature of Utilitarianism is that the morally right action is the one which tends to maximize utility among the available alternatives. If there is a choice between two actions when both increase utility, the morally right action is the one producing the greater good.

Duty Ethics – (Immanuel Kant)

Martin and Schinzinger[2] indicate that using Kant's Theory of Good Intentions, ethical actions are those actions that could be written down on a list of duties, such as be honest, don't cause suffering, be fair to others, etc. Such actions are considered duties because they meet three interwoven conditions:

- Each duty on the list expresses respect for persons
  People deserve respect because they have inherent worth as rational beings who have the capacity for autonomy and for exercising good will. To respect people is to respect their autonomy and their attempts to meet their duties. Immorality occurs when we use people as mere means to an end, nothing more. Equally important is respect for oneself. Kant argued that duties to oneself are fundamental.

- Each duty on the list is a universal principle
  Duties are binding on us only if they are also applicable to everyone. Generally, everyday moral rules pass this test. For example, let us consider "keep your promises." If people only kept promises when they felt like it, promises would not longer hold any meaning for us.

- Each duty on the list expresses an unqualified command for autonomous moral agents, without qualifications or conditions attached
  These moral imperatives require us to do certain things whether we want to or not. These are our duties, independent of whether or not performing them will make us happy.

Kant's theory is very successful in capturing "good intentions" as a feature of morally right actions but Kant thought these principles were absolute in the sense of never having justifiable exceptions. He failed to be sensitive to how principles of duty can conflict with each other, thereby creating moral dilemmas.[2]

Contemporary duty ethicists recognize that many moral dilemmas are resolvable only by making exceptions to simple principles of duty. Some recent duty ethicists emphasize the importance of careful reflection on each situation, weighing all relevant duties in light of all the facts, and trying to arrive at a sound judgment or intuition. They also stress that some principles, such as "Do not kill" or "Protect innocent life" clearly involve more pressing kinds of respects for persons than "keep your promises."[2]

Virtue Ethics - (Aristotle)

Virtue ethics is closely tied to personal character. Virtue ethics seems less concrete because it is harder to describe nonhuman entities such as a corporation in terms of virtue and so this type of ethical theory is somewhat trickier to apply to engineering problems.

We can use virtue ethics by answering questions such as "Is this action honest?" "Will this action demonstrate loyalty to my community and/or my employer?" "Have I acted in a responsible fashion?"

Fledderman[1] contends that as with any ethical theory, it is important to ensure that the traits you identify as virtues are indeed virtuous and will not lead to negative consequences. For example, honor can be a code of dignity, integrity and pride, but there are many examples in history of wars that have been fought in order to preserve the honor of an individual or a nation.

Applying Ethical Theories

Fledderman[1] makes the point that in solving ethical problems, we are not limited to choosing one theory, but use all of them to analyze a problem from different angles and see what result each of the theories gives us. Generally, the rights of individuals should receive relatively stronger weight than the needs of society as a whole, so rights and duty ethics should take precedence over utilitarian considerations.

Much of engineering ethics, according to Martin and Schinzinger[2] can be viewed as a part of applied (or practical) philosophical ethics. Applied ethics necessarily makes contact in many places with philosophical ethical theory because applied ethics is concerned with uncovering rational moral reasons for beliefs and actions as opposed to accepting uncritically whatever beliefs or actions might happen to strike one's fancy as being correct in a given situation. Applied ethics focuses on concrete problems for their own sake, and invokes general theory where helpful in dealing with those problems.

Evaluating Actions

According to Davis,[4] actions can also be evaluated using the following tests:

- **Harms test**: Do the benefits outweigh the harms, short-term and long-term?
- **Reversibility test**: Would I think this was a good choice if I traded places?
- **Colleague test**: What would professional colleagues say?
- **Legality test**: Would this choice violate a law or a policy of my employer?
- **Publicity test**: How would this choice look on the front page of a newspaper?
- **Common practice test**: What if everyone behaved in this way?
- **Wise relative test**: What would my wise old relative do?

References:

[1] Fleddermann, Charles B, "Engineering Ethics", Prentice Hall Engineering Source, Upper Saddle River, NJ, 2004, 134pp

[2] Martin, Mike & Schinzinger, Roland, "Ethics in Engineering", McGraw-Hill, New York, NY, 1996, 417pp

[3] Newhard, Jay, "A Discussion of Three Moral Theories", 2002

[4] Davis, Michael, "Better Communications Between Engineers and Managers: Some Ways to Prevent Ethically Hard Choices", Science and Engineering Ethics 3 (April 1997): 171-213.

No man can always be right.

So the struggle is to do one's best

To keep the brain and conscience clear;

Never to be swayed by unworthy motives

Or inconsequential reasons,

But to strive to unearth the basic

factors involved

And then do one's duty.

Dwight D. Eisenhower

Ethical Leadership
A Vital Engineering Skill

by
Jimmy H. Smith, Ph.D., P.E.
Professor of Civil Engineering and Director, National Institute for Engineering Ethics,
Murdough Center for Engineering Professionalism, College of Enginering
Texas Tech University, Lubbock, Texas

Choosing between good and evil is easy, according to some. They say the real difficulty comes when choosing between competing goods. Design, construction, and maintenance operations, including related functions of accounting, purchasing, and legal, are fundamental to the engineering profession. They contain both risks and benefits...and competing goods. Drawing first from a paper I presented at the Gulf Southwest Conference of ASEE a while back, let me share with you a few thoughts about professional ethics as it relates to the professions in both education and practice.

BASICS
Choosing between good and evil
is usually easy.
It is choosing between
competing goods
...that becomes difficult

Professionals, and in particular, professional engineers, conceive ideas, develop concepts, design products, and implement the construction of a wide variety of systems or products ranging from the tiniest micro-electronics to transportation control systems; from simple systems to highly complex systems. Such projects have both risk and benefit. If we were to plot the number of engineering developments over the past few centuries, or even the past 50 years, the curve would be exponential. Remember that word 'exponential,' for we will return to it later in reference to professional ethics.

Ethics is sometimes a difficult subject to communicate. Like design, situations involving ethical questions do not always have unique answers. When confronted with ethical questions, professionals may be reluctant to discuss them, or ill-prepared to objectively handle them, or both. Open communication, difficult or not, will be beneficial to the profession. Better preparation to handle difficult situations before they arise will come out of higher levels of communication on and understanding of the subject.

Let's first consider the entire spectrum of engineering endeavors from a materials viewpoint. As on many professional undertakings, we make note of what materials are available. For purposes of our discussion today, let's focus on engineering design, and compare it to making of a fabric, and explore the use we make of materials used to weave the 'fabric' of our state agency.

What type of threads have we used to weave the fabric of engineering accomplishments? Certainly, physics and chemistry have played a very major role;

mathematics has been a thread without which our fabric could not have been woven. Also, accounting, legal, and other support groups are a major thread. But creative minds using these threads have played the most important role in our increasing ability to progress from simple wheels and wedges to where we are today. It is our designers and researchers who discover new ways of weaving the threads into different patterns and in doing so, create yet another quantum jump in technology and its use. These are the people who ultimately are instrumental in the selection of another important thread fundamental to all our projects, big and small — the 'ethical' thread. And finally, it is also these individuals who introduce a more personal and subtle aspect to the process, a healthy ego. These two "e" words, ethics and ego, combined with a third, economics, frequently are responsible for the success or failure of new ideas and new concepts.

Continuing our analogy. The thread shuttles carrying the different threads move through the fabric at astonishing speed during the weaving process. Seldom, if ever, do we encounter any conflict among the threads of physics, chemistry, mathematics and engineering theories. And while economics may be a problem, it is reasonably well defined. When all these are used correctly, a smooth fabric results. And when not used correctly, the error is generally easy to detect when we recalculate or ask a colleague to check our work. These threads are very visible to the educated observer.

However, something entirely different happens when ethics, economics and egos collide. These threads are initially either invisible or transparent. And how they fit into the fabric is not as easily determined as are physics, mathematics, chemistry, and engineering theories. And worst of all, when faced with ethical/ego conflict, it is much more difficult to discuss them with colleagues. When there is a collision between ethics and ego, and when ego wins, the fabric, however deceptively attractive, will contain flaws that will propagate throughout the material and may ultimately render it useless. More importantly, left unchecked, it will become dangerous ... not only to the particular project, but also to the company or agency ... and the fabric can become a *shroud*.

Let's pursue a key word just used — propagate; and couple it with the word stressed earlier — exponential. Blend those with some mathematics. And explore together what could happen to the fabric of our workplace.

Before his death, Carl Sagan illustrated how ego and role models can be viewed. The story is about an advisor to a King of long ago who had invented a new game in which the King was the most important piece and must be protected by all the other pieces. The object of the game was to capture the enemy King; and it later became known as chess.

The King was so pleased with the game that he told his advisor to name his own reward and offered jewels, palaces, or land. But the advisor was a humble man and requested only a humble reward: he asked only for a few piles of wheat, one pile for each square on the playing board. A single grain of wheat for the first square, two for the second, and four for the third, and so forth, so that each subsequent square would have twice the last until all 64 squares were filled. While secretly marveling at the unselfishness of his advisor, the King consented.

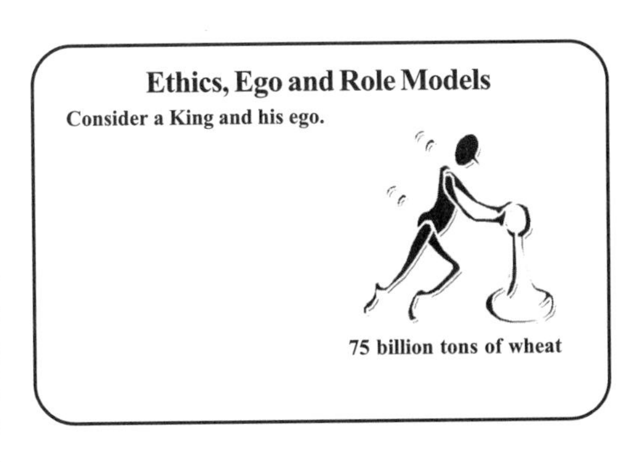

At first the wish was simple to grant. One grain, then 2, 4, 8, 16, 32, 64, 128, 256 ...only a hand full or so. But it was not long before it was discovered that the King did not, and could not, possess enough grain to grant the wish. In fact, by the time the 64th square could have been filled, a total of 75 billion tons of wheat would be required. Based on the current world's production of wheat, it would take 150 years to produce the requested reward. If the game had been invented with a 10x10 grid (100 squares) instead of 8x8, the requested reward would have weighed as much as the Earth.

This is called a geometric progression and the process is called an exponential increase. Carl Sagan relates this to population growth, AIDS, nuclear weapons, and much more. But today, let's think about this in the context of what could happen to our profession if, as mentioned earlier, our ego and ethics collide, and ego wins. Left unchecked, unethical decision-making could propagate exponentially.

Conversely, this same thinking should be cast in the positive sense. Ethical decision-making and actions will propagate exponentially if each professional communicates a clear conviction of the importance of ethics to only a few others who, in turn, do the same.

Educators, as well as professionals in industry and government, are influential to many young minds and in many cases, serve as role models to them. If each professional were to plant (knowingly or unknowingly) a single unethical seed in only two students or young professionals, and those two convey it to only two, and so on, it would not be very long at all before we faced the same dilemma as the King.

Perhaps the analogies presented are too simplistic to relate to the complexities of our profession and the work we do. Perhaps not. Industry and governmental agencies want to attract young aggressive minds, and help them develop an understanding of the basics of their work. We enjoy seeing their confidence level build, and become excited when we see that unmistakable gleam in their eye when a complex theory is truly understood. Yes, we expect them to develop a healthy ego and to take a noticeable pride in their ability and their work.

An entrepreneurial spirit frequently is fostered, and we like that. But with this entrepreneurial spirit, there is no substitute for *honesty* in both education and professional practice. We must make that message clear to the young professional! We must develop effective ways of communicating the importance of ethics.

It has been said that "ethics" starts in the sandbox, not in graduate school or on the first job. It is clear when we read or hear in the news about the O-rings of the space shuttle, computer viruses designed to infect our computer systems, etc., that in some cases at least, the sandbox was a playground for learning how to win, with too little regard for our fellow man.

Let's elaborate on a point made earlier: There exists a delicate balance between ethical and unethical actions within the professions. Which way it will swing in the future will depend in large part on:

1) our commitment to professional integrity
2) our willingness to <u>communicate</u> ethics with each other,
3) the level and quality of the communication,
4) the environment to develop effective communications.

Communication of ethics is the first step, and an important one. We must adhere to our ethics with dedication and conviction. With necessary and sufficient professional ethics in education, industry, and government, we can swing the balance in the positive direction; and in so doing, retain and improve the high standing that our profession enjoys among those we have pledged to protect...the public.

Now let me combine these thoughts with the subject of professional ethics and the responsibilities of both the young professional and others in companies or agencies. Carl Skooglund, former Texas Instruments Vice President and Director of Ethics has some compelling thoughts on industrial ethics. I draw on Carl's paper presented recently to a group of industry leaders throughout Texas. Carl quotes several writers on the subject that include the following.

Price Prichett: In his booklet entitled, *"The Ethics of Excellence,"* states that: *"The ethics you live out as you go about your work can provide the foundation of excellence. The high caliber organization is, after all, merely a reflection of its people."*

Ed Locke: *"If I am being told to do something by the leader, what they are telling me they have a right to tell me and they are probably right in their estimate of the facts. But in the field of ethics, I have to reserve ultimately that little piece of ethics for myself and say...'I must have the last say.'"*

And from Jerry Junkins, former TI Chief Executive Officer, President and Chairman of the Board: *"Resolving a conflict concerning ethical conduct is a matter of personal responsibility. No one should be expected — or permitted — to act in a way that violates his or her personal integrity. Each of you has not only the right, but the obligation, to question and seek clarification of ethical issues arising in the workplace."*

Behavior deemed as unethical can more than negate the rewards of high performance, nevertheless, an employee may face a tremendous dilemma. **<u>Good and decent people can feel the squeeze.</u>**

Carl Skooglund continues:

"What is the answer for the organization and its employees, especially the new employee who is first entering the workplace? I believe that the solution lies in two vital areas. We must hire principled people and train them to understand what is required of them. Equally important, we must provide them the necessary tools for fostering and supporting integrity and ethics in the day-to-day work environment. One of these tools is a clearly communicated set of shared corporate values.

It is essential to understand that there is a big difference between shared corporate goals and shared corporate values. All organizations have goals, many of them appropriately aggressive, and achieving them is often the difference between success and failure. But the organization must also have shared values, because without them it may not recognize where to apply the ethical brakes in the pursuit of those goals. For example, the drive to achieve a sales forecast should never override the obligation to commit to a customer only those product performances and delivery promises that can in good conscience be achieved.

A supportive organization with strong shared values understands the following three principles:

1. Ethical pressures will ultimately arise in the workplace. They must be anticipated and recognized.
2. Left unattended, organizational ethics will ultimately be impacted adversely by these pressures and temptations.
3. Structured, positive, proactive resources must be in place to support and enhance ethical conduct even to the point of strong intervention.

Engineers in engineering organizations carry a particularly important responsibility because their commitments and judgments have a major impact on product and service performance, and, of course, on public safety. Examples of ethical dilemmas facing the professional today are:

- A professional on a restricted budget needs a particular computer software program that will greatly enhance her ability to complete a project on time. But she cannot get approval to purchase it. Her friend has one and copying it would be very simple.
- An engineer responsible for completing the development of a critical system is asked to accelerate the work so that the production phase can be pulled forward. But he is concerned that the system design integrity will be compromised.
- On an airplane trip an accountant found a critical document accidentally left behind by a key competitor. It contained very valuable design and pricing information. It is not illegal for her to have the document but what are her ethical obligations?

Let's look at some specific actions that those of us responsible for establishing the organization's working environment must take to help employees resolve these dilemmas.

- First, we must create candor in the workplace, a safe atmosphere, and an empowering environment where the employee feels secure in bringing up and resolving the touchy issues, where there is trust between employees, where there is no fear of retribution.

- We must ensure that employees have the resources, training, skills, and knowledge to get the job done. The right way.

- Everyone must clearly understand the organization's ethical obligations and expectations.

- Supervisors and managers must set the highest ethical standards through their own style of leadership.

- Employees must feel that they are empowered to influence the work processes and be deeply involved in goal setting and other organizational decisions."

Finally, from a TI document called *"Ethical Leadership:"*

- **"Trust, candor, fairness, and integrity make for a better place to work.**

- An ethical reputation is a positive attribute for organizations. Ethics and integrity define the quality of relationships both inside and outside of the organization. And those relationships determine our success.

- You must get involved and set the example. Those around you will respond in a like manner. Accept responsibility for your part of the organization's ethical reputation.

- Being an engineer, accountant, manager, or other professional in today's complex business environment is a tough job.

- The pressures are tremendous, the demands are continuous.

- Keeping in step with the organization's ethical expectations is not just important, it's absolutely *vital*."

Professionalism, The Golden Years

By William D. Lawson, P.E.

ABSTRACT: Engineering literature throughout the twentieth century typically defines professionalism in terms of a group of attributes, namely, *knowledge*, *organization*, and *the ethic of professional service*. The approach – well known to sociologists – is called the functionalist model, because it presumes a view of society characterized by shared norms and values, stability, and the tendency to maintain equilibrium (functionality) in the presence of social change. This paper presents a detailed exposition of the ideal functionalist model of professionalism, carefully situated within its social, theoretical, and temporal context. This is done in the belief that to clearly understand professionalism during its golden years is the first step toward helping engineers more fully understand professionalism today.

Introduction

There was a time – the middle years of the twentieth century – when professionalism had a relatively clear meaning. Known in sociological literature as the "functionalist" model, a profession in the decades following World War II was commonly described in terms of its quintessential characteristics, or attributes. Although scholars of the professions differ on the number of attributes, following the approach of Kimball and others (Kimball 1992: 323-4), I will use *three* attributes to depict the professional ideal. These are:

- ***Knowledge***, as expressed in a body of theory, professional authority, and higher education;
- ***Organization***, as expressed in professional associations, monopoly and licensing, and professional autonomy; and
- ***The Ethic of Professional Service***, as expressed in the service ideal, codes of ethics, and the career concept.

Although most engineers will agree that times have changed in the past 40-50 years, the reason for presenting this age-old ideal of professionalism is my belief that to clearly understand professionalism in its golden years is the first step toward more fully understanding professionalism today.

The Attributes of a Profession

Knowledge

Body of Theory

Think, for a moment, about the nature of *professional work*. The tasks of professions are *human problems* amenable to expert *service*, and functionalist theory requires that these be problems of universal, or at least widely experienced, social concern. A profession's claims to jurisdiction over these problems involve three parts: claims to classify a problem, to reason about it, and to take action on it. Theoretically, these are the three acts of professional practice, and such practice is tied directly to a system of knowledge that formalizes the skills on which the work proceeds (Abbott 1988: 35-58). This expert knowledge or *body of theory* amounts to an internally consistent system of abstract propositions that describes the classes of phenomena comprising the profession's focus of interest. Theory serves as a base in terms of which the professional rationalizes his or her operations in concrete situations (Greenwood 1957: 68-9). It is the linking of professional skill with the prior or coincidental mastery of the underlying theory that is the true distinction of a professional (Harries-Jenkins 1970: 74).

A professional, then, *by definition*, is knowledgeable (Barber 1965: 18) (Ritzer 1972: 56) (Freidson 1973: 30). They *own* expertise, so much so that one analyst (rather tersely) defines professionalization as "... an attempt to translate one order of scarce resources – special knowledge and skills – into another – social and economic rewards" (Larson 1977: *xvii*). Professionals work with an unstandardized product (Hall 1969: 75-76): their knowledge focuses around areas which, although capable of being classified by science, retain mysteries –hidden elements access to which is privileged and can only be penetrated (if at all) by the profession's esoteric skills. Such skills afford the view of professions as being a class apart (Perrucci and Gerstl 1969: 10-11). In the words of Bledstein, "...the professional person penetrated beyond the rich confusion of ordinary experience, as he isolated and controlled the factors, *hidden to the untrained eye*, which made an elaborate system workable or impracticable, successful or unattainable" (Bledstein 1976: 88-89, emphasis mine).

Professional expertise is a *mixture* of several kinds of practical and theoretical knowledge, both the standardized variety (that which can be *taught*) and the "cognitively indeterminate" (that which is *caught*) (Larson 1977: 41). The optimal base of knowledge for a profession has been described as "...a combination of intellectual and practical knowing, some of which is explicit (classifications and generalizations learned from books, lectures, and demonstrations), some implicit ("understanding" acquired from supervised practice and observation)" (Wilensky 1964: 149-50). The profession's body of theory is like a reservoir from which the professional draws as needed. Although professionals do not directly use much of their abstract knowledge in normal day-to-day practice, society expects professionals to possess a high degree of knowledge and be able to muster *all* of their knowledge for a crisis – or at least have it on call (Goode 1969: 282-3).

Professional Authority

Professional *authority* is intimately linked to the profession's body of theory. "The authoritative air of the professional is a principle source of the client's faith that the relationship he is about to enter contains the potentials for meeting his needs" (Greenwood 1957: 70). Stated another way, the client finds a sense of security in the professional's assumption of authority. We do well to ask, "From whence comes *professional authority*?"

In the classical sense, authority signifies the "rightful, actual and unimpeded power to act;" it is power that is in some sense legitimate and justified and therefore compels trust or obedience (Packer 1982: 108). In the professional-client relationship of the functionalist era, it is the professional who possesses authority with the client being in a subordinate position; thus, professional authority is an expression of social control. Historically professionals drew authority from their high social position – in fact, gentlemanly status and dignity enhanced by a classic, liberal education was perhaps the dominant source of authority and honor for the professions from the seventeenth up until the mid-nineteenth century (Haber 1991: 5-6). But about this time the cultural ideal, the fundamental source of cultural inspiration and legitimacy, shifted to *science*, and scientific knowledge increasingly has dominated intellectual and professional life (Kimball 1992: 200-211).

Although professional authority retains some vestiges of *status* authority, in the functionalist era the principal basis for professional authority was its scientific expertise. This *scientific* authority rests on two sources of social control: legitimacy and dependence. As to legitimacy, academic knowledge legitimizes professional work by clarifying its foundations and tracing them to major cultural values, typically those of rationality, logic and science (Abbott 1988: 54). Thus, through their expertise the professions appropriate the cultural validity and authority of scientific knowledge. With regard to dependence, the kind of authority claimed by professions here involves not only skill in performing services, but also the capacity to judge the experience and needs of clients. Control is expressed, in part, by the client's dependence on the professional's superior competence, with the client's acceptance of authority signifying "a surrender of private judgment" (Starr 1982: 10-11). The distinction between *customers* and *clients* illustrates this point:

> A nonprofessional occupation has customers; a professional occupation has clients. What is the difference? A customer determines what services and/or commodities he wants, and he shops around until he finds them. His freedom of decision rests upon the premise that he has the capacity to appraise his own needs and to judge the potential of the service or of the commodity to satisfy them. The infallibility of his decisions is epitomized in the slogan: "The customer is always right!" In a professional relationship, however, the professional dictates what is good or evil for the client, who has no choice but to accede to professional judgment. Here the premise is that, because he lacks the requisite theoretical background, the client cannot diagnose his own needs or discriminate among the range of possibilities for meeting them (Greenwood 1957: 70).

Professional authority calls for *voluntary* obedience—it is usually enough that the client recognizes they *ought* to follow their professional's advice in awareness of the foul consequences that will befall them if they do not. Stated another way, the authority of the professional is " …more than advice and less than a command, an advice which one may not safely ignore " (Starr 1982: 9-10, 14). Symbols of professional authority – diplomas and certificates, the number of technical aids in an office, the number of articles and books on the professional's résumé, and the like – serve to inform and reinforce the client's awareness of his or her dependency (Bledstein 1976: 96).

Professional authority suggests a relatively wide knowledge gap between the client and the professional – it highlights the layman's comparative ignorance and need in contrast with the professional's knowledge and competence (Parsons 1951: 439) (Greenwood 1957: 70) (Ritzer 1972: 57). This gap between professional and lay knowledge constitutes the historical and logical basis for two well-known professional taboos: advertising and "fee-bidding" for professional services. In both these instances, professional ideology holds that clients lack the ability – the discriminating capacity – to capably select from among competing practitioners or forms of service based on such information (Greenwood 1957: 70). In fact, this ideology views advertising and fee-bidding as contrary to the client's best interests, if not potentially reckless – even *dangerous* – since by their very nature these actions tend to obscure issues and reduce the complexities of professional work to secondary if not surface-level concerns, yet while giving the *appearance* of substance and insight. These tricky practices can dupe clients into *assuming* they know the relevant factors that ought to be considered in a decision (a false sense of security) when in fact they do not, and possibly *cannot.* Such judgments should properly be left to the professional who possesses the requisite knowledge to advise the client as to "what ought to be done." *"Don't try this at home, I'm a professional!"*

Education

Prior to the functionalist era, from the mid-nineteenth century through the first decades of the twentieth century, engineering experienced a gradual shift from "rule of thumb" to "rule of science" (Haber 1991: 296). During this period education became the institutional locus for the cultural ideal of science, but engineers only reluctantly embraced it – ostensibly because of the value they placed on "practical know-how" as opposed to "book-learning." Nevertheless, by the beginning of the functionalist era (1930s –1940s) higher education and professionalism for engineers were firmly emplaced.

Inasmuch as universities and graduate schools are the *producers of professionals* and also the *producers of professional knowledge* (Larson 1977: 50), the link between higher education and professionalism cannot be stated too emphatically. The primary purpose of the professional schools in American universities is to transmit formally a body of expert knowledge that will enable the professional to practice his or her skills at an acceptable level of competence (Freidson 1994: Par20). A secondary purpose of higher education — not always acknowledged because it is achieved not through formal instruction but through contact with faculty and peers — is the transmission of values, attitudes, and commitments that serve to assimilate the novice to a set of professional attitudes and controls, professional conscience and solidarity (Perrucci and Gerstl 1969: 55-56) (Hughes 1958: 33).

In addition to transmitting professional knowledge and values to students, an equally important responsibility of the university is to develop new and better knowledge and theory on which professional practice is based (Mok 1973: 108) (Barber 1965: 20). Thus a division of labor exists between the practice-oriented and the theory-oriented person, the latter devoting his or her professional career to scientific investigation and theoretical systematization. This spawns accelerated expansion of the body of theory and increased specialization (Greenwood 1957).

By way of application, it is appropriate to note that, unlike the classic professions of medicine, law and the clergy, educational requirements for professional status in engineering are met by an undergraduate degree. From the first decades of the twentieth century when engineering education became normative, on to the functionalist era at mid-century, and even now at the dawn of the twenty-first century, the baccalaureate degree alone has been viewed as sufficient for professional status in engineering. This said, the on-going debate regarding the *First Professional Degree* (Russell, et al. 2000: 54-63) can only be favorably impacted by a more informed understanding of what professionalism means. In fact, given the historical and logical precedence of the profession to the university (*i.e.*, the university as servant of the profession and not the other way around), by all rights the more fundamental issue of professional identity should be addressed *first:* engineers must agree on "what they want" before they decide "how to get there." These and other professional issues depend to a large degree on the second attribute of a profession, its distinctive *organization*.

Organization

Professional Associations

Professional occupations have their own unique forms of *organization* and control, with the professional association, or society, being one obvious example. The professional associations constitute an expression of group consciousness and solidarity borne of members' common vocational experiences, interests, and aims. Their broader purpose is to strengthen and elevate the profession's status which they do through defining professional issues and priorities, maintaining standards of performance, and controlling access to the group. Associations seek to serve the internal needs of their professional members while also offering a united front to the various external interests and public entities who interface with the profession.

Typical association activities internal to the professional community promote *communication*: calling meetings, holding conferences and conventions, presenting papers on intellectual and professional concerns, taking unified action on matters of common interest, publishing technical journals – each of these serves to enhance communication among professionals. For "whatever else a professional association does as a collectivity, it provides formal (and persistently informal) means of communication among its constituents" (Moore 1970: 158). The professional associations seek strong —ideally, *full* — membership from the practitioners such that, externally, the associations may profess to speak for the collective interests of the group. By incorporating all or most practitioners in a common organization, professional associations are able to exercise influence over legislatures, training centers, and other components of the professional environment. Such association activities include public relations, publicity campaigns, educational efforts, and lobbying (Moore 1970: 158, 164-66).

Because the professional person is strongly oriented toward his or her peers, *even when he or she competes with them for clients and for public and professional prestige*, there exists a high potential for networking and establishing professional relationships, and the professional associations offer excellent opportunities for this to occur. Younger members have the opportunity to see the acknowledged leaders in their chosen field, and perhaps meet them informally. Rank and file members, elite practitioners, the isolated professional who has no colleagues in his or her department – all may participate in the business of their society through presenting, listening to, or discussing technical papers, sitting on committees, and the like, and as such may affirm their identity as representatives of their chosen profession (Moore 1970: 158-9).

The professional association offers a clear expression of the *professional culture,* which consists of its values, norms, and symbols (Greenwood 1957: 74-5). According to Greenwood, the values of the group are its "basic and fundamental beliefs, the unquestioned premises upon which its very existence rests." Commonly-held professional values include the essential worth of the profession's service to society, authority over clients, self-control, and the values of rationality and neutrality in the professional's science. Professional norms are its guides to behavior in social situations: professions have a system of role definitions that covers "every standard interpersonal situation likely to recur in professional life." The symbols of a profession reflect the culture and include such meaning-laden items as its "insignias, emblems and distinctive dress; its history, folklore, and argot *[i.e., technical jargon]*; its heroes and its villains; and its stereotypes of the professional, the client, and the layman." The idea of a professional culture can be expanded to view the profession as a *community* because its members are bound by a common identity (Ritzer 1977: 63). One of the association's more gratifying community functions is to recognize and honor its members for distinguished service and contributions to the profession and to society.

Apart from the preceding general discussion, the history and development of professional associations in engineering warrants specific mention. It is fair to say, given the nature of professional work, that the engineer who practices in the marketplace will face conflicts between "professional" and "business" interests. Edwin Layton's *The Revolt of the Engineers* (Layton 1971) masterfully exposes the inherent tensions between business and professionalism for engineers and describes how engineering professional associations have taken different views on this fundamental issue. He states that "The balance between business and professionalism has been one of the most important forces in the formation and evolution of engineering societies in America. Most American engineering societies represent compromises between business and professionalism" (Layton 1971: 25).

Monopoly and Licensing

If professional associations are the obvious case of professional organization and control, only slightly less obvious are the monopolization aspects of professional work. Professions seek the formal sanction of society – a *de facto* monopoly over their particular area of jurisdiction (Greenwood 1957: 71-2). Pointing to the legitimate authority of their knowledge and expertise as well as their ethic of professional service, the professions establish institutions that make society's judgments of secondary importance and the profession's judgments paramount (Hughes 1958: 141). Society grants this degree of power to professionals because they are persuaded that "…no one else can do the job and that it is dangerous to let anyone else try" (Goode 1969: 279). The profession's "official" monopoly exists through the profession's control over its training centers, use of the professional title, the professional license, and other privileges and powers.

I have already noted that the university, as both the producer of professionals and the producer of professional knowledge, is inextricably linked to professionalism. Professions are prominently involved in the process of higher education through the customary practice that both professional associations and individual practitioners participate directly in the accreditation process, thus informing the curriculum, instruction, and overall standards for professional education and training. This is a distinctive characteristic of the professions: "only one professional can train or judge another" (Ritzer 1972: 60). Another aspect of the profession's monopoly over knowledge and education is its use of *technical jargon*. The fact that professionals communicate in a sometimes "closed and esoteric vocabulary" perpetuates the art and mystery of their knowledge base and supports their monopoly of skill (Brown 1992: 21-22).

Legal controls also figure in the professions' monopoly. As a means of maintaining credibility and control, the professions hold that no one should be allowed to wear the professional title who has not been conferred it by the appropriate authorities (Greenwood 1957: 71). They seek legal protection of the professional title, particularly when the area of jurisdiction is not clearly exclusive (Wilensky 1964: 145). "Engineer" is a case in point: it is explicitly defined and can only be used as specified by law.

Professions seek monopoly over their jurisdiction through state licensing laws that regulate professional practice. Interestingly, this is one of the weaker forms of control since of necessity, the state only distinguishes the "…qualified from the unqualified, and it has not as a rule concerned itself with nice refinements or degrees of professional skill which might blur this fundamental distinction" (Carr-Saunders and Wilson 1933: 352). The profession's *standard of care* is a similarly defined legal matter:

> The law regards as a civil wrong, for which damages can be recovered by the injured party, the incompetent execution of any act where the relationship between the parties is such as to give rise to a duty to exercise skill and care; and the relationship of professional and client is one to which such a duty is attached… Moreover, … the degree of skill that must be shown is 'such as may be expected, in the circumstances of time and place, from an average person in the profession – one neither specially gifted nor extraordinarily dull' (Carr-Saunders & Wilson 1933: 394-5).

The point is that, given the esoteric nature of scientific expertise, it is important not to penalize *errors of opinion*, or to frighten practitioners into always "playing for safety," but instead to encourage professionals to apply their knowledge and skill toward solving admittedly difficult if not sometimes unsolvable problems (Carr-Saunders and Wilson 1933: 400). Thus, legal standards for professional practice only address the minimal level of competence. And although this is appropriate as a *legal* standard, practitioners *(in the functionalist era, anyway)* looked to norms developed by the professional group that were more stringent than those with a legal basis. "Thus, the real source of control over an individual professional lies in the hands of the profession, with society's (legal) control being weaker. This mechanism allows the profession to maintain its autonomy" (Hall 1969: 77). As we shall see, *autonomy* is the key indicator of professional control.

Professional Autonomy

The normal definition of autonomy is "…the quality or state of being self-governing; *especially* : the right of self-government" (Merriam-Webster 2000). For the professions this means that the profession considers *itself* the proper body to set the terms in which its particular aspect of society, life or nature is to be thought of, and to define the general lines and even the details, of public policy concerning it (Hughes 1965: 3). On the personal (attitudinal) level, autonomy is related to the feeling that the professional is free to exercise his or her judgment and discretion in professional practice (Hall 1969: 81).

Autonomy must not be confused with professional *authority*. Recall that authority is the professional's legitimate and justified power to act in the affairs of the client on matters within the professional's jurisdiction and competence. This authority derives both from the professional's honorable character and status and, more directly, the professional's mastery of and ability to apply scientific knowledge. Professional authority compels the client (who is in a state of dependency) to trust the professional in the belief that the professional can do the job and will look out for his or her interests. Therefore, "authority and trust go hand in hand" (Marquand 1997: 146) — the professional *has* to be trusted if he or she is to do their work (Goode 1969: 296). It is here that *autonomy* enters the scene: autonomy corresponds to the degree which trust *actually occurs* in the professional-client relationship. In other words, autonomy indicates how much the client (or society) actually believes the professions' authoritative claims. For example, theoretically at least, the client might completely trust the professional's skill and competence (at the individual level) as well as the larger professional structure including its development of theory, educational and training centers, licensing process, professional ideals, and the like (the occupational level). Such complete trust will translate into *full autonomy* for both the individual practitioner and the profession – they are *free to act*, with the only proviso being the expectation that they do what *they think* is best. But, for a number of reasons, the client might not fully trust in the ability of the profession or a particular practitioner to capably handle his or her affairs. In this case the client will grant only limited autonomy and will require other reassurances besides the word and acts of the professional. Of course, our concern here is with the ideal-typical profession where authority will be completely *valid*, trust will be *high* and, correspondingly, autonomy will be *full*.

One illustration of autonomy is that a true professional, according to the ideology of professions, is never "hired." Rather, he or she is retained, engaged, consulted, etc., by someone who has need of his or her services. Thus the professional has, or should have, almost complete control over what he or she does for the client (Hughes 1965: 9). Professional autonomy means *freedom* – freedom for the professions "to regulate themselves and act within their spheres of competence" (Wilensky 1964: 146), otherwise known as *financial* autonomy and *technical* autonomy, respectively. More explicitly, this is "...freedom *from* superordinate power, the right to make decisions and judgments about practice without instructions or interferences from above ...[and also] ...freedom *to* exercise power over subordinates" (Haug 1975: 207, emphasis in the original). The complexities of autonomy become apparent when one realizes that autonomy exists at both the occupational (structural) and the individual (attitudinal) levels, as well as both internally and externally to the profession. Different analysts have studied different aspects of this issue (Hall 1969: 81) (Perrucci and Gerstl 1969: 12) (Daniels 1973: 41, 52) (Freidson 1973: 33), but whatever the forum, freedom to self-regulate is the key:

> Students of the professions have pointed out that the autonomy granted to professionals who are basically responsible to their consciences (though they may be censured by their peers and in extreme cases by the courts) is necessary for effective professional work. Only if immune from ordinary social pressures and free to innovate, to experiment, to take risks without the usual social repercussions of failure, can a professional carry out his work effectively. ...The ultimate justification for a professional act is that it is, to the best of the professional's knowledge, the right act. He might consult his colleagues before he acts but the decision is his (Etzioni 1969: *x*).

Such freedom is not without obligations. The professions exercise great care to explain why professional autonomy is not a matter of *self-interest*, but is a requirement for offering the best possible service in the *public's interest*. These explanations reside in the professions' ethical codes which "...stand as the sign of the type of self-policing the professional group offers when justifying its desire for autonomy" (Daniels 1973: 45-46). Thus we now direct our attention to the third attribute of a profession, the *ethic of professional service*.

Ethic of Professional Service

The Service Ideal

As we consider the ethic of professional service, it will be helpful to briefly review the professional attributes discussed thus far. To wit, a professional is a person who owns expertise, who has mastered a body of knowledge including both practical know-how and esoteric theory. He or she usually gains this mastery through formal, university education and extended training. Such knowledge is the primary source of professional authority which legitimately places the professional in a position of control over the lay client. We noted that distinctive forms of organization and control characterize the professions, such organizational forms also reflecting the profession's unique expertise. Professions establish professional associations, they seek a monopoly to practice in their area of jurisdiction through licensing and related means, and most importantly, professions possess a high level of autonomy – freedom of self-regulation. We also introduced the key point that the nexus of authority and autonomy is *trust*, for autonomy exists only to the extent that vulnerable clients *believe* the authoritative claims of individual practitioners and the professions. The trust relationship places *moral obligations* on the professional, which brings us to our third attribute of a profession: the embodiment of these moral obligations has come to be known as "the ethic of professional service" or alternatively, "the service ideal."

The service ideal expresses the notion that "the technical solutions which the professional arrives at should be based on the *client's needs*, not necessarily the best material interests or needs of the professional" (Goode 1969: 278, emphasis mine). In other words, the service ideal obligates professionals to place their client's needs above their own and to perform well in arenas where they generally are immune to their client's oversight (Bachner 1991: *xii*) (Hall 1969: 75). In contrast to the norms of business – where commerce is among traders and *caveat emptor* (let the buyer beware) is the rule – the service ideal recognizes and accounts for the dependent and vulnerable position of the lay client and embodies a different, more genteel, rule: *credat emptor* (let the buyer trust) (Hughes 1965: 3).

We may ask, How did the service ideal come into being? Have professionals always had this obligation? In response, apart from much recent scholarship which takes a skeptical if not cynical view of the service ideal, two explanations have merit. One, that of Samuel Haber, places the origins of the service ideal in the honorable and dignified bearing of the landed gentry of eighteenth-century England (Haber 1991: *ix*). The second, that of Bruce Kimball, identifies the service ideal as a nuance derived from the first three of six rhetorical moments in the long history and usage of the term "profession." Thus, he traces the service ideal to the clergy of colonial America and what has been called the *Protestant ethic*, this in turn deriving from Christ's dialectical servant claims as recorded in the Gospels of Mark and Matthew (*see Mark 10:42-45, esp. vv.43-44: "...but whoever wishes to be great among you shall be your servant; and whoever wishes to be first among you shall be slave of all"* (Kimball 1992: 32-33, 103). The point is that the ethic of placing someone else's interests ahead of one's own, *especially* when that someone else is in a weaker or subordinate position, has been around for hundreds if not thousands of years. As regards the professions, this was viewed as fully consistent with the notion of professionalism and became subsumed by it, so much so that the terms "professional" and "ethical" have been used interchangeably (Greenwood 1957: 72).

As a norm of behavior, the service ideal provides the basis on which trust is erected and thus enables the *institutionalization of trust* between the client and the professional. This formalized protocol becomes particularly important in times of crisis when, as is often the case, the relationship between practitioner and client is among strangers (Perrucci and Gerstl 1969: 15). The service ideal is thus the preferred means to regulate professional-client relationships. Consider, for example, what would happen if the service ideal did not exist:

> The client is peculiarly vulnerable; he is both in trouble and ignorant of how to help himself out of it. If he did *not* believe that the service ideal were operative, if he thought that the income of the professional were a commanding motive, he would be forced to approach the professional as he does a car dealer – demanding a specific result in a specific time and a guaranty of restitution should mistakes be made. He would also refuse to give confidences or reveal potentially embarrassing facts. The service ideal is the pivot around which the moral claim to professional status revolves (Wilensky 1964: 140).

Aspiring practitioners internalize the attitudes and values embodied in the service ideal during professional training, this so that appropriate behavior becomes "natural" and external social controls will not be required later: "The professional is taught to monitor himself" (Daniels 1973: 43). Internalization of the service ideal at the individual level also sees expression as a group phenomenon: "…members care very much about each other's good opinion" (Daniels 1973: 43-44).

Although the service ideal is complex and has been viewed in different, not necessarily compatible, ways (Kimball 1992: 316), the key point for our purposes is that the service ideal embodies acceptable norms of behavior which the professions see as a form of internalized self-regulation. For many years *internalized* was indeed the operative word, for not until the dawn of the twentieth century did the professions expressly write out their ideals of service in the form of *codes of ethics*.

Regulative Code of Ethics

Between 1904 and 1922, practically every established profession in existence developed a code of ethics (Adams 1993) (Wilensky 1964: 143), civil engineering adopting theirs in 1914. Interestingly, even though the service ideal was clearly accepted as part of the professional ideology, codes of ethics did not follow as an obvious necessity and engineers at that time viewed them with deep ambivalence:

> There has been much discussion by engineers of the need of adopting a comprehensive code in order that the ideals of the profession may be presented clearly to the young engineer. On the one hand, these efforts have been scoffed at; indeed, in the case of one of the national engineering societies, it was "decided that no gentleman needed a code of ethics, and that no code of ethics would make a gentleman out of a crook" (Newell 1922: 133).

But beginning with electrical (in 1912), followed by mechanical and civil (both in 1914), these founder engineering societies adopted ethics codes (Layton 1971: 70), the goal being to outline generally approved ways of accomplishing the "universal" good (MacIver 1955: 52). "Through its ethical code the profession's commitment to the social welfare becomes a matter of public record" (Greenwood 1957: 72), and while specific details vary among the professions, the essential elements of codes of ethics are fairly uniform: they provide guidance and support for responsible engineers, establish shared minimum standards, and describe relationship norms between the professional and his or her colleagues, clients, and society (Martin and Schinzinger 1996: 106-7). Ethics codes are commonly understood as *not* self-enforcing. "The professional association, being nominally a society of equals, must adopt other procedures [to enforce discipline]; normally, these will rest primarily on an internal, quasi-judicial body commonly known as a "committee on ethics," which will review complaints and, if necessary, recommend disciplinary action" (Moore 1970: 116).

Among other things, codes of ethics place significant restrictions on the financial and competitive aspects of professional practice: they hold service to humanity as paramount and relegate financial gains or rewards to a subordinate consideration (MacIver 1955: 51). For example, the codes forbid specific forms of indirect remuneration because they might lead to a conflict between duty and self-interest (Carr-Saunders and Wilson 1933: 432). These and other restrictions affect the professional's relationships with both clients and colleagues. In the case of clients, the professional ought to give maximum caliber service: "The nonprofessional can dilute the quality of his commodity or service to fit the size of the client's fee; not so the professional" (Greenwood 1957: 73). Colleague relationships, in a similar vein, are to be "cooperative, equalitarian, and supportive" with intraprofessional competition being "...a highly regulated competition, diluted with cooperative ingredients which impart to it its characteristically restrained quality" (Greenwood 1957: 73). Furthermore, professional ideology holds that these types of ethical behaviors are "good business" and inherently practical: "A reputation for honesty and competence enhances the desirability of a practitioner to his clients and thus the rewards available to him in his professional career" (Daniels 1973: 44).

As would also be expected given the professions' organic tie to expertise, *competence* is not only a technical but an *ethical* issue. Competence goes beyond the minimum standards for admission to a profession and incorporates the maintenance and improvement of both personal and collective skills and practices. "Despite the patent difficulty of doing so in the contemporary world, the professional is supposed to keep current with developments in his field, so that his clients do not seriously suffer relative harm from his failure to do so... Competence is for a purpose: *conscientious performance"* (Moore 1970: 13-14, emphasis added). Such high demands and intense commitment to the profession has resulted in the professional career being termed a "calling."

Professional Career

In a manner similar to our discussion of the ethic of professional service, the idea of the professional career being a "calling" also has theological roots in the Protestant ethic of the seventeenth and eighteenth centuries. Simply put, a "calling" carries the idea that one's worldly vocation is the realization and fulfillment of one's spiritual vocation (Kimball 1992: 33-34). It enshrines "an avowal to a higher purpose" (Wittlin 1965: 91-92): "The very idea of professional *callings*, what Weber called a *beruf*, suggests that individuals who enter professions are called by inner promptings to provide some service to humanity, their country, or God" (Daniels 1973: 42-43).

That the professional career can be viewed as a "calling" carries significant implications. One is the intensive level of commitment involved, this being reflected in how the profession constitutes a lifelong career and elicits strong identification with the work (Perrucci and Gerstl 1969: 12-13). A profession is typically the terminal occupation for its members – the financial and temporal investment in the occupation is such that the trained professional typically does not leave the profession (Hall 1969: 77). Another implication of "calling" is the professional's dedication to the work and the feeling that he or she would probably want to continue in the occupation even if fewer rewards were available (Hall 1969: 81-82). Long years of preparation to enter the profession and long hours of practice to do the work of a profession are the norm (Moore 1970: 7-9). Identity is yet another: a person's work is one of the more important parts of his or her social identity and of the self (Hughes 1958: 43). Thus, "To the professional person his work becomes his life" (Greenwood 1957: 75).

Summary And Conclusions

Knowledge, organization, the ethic of professional service – these, then, are the attributes of a profession. As an ideal, they set the standard for what professionalism meant in the middle years of the twentieth century. I find this to be absolutely *fascinating:* in contrast with the ambiguity, tension and change inherent in today's professional world, that an era actually existed where professionalism could be defined in more or less certain terms is the sole reason I choose to call this period "the golden years." For there is much value in being able to understand and live out the practice and the ideals of professionalism. Perhaps you, like me, resonated with recognition and agreement as you encountered and thought about the functionalist attributes of a profession. Even today, these ideals can inform much of professional education and practice, for vestiges of them remain. But we must all be careful to realize that, indeed, times *have* changed. The professional ideals described herein apply to concepts of society, clients, and practitioners that no longer match reality, and in the case of engineering, perhaps some of them never did.

Thus the professional ideal presented in this paper is most properly understood as a classic – like the Glenn Miller Orchestra, the 55' Chevy, and Leave it To Beaver, it belongs to its own era. Notice, however, that I have nowhere said that the classic view of professionalism is bad. Far from it, I cannot think of a better way to introduce professional values to young engineers so that they may draw on this wealth of ideology. But this should only be the beginning; fresh and relevant concepts of professionalism must be articulated for our times – concepts that address the changes that have characterized society and engineering in the past fifty years. Of course, this paper does not deal with these developments – they will be a future topic. For now I will be satisfied that I have made a contribution if the engineer of the twenty-first century clearly understands professionalism in its golden years, and equally important, he or she realizes that times have changed.

ACKNOWLEDGMENTS

I thank Jimmy Smith and Bob Lockhart for their review and helpful comments on earlier drafts of this paper.

References Cited

Books and Monographs

Abbott, Andrew, *The System of Professions: An Essay on the Division of Expert Labor*, The University of Chicago Press, Chicago, IL, 1988, 435pp.

Bachner, John Phillip, *Practice Management for Design Professionals, A Practical Guide to Avoiding Liability and Enhancing Profitability*, John Wiley & Sons, Inc., New York, NY, 1991, 371pp.

Bledstein, Burton J., *The Culture of Professionalism, The Middle Class and the Development of Higher Education in America*, W. W. Norton & Company, Inc., New York, NY, 1976, 354pp.

Broadbent, Jane; Dietrich, Michael; and Roberts, Jennifer (editors), *The End of Professions? The Restructuring of Professional Work*, Routledge, London, 1997, 150pp.

Brown, JoAnne, *The Definition of a Profession: The Authority of Metaphor in the History of Intelligence Testing, 1890-1930*, Princeton University Press, Princeton, NJ, 1992, 214pp.

Carr-Saunders, A. M. and Wilson, P. A., *The Professions*, Oxford University Press, Oxford, Great Britain, 1933, 536pp.

Etzioni, Amitai (editor*), The Semi-Professions and Their Organization: Teachers, Nurses, Social Workers*, The Free Press, New York, NY, 1969, 328pp.

Freidson, Eliot (editor), *The Professions and Their Prospects*, Sage Publications, Beverly Hills, CA, 1973, 333pp.

Haber, Samuel, *The Quest for Authority and Honor in the American Professions, 1750-1900*, University of Chicago Press, Chicago, IL, 1991.

Hall, Richard H., *Occupations and the Social Structure*, Prentice-Hall, Inc., Englewood Cliffs, NJ, 1969, 393pp.

Hughes, Everett C., *Men and Their Work*, The Free Press, Glencoe, IL, 1958, 184pp.

Jackson, J. A. (editor), *Professions and Professionalization*, Cambridge University Press: London, 1970, 226 pp.

Kimball, Bruce A., *The "True Professional Ideal" in America, A History*, Blackwell Publishers, Cambridge MA, 1992, 429pp.

Larson, Magali Sarfatti, *The Rise of Professionalism, A Sociological Analysis*, University of California Press, Berkeley, CA, 1977, 309pp.

Layton, Edwin T., Jr., *The Revolt of the Engineers: Social Responsibility and the American Engineering Profession*, Press of Case Western Reserve University, Cleveland, OH, 1971, 286pp.

Lynn, Kenneth S. (editor), *The Professions in America*, Houghton Mifflin Company, Boston, MA, 1965, 273pp.

Martin, Mike W. and Schinzinger, Roland, *Ethics in Engineering, Third Edition*, The McGraw-Hill Companies, Inc., New York, NY, 1996, 439pp.

Moore, Wilbert E., *The Professions: Roles and Rules*, Russel Sage Foundation, 1970, 301pp.

Parsons, Talcott, *The Social System*, The Free Press, Glencoe, IL, 1951, 575pp.

Perrucci, Robert and Gerstl, Joel E., *Profession Without Community: Engineers in American Society*, Random House, New York, 1969, 194 pp.

Perrucci, Robert and Gerstl, Joel E. (editors), *The Engineers and the Social System*, John Wiley and Sons, Inc., New York, 1969, 344 pp.

Ritzer, George, *Man and His Work: Conflict and Change*, Meredith Corporation, New York, NY, 1972, 413pp.

Starr, Paul, *The Social Transformation of American Medicine*, Basic Books, 1982, 514pp.

Vollmer, Howard M. and Mills, Donald L. (editors), Professionalization, Prentice-Hall, Inc., Englewood Cliffs, NJ, 1966, 365pp.

Periodical and Journal Articles

Adams, Guy B., "Ethics and The Chimera of Professionalism: The Historical Context of an Oxymoronic Relationship," *American Review of Public Administration*, June 1993, Vol. 23, No. 2, pp. 117-139.

Barbar, Bernard, "Some Problems in the Sociology of the Professions", in *The Professions in America*, Kenneth S. Lynn (ed.), Houghton Mifflin Company, Boston, MA, 1965, pp. 15-34.

Broadbent, Jane, Dietrich, Michael and Roberts, Jennifer, "The End of the Professions?" in *The End of the Professions? The Restructuring of Professional Work*, Jane Broadbent, Michael Dietrich and Jennifer Roberts (eds.), Routledge, London, 1997, pp. 1-13.

Daniels, Arlene Kaplan, "How Free Should Professions Be?" in *The Professions and Their Prospects*, Eliot Freidson (ed.), Sage Publications, Beverly Hills, CA, 1973, pp. 39-57.

Etzioni, Amitai, "Preface," in *The Semi-Professions and Their Organization: Teachers, Nurses, Social Workers*, Amitai Etzioni (ed.), The Free Press: MacMillan Publishing Co., 1969, pp. *v-xviii*.

Freidson, Eliot, "Professions and the Occupational Principle," in *The Professions and Their Prospects*, Eliot Freidson (ed.), Sage Publications, Beverly Hills, CA, 1973, pp. 19-38.

Freidson, Eliot, "Method and Substance in the Comparative Study of Professions," Plenary Address by Eliot Freidson, Conference on Regulating Expertise, Paris, April 4, 1994, full text available on-line, http://itsa.ucsf.edu/~eliotf , retrieved 4/29/00.

Greenwood, Ernest, "Attributes of a Profession," *Social Work*, July 1957, pp.45-55, reprinted in *Ethical Issues in Engineering*, Deborah G. Johnson (ed.), Prentice Hall, Englewood Cliffs, NJ, 1991, pp. 67-77 (pagination in the text is from this reprint).

Goode, William J., "The Theoretical Limits of Professionalization," in *The Semi-Professions and Their Organization: Teachers, Nurses, Social Workers*, Amitai Etzioni (ed.), The Free Press: MacMillan Publishing Co., 1969, pp. 266-313.

Haug, Marie R. , "The Deprofessionalization of Everyone?" *Sociological Focus*, Vol. 8, No. 3., August 1975, pp. 197-213.

Harries-Jenkins, G., "Professionals in Organizations," in *Professions and Professionalization*, J. A. Jackson (ed.), Cambridge University Press: London, 1970, pp. 53-107.

Hughes, Everett C., "Professions," in *The Professions in America*, Kenneth S. Lynn (ed.), Houghton Mifflin Company, Boston, MA, 1965, pp. 1-14.

Jackson, John A., "Professions and Professionalization – Editorial Introduction," in *Professions and Professionalization*, J. A. Jackson (ed.), Cambridge University Press: London, 1970, pp. 1-15.

Marquand, David, "Professionalism and Politics: Towards a New Mentality?" in *The End of the Professions? The Restructuring of Professional Work*, Jane Broadbent, Michael Dietrich and Jennifer Roberts (eds.), Routledge, London, 1997, pp. 140-147.

MacIver, Robert, "The Social Significance of Professional Ethics," 1955 article reprinted in *Professionalization*, Howard M. Vollmer and Donald L. Mills (eds.), Prentice Hall, Englewood Cliffs, NJ, 1966, pp. 50-55.

Merriam-Webster's Collegiate Dictionary, s.v. "autonomy," www.m-w.com/cgi-bin/dictionary, accessed July 3, 2000.

Mok, Albert L., "Professional Innovation in Post-Industrial Society," in *The Professions and Their Prospects*, Eliot Freidson (ed.), Sage Publications, Beverly Hills, CA, 1973, pp. 105-116.

Newell, Frederick Haynes, "Ethics of the Engineering Profession," 1922 article reprinted in *Professionalization*, Howard M. Vollmer and Donald L. Mills (eds.), Prentice Hall, Englewood Cliffs, NJ, 1966, pp. 133-137.

Packer, James I., "Authority," in *New Bible Dictionary, Second Edition*, J. D. Douglas (ed.), Inter-Varsity Press, Leicester, England, 1982, pp. 108-109.

Russell, Jeffrey S., Stouffer, Brewer, and Walesh, Stuart G., "The First Professional Degree: A Historic Opportunity," *Journal of Professional Issues in Engineering Education and Practice*, ASCE, Vol. 126, No. 2, April, 2000, pp. 54-63.

Wilensky, Harold, "The Professionalization of Everyone?" *American Journal of Sociology*, Vol. 70, 1964, pp. 137-158.

Wittlin, Alma S. "The Teacher," in *The Professions in America*, Kenneth S. Lynn (ed.), Houghton Mifflin Company, Boston, MA, 1965, pp. 91-109.

PROFESSIONAL RESPONSIBILITY: THE ROLE OF ENGINEERING IN SOCIETY[1]

Steven P. Nichols, PhD, JD, PE
Professor of Mechanical Engineering
Director, Clint W. Murchison Chair for Free Enterprise
Associate Vice President for Research
The University of Texas at Austin
and
William F. Weldon, MSME, PE
Josey Centennial Professor Emeritus in Energy Resources
Department of Mechanical Engineering
The University of Texas at Austin

Abstract

We argue that the practice of engineering does not exist outside the domain of societal interests. That is, the practice of engineering has an inherent (and unavoidable) impact on society. Engineering is based upon relationships (inter alia).

An engineer's conduct (as captured in professional codes of conduct) towards other engineers, towards employers, towards clients, and towards the public is an essential part of the life of a professional engineer, yet the education process and professional societies ostensibly pay inadequate attention to the area. If one adopts Skooglund's definition of professional ethics (how we agree to relate to one another), then the codes of professional conduct lay out a road map for professional relationships. As professionals, engineers need to embrace their codes and to realize that they have personal stake in the codes as well as the process of developing the codes. Yet, most engineers view professional codes as static statements developed by "others" with little (or no) input from the individual engineer. Complicating the problem, questions of professionalism (such as ethics) are frequently viewed as topics outside the normal realm of engineering analysis and design. In reality, professional responsibility is an integral part of the engineering process.

[1]Originally published by Opragen Publications in *Science and Engineering Ethics* **3**, #3, pp 327-337, 1997. Repreinted here with permission.

I. Introduction

The essence of engineering is design.

B. V. Koen [1]

This article examines the relationship between engineers and society and engineer's professional responsibilities given that relationship. This examination is particularly important for engineers in the execution of their professional responsibilities and for students preparing to enter fields of engineering.

A review of literature yields a series of discussions on the definition of an engineer, descriptions of the design process, and of "what engineers do".[2] Articles and books also explore topics such as engineering professional ethics and legal aspects of engineering.[3] Recently, Davis presented a useful historical perspective of engineering in "Preface to Engineering Ethics".[4] Yet, differences exist as to what engineers do.[5] This article examines relationships between engineers and society but does not attempt to develop a new definition of engineering. The article does, however, touch on topics relevant to the definition of engineering.

---

[1] Koen, Billy V., "Toward a Strategy for Teaching Engineering Design", Journal of Engineering Education, July, 1994.

[2] See, for example, the following:
Koen, B.V., Definition of the Engineering Method, American Society or Engineering Education, Washington, D. C., 1985.
Walton, J. W., Engineering Design: From Art to Practice, West Publishing Company, 1991.
Ferguson, E. S., "How Engineers Loose Touch," Invention & Technology, Winter 1993.
Petroski, Henry, "Failed Promises," American Scientist, Vol. 82, January 1994.

[3] See, for example, the following:
Rabins, M., and Harris, E., "Controls, Risk, & Educational Responsibility: The Ethical/ Professional Links," Advances in Control Education, Tokyo, Japan, August 1, 1994.
Harris, E., Pritchard, M. Rabins, M. Engineering Ethics Concepts and Cases, Wadsworth Publishing Company, 1995.
Martin, M. and Schinzinger, R. Engineering Ethics, 2nd Ed., McGraw-Hill, New York, 1988.

[4] Davis, Michael, "An Historical Preface to Engineering Ethics," Science and Engineering Ethics, Vol. 1, No. 1.

[5] See, for example, the definitions for an engineer given in Spier (Spier, R. Science, Engineering and Ethics: Running Definitions, Science and Engineering Ethics, Volume 1, Issue 1, 1995, pg. 7 et. seq.) vs. the definition given by Koen (Koen, B.V., Definition of the Engineering Method, American Society or Engineering Education, Washington, D. C., 1985.)

II. The Role of Engineering in Society: Engineering Design

Some will say that I'm an academic and that I'm supposed to be a scientist, but I have this craving to be an engineer. Waldron.[6]

The National Research Council recently recognized the need for improved engineering design and engineering design education.[7] Although the literature holds numerous articles on engineering design,[8] the authors wish to concentrate on the interaction between engineers and society. In the course of exploring the relationship of engineering design to the balance of the engineering discipline, the authors developed a series of diagrams ostensibly representing interactions between engineering and society. The authors have found these diagrams useful in discussions with students and colleagues and hope that it will prove useful in continuing discussions.

One of the first sources of confusion, particularly with those that are not engineers or scientists, is the distinction between science and engineering.[9] The primary role of science is to develop knowledge and understanding of the physical universe.[10] As pointed out by Davis and others, an important distinction is that this pursuit of knowledge (science) may occur largely without regard to societal need (or to societal

---

[6] Waldron, K.J., "Secret Confessions of a Designer." Mechanical Engineering, November, 1992.

[7] National Research Council, Improving Engineering Design: Designing for Competitive Advantage, National Academy Press, Washington, D. C., 1991.

[8] For discussions on engineering design with an emphasis on engineering design education, see the following articles:

American Society of Mechanical Engineering, Innovations in Engineering Education: Resource Guide, ASME, 1993.

Chaplin, C., "Creativity in Engineering Design—The Educational Function," The Education and Training of Charted Engineers for the 21st Century, A Study Undertaken for the Fellowship of Engineering 2 Little Smith Street, Westminster, London, November, 1989.

Efatpenah, K. Nichols, S., Weldon, W., "Design in the Engineering Curricula: A Changing Environment, Advances in Capstone Education: Fostering Industrial Partnerships, August 3-5, 1994.

Nichols, S., "The Mechanical Engineering Design Projects Program: An Experience in Industrial/University Cooperation," Innovations in Engineering Design Education: Resource Guide, ASME, 1993.

Pugh, S., Total Design, Addison Wesley, 1991.

[9] For an interesting discussion of the distinction between scientist and engineers, see Davis, Michael, "An Historical Preface to Engineering Ethics", Science and Engineering Ethics, Volume 1, No. 1, 1995.

[10] This definition is consistent with definitions by other authors such as Spier, Raymond, "Science and Engineering Ethics: Running Definitions", Science and Engineering Ethics, Volume 1, No. 1, 1995.

implications).[11] The direction of science research as been described by some as curiosity based research and is not necessarily driven by the values of society. Societal values (and resulting priorities) do not necessarily define the bounds or direction or scope of scientific curiosity.[12] This is not a criticism of science, for such is the nature of "inquiring". Furthermore, it is often not possible to determine relevance of a particular field of scientific inquiry to the future needs of society.[13] Given this curiosity driven process, the base of scientific knowledge about the physical universe may be represented by an amoebae-like structure uneven in its extent in the various directions with current scientific research efforts acting to extend its coverage (see Figure 1).

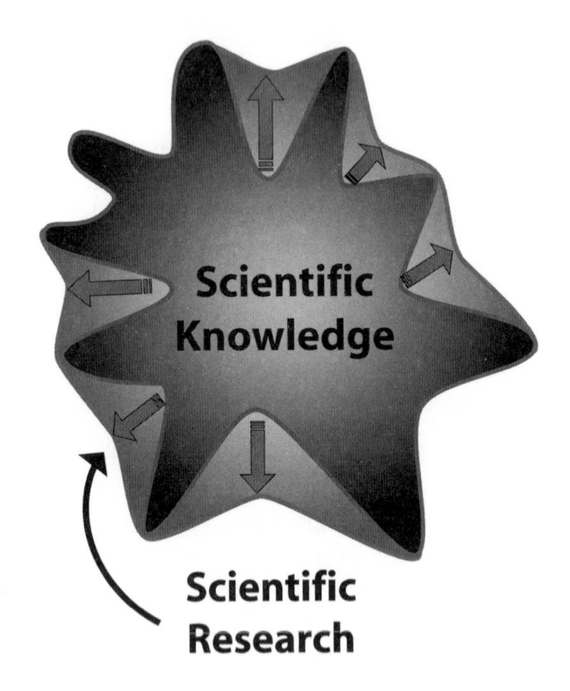

Figure 1.

[11] Davis supports this with a demonstration of the interests of scientists vs. engineers. See Davis, "An Historical Preface to Engineering Ethics, Science and Engineering Ethics", Vol. 1, No. 1, Pg. 42-43.

[12] One could pursue many areas of science without regard to societal implications. One may contrast the pursuit of science with the pursuit of law which is entirely value based. Free of the values of society, one could argue that there can be no legal profession.

[13] This does not mean, of course, that professionals in the pursuit of sciences do not encounter questions of ethics. Much of science, indeed, does have close ties to society and society. We argue that engineering, by its very nature, <u>must</u> address topics associated with societal values.

The utilization of scientific knowledge over time establishes that some of the knowledge is immediately relevant to societal needs while other parts are less immediately relevant.[14] While the congruence of societal need with scientific knowledge is much more complex than indicated in this article, it may be represented for the purpose of this discussion by a Venn Diagram as seen in Figure 2. The authors maintain that it is this overlap of scientific knowledge with societal need, more specifically, the application of scientific knowledge to the needs of society, that is the domain of engineering (inter alia).[15] Clearly, the extent of human enterprise is much more complex than is represented here. If, for example, it is in the interest of society to increase our store of scientific knowledge, then engineers and scientists who ply their trade in the frontiers of scientific research are both serving societal need. Nevertheless, our contention is that the central focus of the engineering profession is the application of scientific knowledge to meet societal needs.

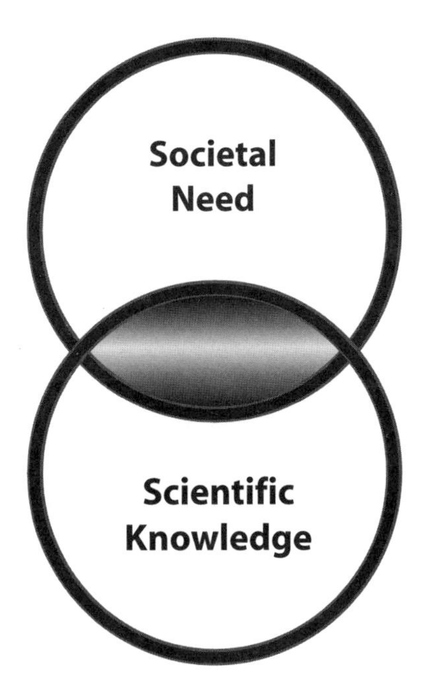

Figure 2.

This analogy can be extended by superimposing the distinction of the creative versus the analytical aspect of the human enterprise.[16] We can represent this aspect of the human intellect by another Venn Diagram shown in Figure 3. As indicated in the diagram, one may pursue efforts without involving analytical skills, and one may

[14] Society may never realize the relevancy of a particular scientific inquiry.

[15] This is more fully discussed below.

[16] See, Left Brain, Right Brain; Springer, S.P., 1993.

apply analytical skills without entering the domain of creativity. For example, as engineers apply commercial software to the solution of an engineering problem, the application of analytical skills, per se,[17] may involve little or no creativity.

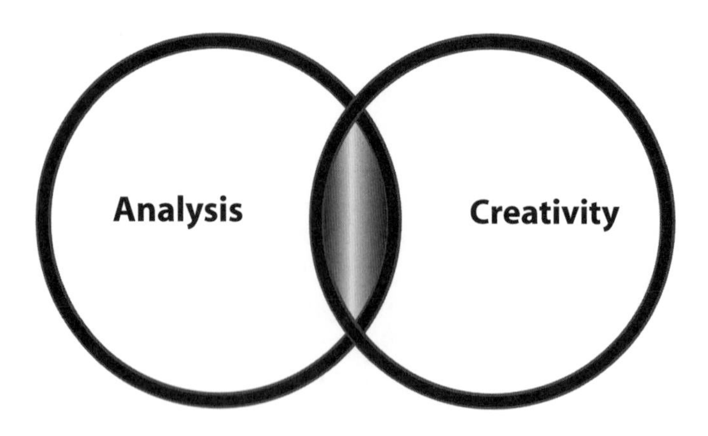

Figure 3

One may superimpose these two Venn Diagrams and use the resulting diagrams to examine engineering enterprise as shown in Figure 4.

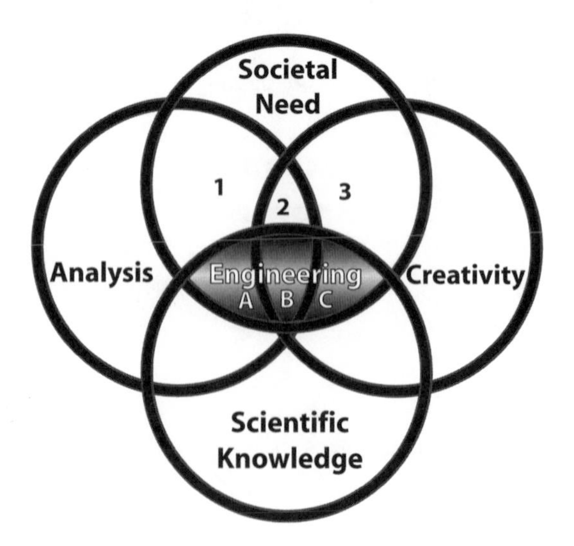

Figure 4

[17] The engineer may use creativity, of course, in how the analytical results are applied or interpreted.

Considering the intersection of scientific knowledge with societal need (designated as the domain of engineering), the authors will discuss three sectors, shown as A, B, and C.

Sector A may be used to represent the intersection of analytical talents with the engineering domain. This may be used to represents engineering science, an ability to model complex systems and predict their response to various inputs under various conditions.[18] This segment of engineering has, of course, been the subject of intense development over the last half century and has benefited most directly from the availability of fast digital computers.

Sector C, the intersection of our creative capacity with the engineering domain, can be viewed as representing those sudden intuitive leaps often responsible for revolutionary advances in technology[19] as well as those aspects of engineering, not yet fully supported by engineering science, that remain more art than science.

The third sector, B (the intersection of knowledge and need with both creative and analytical capability) can be used to represents engineering design and much "real world" problem solving. This sector could include activities ranging from developing innovative products and processes, to creating an innovative bridge design, to developing a new control process for petrochemical production. This vision of engineering design as the intersection represented by Section B is consistent with statements expressed by Pahl and Beitz,[20] Dixon,[21] and Penny.[22]

Current approaches to teaching used in engineering schools have been designed more for developing analytical skills (Sector A) than creative skills.[23] The Accreditation Board for Engineering and Technology (ABET) identifies engineering as "that profession in which knowledge of the mathematical and natural sciences gained by study, experience, and practice is applied with judgment to develop ways

---

[18] The Accreditation Board for Engineering and Technology states that "engineering sciences have their roots in mathematics and basic sciences but carry knowledge further toward creative application. These studies provide a bridge between mathematics and basic sciences on the one hand and engineering practice on the other." Criteria for Accrediting Programs in Engineering in the United States, Section IV.C.3.d.(3), 1995-96.

[19] Called "significant novelty" by Spier, Spier, R. Science, Engineering and Ethics: "Running Definitions, Science and Engineering Ethics", Volume 1, Issue 1, 1995, pg. 7.

[20] Pahl, G., and Beitz, W., Engineering Design: A Systematic Approach, (Edited by Wallace, K.), The Design Council, London, Springer-Verlag, London, Paris, Tokyo, 1988.

[21] Dixon, J.R., Design Engineering: Inventiveness, Analysis and Decision Making, McGraw-Hill, 1966.

[22] Penny, R.K., Principles of Engineering Design, Postgraduate 46, 344-349, 1970

[23] National Research Council, Improving Engineering Design: Designing for Competitive Advantage, National Academy Press, Washington, D. C., 1991.

to utilize, economically, the materials and forces of nature for the benefit of mankind " (emphasis added). ABET further recognizes that "a significant measure of an engineering education is the degree to which it has prepared the graduate to pursue a productive engineering career that is characterized by continued professional growth".[24] One can conclude that analytical skills are essential tools for engineers[25] but are not sufficient for a complete engineering education. An education which only uses classroom problems in which all variables are accurately known and only one correct answer exists not only misrepresent the situations engineers encounter in their jobs, but also do little to stimulate creativity. A trend toward using open-ended problems in the engineering science and analysis classroom is a healthy step in the direction of more complete and relevant engineering education.

This four-circle representation of human endeavor (Figure 4) is useful in the perspective it offers for other enterprises as well. Sector 1, the intersection of analytical skills with societal needs outside the bounds of scientific knowledge might include economics and philosophy while sector 2 may encompass the arts. Sector 3 may be used to represent those societal needs outside the bounds of scientific knowledge that required both analytical and creative skills, perhaps including public policy, business administration and music.

The view of engineering presented in this paper differs from the view of "method" presented by Koen, and the notion of "significant novelty" presented by Spier. Spier argues, "There is a product that results from the activity of an engineer" and interprets the term "product" broadly (we presume to include processes). Our emphasis, however, is not on the product, but the engineer's interaction with society.

III. Professional Responsibility and Engineering Ethics

"The rationale for teaching ethics to engineers and computer scientists seems fairly obvious. Their work (developing, designing and implementing technologies) has an enormous impact on the world."

Johnson[26]

Given an engineer's inherent interaction with society and societal needs, one may understand the importance of an engineer's responsibility to society. Since the Grinter report, engineering education has made significant progress in strengthening the basic sciences in engineering, including mathematics, chemistry, and physics.[27,28]

---

[24] Section IV.C.2, Criteria for Accrediting Programs in Engineering in the United States, Engineering Accreditation Commission, ABET, 1995-96.

[25] One could argue, that it is the use of these mathematical and natural sciences and other analytical tools which partially differentiate engineers for other problem solving professions.

[26] Johnson, Deborah G., "Why Teach Ethics in Science and Engineering," Annual Meeting of the American Association for the Advancement of Science Seminar "Teaching Ethics in Science and Engineering 10-11 February, 1993. The comments also appeared in Science and Engineering Ethics, Vol. 1., No. 1.

Recent trends toward increasing discussion of professionalism in the classroom notwithstanding, topics of professional responsibility (as compared to sciences, engineering sciences, and engineering analysis) have received surprisingly little attention in engineering education over the last several decades.[29] The authors fear that professional responsibility may also have been underemphasized in the practice of engineering.

The authors include in the area of engineering "professional responsibility" the following topics:

-Safety and Welfare of the Public and of Clients,

-Professional Ethics,

-Legal Liabilities of Engineers,

-Environmental Responsibilities,

-Quality, and

-Communications

Each of these topics relates to the interaction of an engineer to clients, society, employers, employees, and to the engineering profession. Regarding engineering ethics, Whitbeck argues that engineers should study engineering ethics from the perspective of a moral agent as opposed to a moral judge.[30] We fully subscribe to this approach not only for teaching engineering ethics, but also for teaching (and practicing) in other areas of professional responsibility. For engineers, engineering ethics is not a topic separate from engineering, it is part of the essence of engineering as it pertains to the professional responsibilities that the engineer has with society. The results of an NSF sponsored workshop on engineering ethics in the classroom observed techniques from engineering design methodology to address ethical dimensions of engineering problems, designs, and interactions.[31] One may consider

---

[27] Grinter, L. E. (ed.), "Report on Evaluation of Engineering Education," Journal of Engineering Education, 1952.

[28] See Efatpenah, K., *et al.*, Design in the Engineering Curricula, Advances in Capstone Education Conference, Provo, Utah, August 1994.

[29] "Engineering Ethics in the Classroom," Engineering Ethics in Engineering Education, Vivian Weil (editor), Center for the Study of Ethics in the Professions, Illinois Institute of Technology, Chicago, Illinois, 1992. Proceeds from an NSF sponsored workshop in professional ethics.

[30] Whitbeck, Caroline, "Teaching Ethics to Scientists and Engineers: Moral Agents and Moral Problems," Science and Engineering Ethics, Vol. 1, No. 3, 1995.

numerous engineering design methodologies which will illustrate the point.[32] Pugh, for example, includes the following elements in the "engineering design core":[33]

-Understanding the Market (problem definition: societal need)
-Design Specification (specifying the needs)
-Concept Design
-Detail Design
-Manufacture
-Sell

Pugh's methodology focuses on product design, but also has applications in process design and general problem solving. Experienced engineers would not logically delay consideration of economic issues until after completion of detail design. That would not allow the engineer to consider economic and performance tradeoffs that are essential in the overall evaluation of alternative designs to be analyzed in the Concept Design element. It is just as important that engineers first approach ethical, safety, liability, environmental, quality, and communications issues in the first step of the design process, rather than allowing the design to proceed without regard to these issues. This allows engineers to address and analyze each element of the problem from the problem statement to the release of the product or service to the customer. This allows engineers to integrate (naturally) the consideration of ethical and other concerns directly into the design process and to expand the alternative designs to potentially eliminate or reduce problems rather than simply react to the problems.

This article started with a quote stating that "the essence of engineering is design." The Accreditation Board for Engineering and Technology (ABET) defines design as follows:

> *Engineering design is the process of devising a system, component, or process to meet desired needs. It is a decision making process (often*

[31] "Engineering Ethics in the Classroom," Engineering Ethics in Engineering Education, Vivian Weil (editor), Center for the Study of Ethics in the Professions, Illinois Institute of Technology, Chicago, Illinois, 1992. Proceeds from an NSF sponsored workshop in professional ethics.

[32] Problem solving and design methodologies may be found in texts such as Prasad, B. Concurrent Engineering Fundamentals, Printice Hall, 1995, Jones, J.V., Engineering Design: Reliability, Maintainability and Testability, TAB Professional and Reference Books, 1988, Walton, J., Engineering Design: From Art to Practice, West Publishing Company, 1991, and many others.

[33] Pugh, S., Total Design, Integrated Methods for Successful Product Engineering, Addison-Wesley Publishing Company, 1991

iterative), in which the basic sciences and mathematics and engineering sciences are applied to convert resources optimally to meet a stated objective. ... ***it is essential to include a variety of realistic constraints, such as economic factors, safety, reliability, aesthetics, ethics and social impact.***[34] (emphasis added)

ABET's definition of design involves engineering activities which include open ended problems. These activities include machine design, product and process engineering, manufacturing engineering, and applications engineering. This broad definition of design includes most of engineering activities involving societal interaction. Due to their interactions with society, engineers assume the responsibility inherent in such interactions. ABET's definition of design acknowledges that relationship of engineering to society in the recognition of "realistic constraints" in the design process (remember that "design is the essence of engineering"[35]) The National Research Council also recognized the importance of engineering in society.[36] Yet engineers frequently give limited attention to the codes which guide their interaction with society.[37] Skooglund proposes that professional ethics describe "how we agree to relate to one another".[38] This pragmatic definition of professional ethics can be useful in examining how engineers view their codes.

Development of course material[39] in the last decade have allowed engineering degree programs to expand course offerings in fields of professional responsibility. Additionally, faculty have developed problems for analytical courses which include issues of professional responsibility. (See Broom and Peirce, "The Heroic

---

[34] Criteria for Accrediting Programs in Engineering in the United States, Criteria iv.C.3.d.(3)(c), Effective for Evaluation During the 1995-96 Accreditation Cycle, Engineering Accreditation Commission, Accreditation Board for Engineering and Technology, Inc. 111 Market Place, Suite 1050, Baltimore, Maryland 21202.

[35] Note that the authors are combining Koen's statement (see footnote 1) with ABET's definition of design. It is not clear that Koen would agree with ABET's definition of design. The author's are comfortable with the combination of the approaches.

[36] See footnote 7.

[37] Vandenburg, W. H., and Khan, N., "How Well is Engineering Education Incorporating Societal Issues", ppg. 357-61, Journal of Engineering Education, Vol. 83, No. 4, October 1994.

[38] Skooglund, C., El Paso Faculty Workshop On Ethics & Professionalism, Texas State Board of Registration for Professional Engineers and the Murdough Center for Engineering Professionalism, April 15-16, 1993. This is one in a series of workshops developed by Professor Jimmy Smith to provide focus and experience to faculty members in the integration of the discussion of ethics into the engineering curriculum.

[39] See for example, Martin, M. and Schinzinger, R. Ethics in Engineering, McGraw Hill, 2nd Edition, 1988, and Harris, C, Pritchard, M., and Rabins, M, Engineering Ethics, Concepts and Cases, Wadsworth Publishing Company, 1995.

Engineer".)[40] These developments supporting an engineer's ability to address areas of professional responsibility are encouraging. The authors still believe that engineering programs currently are releasing far too many engineers (who do not understand their professional responsibilities to society) on an unsuspecting public. Just as troublesome, we are releasing the public upon (unnecessarily) unsuspecting engineers and engineering graduates. Observations by Vandenburg and Khan[41] support these concerns. As stated in Engineering Education for a Changing World, "... engineering colleges must not only provide their graduates with the intellectual development and superb technical capabilities, but following industry's lead, those colleges must educate their students to work as part of teams, communicate well, and understand the economic, social, environmental and international context of their professional activities."[42]

Engineers need to develop a fundamental understanding of their professional responsibilities. Few engineers have an opportunity, however, to develop or contribute to the development of a professional code of ethics. As a result, engineers are in danger of viewing codes of ethics as static codes dictated by "others" for engineering applications. Compare this to the process by which attorneys in the United States develop professional codes regulating their conduct. State bars and their members develop and periodically review their professional codes of conduct. Statewide debate about the codes can be heated and can produce significant discrepancies from state to state in rules of professional conduct. One should expect these discussions to become heated, since these codes describe how professionals (attorneys) will relate to clients, courts, the public, and other attorneys. At the end of the review process, the code describes how the parties will "relate to one another" (using Skooglund's terminology). Partially due to the process used to develop and review their codes of professional conduct, attorneys tend to internalize these codes.

The authors do not suggest that the engineering profession model itself after the legal profession,[43] but rather we suggest that engineers examine and adopt "best practices" in development of rules of professional conduct which encourage engineers to understand and internalize their professional codes. Engineers need to develop a fundamental understanding of their professional responsibilities. Students at (at

---

[40]Broome, T. H. and Pierce, J., "The Heroic Engineer", p. 51 et seq., Journal of Engineering Education, Vol. 86, No. 1 January, 1997.

[41] "Given current economic, social and environmental trends and policies, the study shows cause for deep concerns ... ", Vandenburg, W. H., and Khan, N., "How Well is Engineering Education Incorporating Societal Issues", ppg. 357-61, Journal of Engineering Education, Vol. 83, No. 4, October 1994.

[42] American Society for Engineering Education, Engineering Education for a Changing World, A Joint Project the Engineering Deans Council, and Corporate Roundtable of the American Society for Engineering Education, 1994.

[43] In fact, substantive differences from state to state have some serious drawbacks.

least) one engineering college have developed their own codes of conduct (how they will relate to one another and the college) for their academic career.[44] This experience gives the students a personal approach to professional codes of conduct necessary in the engineering profession. These students had an opportunity to integrate their "professional code" into their daily work as engineering students. This allows students to internalize their professional responsibilities and to develop a fundamental understanding of their obligations and resulting consequences. Students at other universities and the engineering profession would be well served to learn from the experiences of these students who developed their own code.

Since it is difficult for every practicing engineers to participate in the development of national professional codes, it may be better to localize this experience for professional engineers. This can be done by developing codes for conduct at company, division, or departmental levels in traditional engineering environments. Texas Instruments and Bell Helicopter have had positive experiences developing company codes which are intended to describe how professionals "agree to relate to one another.[45]

We suggest, however, that the most effective mechanism is the personal involvement of each engineer in integrating the topics of safety and welfare of the public, professional ethics, legal aspects, environmental responsibilities, quality, and communications into the methodologies which that engineers use to approach and solve problems in the ordinary course of practice.[46,47] The healthy debate among engineers (as well as clients and employers) which should naturally arise in the integration and the application of the methodologies will serve to underscore the nature and importance of the role that the engineer has in society (health, safety, and welfare of the public), the role the client has in engineering design (economics, reliability, maintainability, and other associated topics of quality), the effects of engineering activity on society, and the relationship of society to engineering activities.[48]

---

[44] See for example, "Student Code of Ethics", Civil Engineering Department, University of New Mexico, Revision 041393, 1993.

[45] Personal conversations with Carl Skooglund, Texas Instruments.

[46] The integration of topics of professional responsibilities into traditional engineering methods has been discussed earlier in this article.

[47] This could be considered a natural extention of "Concurrent Engineering" in which the elements of design, manufacturing, and other issues are considered concurrently in engineering methodology. The concurrent methodology would include design for manufacturing, design for reliability, design for maintainability, design for assembly, design for environment, design for safety, design for economics, etc. This supports the apporach taken by Pugh in his concept of "Total Design". Some elements of integration has been imposed by regulation (such as environmental regulation).

[48] This includes product liability, protection of intellectual property, environmental regulations, etc.

Some Thoughts About Good Engineering

by
Taft H. Broome, Jr. Sc.D., Fellow AAAS
Professor, College of Engineering, Howard University
25 May 2004

Let us think about the term "good engineering," and let us think about it in the company of engineers. This is to presume that we agree to think about good engineering in anticipation of talking about it with engineers.

Perhaps first in mind is the noun "engineering," which for engineers may refer to the theory or to the practice of engineering. For us, engineering theory connotes the learned discipline of engineering that we associate with university teaching and research. For philosophers, it is engineering praxis or, more precisely, engineering praxiology. For us, engineering practice connotes activity that is informed by principles and methods of the learned discipline of engineering.

Perhaps second is the adjective "good," which has meaning only when taken together with "bad." For engineers it may suggest a technical sense of the term "good engineering," or it may suggest a non-technical sense, or both senses. For us, a technical sense refers to the tradition of engaging the learned discipline of engineering separately from ethics, or without respecting a primacy of ethics:

> The safety of the public shall be the highest law. (Cicero)[1]

Engineering analysis can be performed with or without conscious notice of ethical constraints, and engineering design can be carried out with ethical, economic and other constraints each enjoying the same status with the designer. A non-technical sense might suggest professional constraints, such as codes of professional ethics, or a broader array of constraints enabled by the learned discipline known as engineering ethics. Can engineering always be good in technical as well as non-technical senses?

> A prince who desires to maintain his position must learn to be not always good, but to do so or not as needs may require.
>
> (Machiavelli, *The Prince*)[2]

On Sunday morning, December 28, 1986, a piece I had written appeared in the Outlook section of the *Washington Post*.[3] The piece was entitled "The Slippery Ethics of Engineering." In it, I maintained that engineers should subordinate the business interests of their employers to the interests of the public in the general health and safety. But, I argued that engineers could not always assure the general health and safety, so they should seek general acceptance of the risks associated with engineering. The piece was syndicated by the *Los Angeles Times News Service*, and was thereby reprinted nationally as well as abroad. Most readers were comforted merely by the revelation that engineers think about ethics. Judging by the copyright privileges requested by Kinko's and others seeking to supply college students with the piece, I concluded that it was generally embraced by the scholarly community. Responses from the engineers, however, were mixed and impassioned. All seemed agreed that *what* I said was correct, but most seemed agreed that I *should not* have said it.[4]

Another category of thinking about good engineering has more to do with the engineer than with his or her engineering. Michael Pritchard would tell us that good engineering is that which is done by the good engineer, and Mike would say that the good engineer is good in the moral sense.[5] Similarly, I argued that good engineering is that which is done by the engineer of good character, the difference being that character is a judgment of others about the engineer. Specifically, I was interested in the kind of character that would inspire others to approve of one's work, even when proven lethal, and that would inspire them to encourage him or her to try again after failure.[6]

Another thought belonging to this category focuses on competence.

> Machiavelli wrote only for petty Princes. (Frederick II-The Great)[7]

Here "petty" does not imply that the Prince has limited powers of forceful action; rather, it is meant in senses that assume both incompetence and character defects in the Prince. Recalling that competence is a condition in the codes of ethics of perhaps most engineering professional societies, we make the point that engineering is as good as it gets whenever the decent, competent engineers are in charge and emulated by the other engineers.

> I never heard any man praised whose courage was not paired with decency.[8]

The good engineer does not devalue parties affected by their work in order to do technically good work, nor turn from technical work simply because of its potential lethality. Rather, he or she is capable of doing engineering while observing the highest standards of technical *and* ethical work. Fallible, the good engineer neither covers up nor blames others for error, but is capable of what is known in basketball as "the good foul":

> Two minutes are left in the game, you are down by one point, and the opponents have possession of the ball. So, you foul the opponent: not in a harmful way; and in clear sight of the referee. If the opponent misses the free throw, you will get the ball with enough time to win the game.

This is the engineer who comes forward when in error, who blows the whistle on himself, so to speak. Thus, we have the engineer who has the power to put others to risk but who can be trusted to use that power wisely.

The worst engineer is easily imagined but not so easily anticipated. Of course, incompetence and bad character make a dangerous combination. ABET policies reflect a growing concern among employers that the workforce suffers a competency problem, and there remains concern that the USA suffers from anti-intellectualism and a measure of moral decay. I have considered the problem of self-interest in the USA, and focused on a character type that can be anticipated. In a hate speech episode, he is neither the taunter nor the taunted, nor an intervener, but an on-looker.[9] His type is the first of the damned that Dante and Virgil encountered after passing through the Gates of Hell:

> There sighs, weeping, loud wailing resounded through the starless air, for which at the outset I shed tears.
>
> Strange languages, horrible tongues, words of pain, accents of anger, voices loud and hoarse, and sounds of blows with them,
>
> made a tumult that turns forever in that air darkened without time, like the sand when a whirlwind blows.
>
> And I, my head girt with horror, said: "Master, what is this I hear? and what people is this who seem so overcome by grief?"
>
> And he to me: "This wretched measure is kept by the miserable souls who lived without infamy and without praise.
>
> They are mixed with that cowardly chorus of angels who were not rebels yet were not faithful to God, but were for themselves…" (Dante)[10]

There exists a remedy for this type of character in educational psychology; it is called character development, and its aim is to prevent adolescence in adulthood. My own prescription was what Joseph Campbell called the hero's journey. And there are ways to incorporate that journey into the engineering problem-solving classroom.[11]

Surely more thinking about good engineering can be done, but enough of it has been done to share an achievable vision of an engineering community that is good inasmuch as its leaders are competent and of good character.

References

[1] Cicero, 45 BC. *De Legibus*, bk. 3, ch. 3, sec. 8. (See: Swainson, ed. 2000. *ENCARTA Book of Quotations*. New York, NY: St. Martin's Press, p. 212.)

[2] Machiavelli, N. 1531. *Discourses on the First Ten Books of Titus Levy*. (See: Swainson, ed. 2000. *ENCARTA Book of Quotations*. New York, NY: St. Martin's Press, p. 592.)

[3] Broome, T.H. 1986. The Slippery Ethics of Engineering. *Washing Post*, "Outlook" section, Sunday, December 28, pp. D2-4.

[4] Broome, T.H. 1995. STS Calling. *AWIS Magazine*, vol. 24, no. 6, pp. 12-13.

[5] Pritchard, M. Unpublished manuscript.

[6] Broome, T.H. 1999. Case Study: The Concrete Sumo. *Science and Engineering Ethics*, vol. 5, issue 4, pp. 542-547. This paper, its Commentaries, and my Reply are introduced by Pritchard, M. p. 541. Commentaries: Weil, V., Understanding the "Concrete Sumo," pp. 548-549; Pritchard, M., Broome's "Concrete Sumo," pp. 550-553; Herkert, J., Concrete Ethics, pp. 554-555; Davis, M., Wrestling With Broome's Concrete Sumo, pp. 556-564. Broome, TH., Reply to Commentaries, pp. 565-567; October.

[7] Frederick II. (P. Sonnino, transl. 1981) 1740. *Anti-Machieval*, 13. (See: Frank, ed. 1999. *Quotenary*. New York, NY: Random House, p. 476.)

[8] Eschenbach, W. 1980. (Hatto, A.T., transl.) *Parzifal*. New York, NY: Penguin, p. 178.

[9] Broome, T.H. 1996. The Heroic Mentorship. *Science Communication: An Interdisciplinary Journal of the Social Sciences*, vol. 17, no. 4, pp. 398-429; June.

[10] Dante, *Inferno*, Canto XIII, seventh circle, second sub-circle: the violent against themselves.

[11] Broome, T.H. and Peirce, J. 1997. The Heroic Engineer. *J. of Engineering Education*, vol. 86, no. 1, pp. 51-55; January.

Great Achievements and Grand Challenges

By Wm. A. Wulf, President of the National Academy of Engineering

This article is a revised version of the talk Dr. Wulf gave 22 October 2000 at the NAE Annual Meeting; Reproduced with permission of Dr. Wulf

Poised as we are between the twentieth and twenty-first centuries, it seems to me that this is the perfect moment to both reflect on the accomplishments of engineers in the last century and ponder the challenges facing them in the next.

Great Engineering Achievements of the 20th Century

This past February, working with the engineering professional societies, the NAE selected the 20 greatest engineering achievements of the twentieth century. The main criterion for selection was *not* technical "gee whiz," but how much an achievement improved people's quality of life. The result is a testament to the power and promise of engineering to improve the quality of human life worldwide.

Reviewing the list, it's clear that if *any* of its elements were removed our world would be a very different place—and a much less hospitable one. The list covers a broad spectrum of human endeavor—from the vast networks of the electric grid (no. 1) to the development of high-performance materials (no. 20). In between are advancements that have revolutionized virtually every aspect of the way people live (safe water, no. 4, and medical technologies, no. 16); the way people work (computers, no. 8, and telephones, no. 9); the way people play (radio and television, no. 6); and the way people travel (automobile, no. 2, and airplane, no. 3).

In announcing the achievements, former astronaut Neil Armstrong noted that, "Almost every part of our lives underwent profound changes during the past 100 years thanks to the effort of engineers, changes impossible to imagine a century ago. People living in the early 1900s would be amazed at the advancements wrought by engineers." He added, "As someone who has experienced firsthand one of engineering's most incredible advancements—space exploration—I have no doubt that the next 100 years will be even more amazing." Given the immediacy of their impact on the public, many of the achievements seem obvious choices, such as the automobile and the airplane. The impact of other achievements are less obvious, but nonetheless introduced changes of staggering proportions. The no. 4 achievement, for example, the mechanisms to supply and distribute safe and abundant water, together with sanitary sewers, literally changed the way Americans lived and died during the last century. In the early 1900s, waterborne diseases like typhoid fever and cholera killed tens of thousands of people annually, and dysentery and diarrhea, the most common waterborne diseases, were the third largest cause of death. By the 1940s, however, water treatment and distribution systems devised by engineers had almost totally eliminated these diseases in America and other developed nations.

Engineering is all around us, so people often take it for granted. Engineering develops consumer goods, builds the networks for highway, air, and rail travel, creates innovations like the Internet, designs artificial heart valves, builds lasers for applications from CD players to surgical tools, and brings us wonders like imaging technologies and conveniences like microwave ovens and compact discs. In short, engineers make our quality of life possible. The NAE's full list of engineering achievements, with an expanded explanation of each item, can be found on the Web at www.greatachievements.org. The short form of the list appears below:

1. Electrification—Vast networks of electricity provide power for the developed world.
2. Automobile—Revolutionary manufacturing practices made cars more reliable and affordable, and the automobile became the world's major mode of transportation.
3. Airplane—Flying made the world accessible, spurring globalization on a grand scale.
4. Water Supply and Distribution—Engineered systems prevent the spread of disease, increasing life expectancy.
5. Electronics—First with vacuum tubes and later with transistors, electronic circuits underlie nearly all modern technologies.
6. Radio and Television—These two devices dramatically changed the way the world receives information and entertainment.
7. Agricultural Mechanization—Numerous agricultural innovations led to a vastly larger, safer, and less costly food supply.
8. Computers—Computers are now at the heart of countless operations and systems that impact our lives.
9. Telephone—The telephone changed the way the world communicates personally and in business.
10. Air Conditioning and Refrigeration—Beyond providing convenience, these innovations extend the shelf-life of food and medicines, protect electronics, and play an important role in health care delivery.
11. Highways— 44,000 miles of U.S. highways enable personal travel and the wide distribution of goods.
12. Spacecraft—Going to outer space vastly expanded humanity's horizons and resulted in the development of more than 60,000 new products on Earth.
13. Internet—The Internet provides a global information and communications system of unparalleled access.
14. Imaging—Numerous imaging tools and technologies have revolutionized medical diagnostics.
15. Household Appliances—These devices have eliminated many strenuous, laborious tasks, especially for women.
16. Health Technologies—From artificial implants to the mass production of antibiotics, these technologies have led to vast health improvements.

17. Petroleum and Petrochemical Technologies—These technologies provided the fuel that energized the twentieth century.
18. Laser and Fiber Optics—Their applications are wide and varied, including almost simultaneous worldwide communications, noninvasive surgery, and point-of-sale scanners.
19. Nuclear Technologies—From splitting the atom came a new source of electric power.
20. High-performance Materials—They are lighter, stronger, and more adaptable than ever before.

Challenges for the 21st Century

So much for the achievements of engineering in the twentieth century; now let's look forward to the challenges of the twenty-first. I am an optimist. I believe 2100 will be "more different" from 2000 than 2000 was from 1900. I believe that the differences will bring further improvements in our quality of life, and that these improvements will be extended to many more of the people on the planet! But that is a belief, not a guarantee—and there are profound challenges twixt here and there. Some of those challenges are reflected in the NAE program initiatives: megacities, Earth systems engineering, technological literacy of the general public, and so on. Rather than talk about all of these challenges, I want to talk in depth about just one. It's a challenge that I haven't written or spoken about yet, that I believe may be the greatest challenge for the twenty-first century, that I want to start an NAE program on, and that I want to begin a dialogue with you about.

The challenge is *engineering ethics!*

Let me start by being clear that I believe engineers are, on the whole, very ethical. Indeed, ethics is a subject of great concern in engineering, reflecting the profession's responsibility to the public. There are ethics courses at many engineering schools. There is a bewildering array of books on the subject. Every engineering society has a code of ethics—most start with something like "… hold paramount the health and welfare of the public."[1] These codes typically go on to elaborate the engineer's responsibility to clients and employers, the engineer's responsibility to report dangerous or illegal acts, the engineer's responsibility with respect to conflicts of interest, and so on.

Beyond the codes are the daily discussions that occur in the work of engineering. I have vivid memories of discussions with my father and uncle, with my professors, and with many colleagues—about everything from design margins to dealing with management pressure to cases where tough choices had to be made.

All of that is still in place. It's part of why I am proud to be an engineer!

So—why do I want to talk about engineering ethics? Why do I believe it may be the greatest challenge of the twenty-first century? Why do I think we need to start an

NAE program activity on the topic? The reason is the confluence of two intertwined issues—briefly: (1) Engineering is changing; specifically, it is changing in ways that raise new ethical issues, and (2) these new issues are, I believe, "macro-ethical" ones that are different in *kind* from those that the profession has dealt with in the past.

The literature on engineering ethics, the professional society codes, and the college ethics courses all focus on the behavior of *individual* engineers; these have been called "micro-ethical" issues. The changes I will discuss pose new questions for the *profession* more than for the individual; such issues are called "macro-ethical" ones.

The challenge is *engineering ethics!*

In medicine the individual ethical issues are very similar to those in engineering. But, in addition, there are many macro-ethical issues. For example, the individual medical doctor cannot and should not make broad policy decisions about "allocation"—who should receive scarce organs for transplant, or doses of a limited stock of medicine, or even the doctor's attention when there are more ill than can be accommodated. The profession, or better, society *guided* by the profession, needs to set these policies.

Several things have changed to create these new macroethical questions in engineering, but I am going to focus on one—complexity. Moreover, I will focus specifically on complexity arising from the use of information technology and biotechnology in an increasing number of products. The key point is that we are increasingly building engineered systems that, because of their inherent complexity, have the potential for behaviors that are impossible to predict in advance.

Let me stress what I just said. It isn't just *hard* to predict the behavior of these systems, it isn't just a matter of taking more into account or thinking more deeply—it is *impossible* to predict all of their behaviors.

There is an extensive literature on engineering failures—the Titanic, Three Mile Island, etc. Engineering has, in fact, advanced and made safer, more reliable products because it has been willing to analyze its failures. I found two books on such failures particularly interesting: *Normal Accident*s, by Charles Perrow (1985, Basic Books[2]) and *Why Things Bite Back,* by Ed Tenner (1997, Vintage Books).

I found them interesting because of the progression in thinking in the 13 years between them about why systems fail and what engineers should do about it. For Perrow, the problem is that we don't think about multiple failures happening at once in "tightly coupled systems"—and the clear implication is that the solution is to think about them! For Tenner, there is a beginning of a glimmer that very complex systems have behaviors that are *really* hard to predict. But one still gets the feeling that if we just thought about it harder, if we just thought in the larger context in which the system is embedded, we would anticipate the problems.

Perrow and Tenner are not engineers—they are a historian and a sociologist—and they use the tools of their disciplines to analyze why failures happen. Mathematics isn't one of those tools, and so they are unlikely to have encountered the technical explanation I am about to give you. And, of course, they are partly right about the earlier failures they analyze—those systems may not yet have crossed the threshold beyond which prediction is impossible.

Over the last several decades a mathematical theory of complex systems has been developing. It's still immature compared to the highly honed mathematical tools that are the heart of modern engineering, but one thing is very solid—*a sufficiently complex system will exhibit properties that are impossible to predict a priori!*

I said the theory was "immature"—unfortunately, it also carries some undeserved baggage. The term used for these unanticipated behaviors is "emergent properties," a term that originally arose in the 1930s in "soft" sociological explanations of group behaviors. Some postmodern critics of science have also tried to use the work to discredit reductionist approaches to scientific research. Despite this baggage, there are solid results, and the impossibility (or "intractability," to use a more technical term) is one of them.

I don't want to get technical, but I need to give you a flavor of why I say "impossible." Consider the question of why software is so unreliable. There are many reasons, but one of them is not "errors" in the sense that we usually use the term. In these cases the software is doing exactly what it was designed to do; it is running "to spec." The problem is that the implications of the specified behavior were not fully understood because there are so many potential circumstances, and the software designers simply couldn't anticipate them all. Not didn't, but *couldn't*! There are simply too many to analyze!

Let me just give you an idea of the magnitude of the numbers. The number of atoms in the universe is around 10^{100}. The number of "states" in my laptop, the configuration of 1s and 0s in its memory, is about $10^{10,000,000,000,000,000,000}$. That's just the number of states in the primary memory, and doesn't count those on the disk.

If every atom in the universe were a computer that could analyze 10^{100} states per second, there hasn't been enough time since the Big Bang to analyze all the states. When I say that predicting the behavior of complex systems is impossible, I don't mean that there isn't a process that, given enough time, could consider all the implications—

So, that's what has changed. We can, and do, build systems not all of whose behaviors we can predict. We do, however, know that there will be some such unpredicted behaviors—we just don't know what they will be. The question then is: How do we *ethically* engineer when we know this—when we know that systems will have behaviors, some with negative or even catastrophic consequences—but we just don't know what those behaviors will be? Note that it wouldn't be an ethical question if we didn't anticipate that systems would have these negative properties. Ethicists and the courts alike have long held that if an engineer couldn't reasonably know the consequences of his or her actions, that's okay. But here we know! So how should we behave? How should we "engineer"?

A concrete example is the programmatic theme the NAE has embarked on—Earth Systems Engineering. Clearly the ecosystem, our planet, is not fully understood, and is a very complex, interconnected system. It's a clear example of a system where "everything is connected to everything." Every action will have an effect on the whole, albeit perhaps not a large one in most cases (but we have many examples where we thought that an action wouldn't have a large negative impact, but it did). It's a system where, even if we did understand all the parts, we would not be able to predict all of its behaviors.

Moreover, we must recognize that the Earth is already a humanly engineered artifact! Whether we consider *big* engineering projects, as in the proposed restoration of the Everglades, or simply paving over a mall parking lot that happens to feed an aquifer vital to a community hundreds of miles away, we have changed the planet.

Consider the case of the Everglades—either we do something or we don't; both are conscious acts. Either way, knowing that we can't predict all of the consequences, how do we proceed ethically? How do we behave? How do we choose? Clearly these are deep issues, and issues for the whole profession, not the individual engineer. The kind of ethics embodied in our professional codes doesn't tell us what to do.

This spring, Bill Joy, cofounder and chief engineer of Sun Microsystems, raised a somewhat related, but different, issue. In what I thought was an irresponsibly alarmist article in *Wired* magazine (8.04), Joy mused that the interaction of information technology, nanotechnology, and biotechnology would lead to self-replicating systems that would "replace" human beings. He then raised the question of whether we should stop research on some or all of these technologies. I abhor the way that Joy raised the question, but I think we have to deal with the fact that something like it is at the root of the public's concerns over cloning, genetically modified organisms, etc. We are meddling with complex systems; how can the public be assured that we know all of the consequences of that meddling?

I am personally repelled, however, by the notion that there is truth that we should not know. I can embrace the notion that there are ways we should not *learn* truth, research methods we should not use—the Nazi experiments on humans, or perhaps even fetal tissue research, for example. I can embrace the notion that there are unethical, immoral, and illegal ways to *use* our knowledge. But I can't embrace the notion that there is truth, knowledge, that we should not know.

It's ironic that the first academies in the seventeenth century were created because science, this new way of knowing truth, was not accepted by the scholastic university establishment. Even more than a hundred years later, Thomas Jefferson was making a radical assertion, when, in founding the University of Virginia, the first secular university in the Americas, he said, "This institution will be based on the illimitable freedom of the human mind. For here we are not afraid to follow the truth wherever it may lead ..."

That's the spirit of the pursuit of knowledge that I teethed on. Yet here I am in the Academy asking whether there is truth we should not know.

Alas, I also have to admit that the history of the *misuse* of knowledge is not encouraging. I do not know the answer to Joy's question, but it is also a macroethical one; it is not an issue for each of us individually. You might reasonably ask why we engineers need to ponder this as *our* ethical question? It's because science is about discovering knowledge; engineering is about *using* knowledge to solve human problems. So, while I can't bring myself to agree with the implied answer in Bill Joy's question, I do believe it raises a deep question for engineers about the use of knowledge.

How should we behave to ensure proper use of knowledge? Again, it's a question for the profession, not the individual. While an individual engineer perhaps should object to improper use of knowledge, such an act by itself will not prevent misuse. We need a guideline.

Conclusion

I could give other examples of new macroethical issues that engineering must face, but let me just summarize.

Engineering — no *engineers* — have made tremendous contributions to the quality of life of citizens of the developed world. There have been missteps, and there is much to be done even to bring the benefit of today's technology to the rest of the world. But I am unabashedly optimistic about the prospects for further increasing our quality of life in the twenty-first century and for spreading that quality of life around the globe.

However, that is not guaranteed. There are significant challenges, and, in fact, those challenges are not a bad operational definition of what the NAE program should be. One of these challenges, and perhaps the greatest one, is a class of macroethical questions that engineers must face. There are many such issues, but I chose two to illustrate the point.

Projects such as the further modification of the Everglades will be done with imperfect knowledge of all of the consequences. They *should* be done with the certainty that some of the consequences will be negative—perhaps even disastrous. At the same time we do not have the luxury of "opting out." Not to act is also an action—so we must address the question of what constitutes ethical behavior under such circumstances. Does the current nature of the engineering process support, or even allow, such behavior?

A separate but related question is how we ethically use the increasing knowledge we have of the natural world, and the power that knowledge gives us to modify nature—which I think is the substantive question raised by Bill Joy's article.

Both of these are questions on which society must give us guidance—our professional codes do not address them. But we must raise the issue and provide society with the information to help it decide, and we had better do it soon!

I happened on a quote from John Ladd, emeritus professor of philosophy at Brown, that captures part of the point I have tried to raise. He said, "Perhaps the most mischievous side effect of [ethical] codes is that they tend to divert attention from the macro-ethical problems of a profession to its micro-ethical ones." Our ethical codes are *very* important, but now we have another set of issues to address. Let's not let our pride in one divert us from thinking hard about the other.

Notes

[1]This particular wording is from the National Society of Professional Engineers code, but many others are derived from it and use similar language.

[2]Charles Perrow released an updated edition of *Normal Accidents* in 1999 (Princeton University Press)

Tomorrow's Engineer — What Do We Really Need?

By Joe Paul Jones, P.E.[1]

Former Vice President, Freese & Nichols, Inc., Ft. Worth, Texas
Past President, National Society of Professional Engineers

I am pleased to have the opportunity to share some thoughts with you on the characteristics that are really needed in tomorrow's engineer.

The subject is timely with me since NSPE has been attempting to develop a vision of needs in the next century with its Task Force 2000. We started this study in order to determine how our professional society should be molded to serve tomorrow's engineers. This, of course, demanded a systems look at our profession and mushroomed into examination of a model that has three sub-models: education, registration, and organization, all of which must be reactive to the individual engineer and the environment in which he or she will work.

...our profession is changing at a dizzying pace.

The one certain fact is that our profession is, and has been throughout my career, changing at a dizzying pace. The obvious answer that I must give to what we really need for tomorrow's engineer is that they must be prepared for change. They must be more than a technical engineer, and they must have a sensitivity for the world that surrounds them.

Let's develop this in an understandable way. I have had a long career in engineering. It has been challenging, interesting, and at times trying, but always rewarding. Rewarding both financially, and most importantly, by giving a feeling of satisfaction only obtained when we know that we've done something worthwhile.

During the past 40 years, I have moved from using a slide rule to a hand-operated desk calculator, then electrical, and on to a $300 electronic calculator that is now surpassed by a pocket size machine that costs less than $10. Finally, a P.C. that was doubled in capacity and reduced in price by another, within a year of its delivery.

Why reflect back? Because these catastrophic changes in our tools and methods of engineering have been accelerating at an exponential, not linear rate. Quite obviously, this change will continue and, if examined properly, may give some clues as to what we really need in tomorrow's engineers.

[1]Presented at the Fourth Annual Murdough Symposium, 1991

But first, another aspect of the engineering profession that has changed drastically during my career is the need for a professional to be far more than a technician. We live today in an era of total quality management, of competition from firms throughout the world, and of clients that have a wide range of choice of engineers and have been able, therefore, to make their selection based on whom they like to work with, and who will most likely make them look good to their clients.

This means that today's, and most certainly tomorrow's, engineer needs to be a much broader person. This should be started during the educational foundation building days, and continually developed throughout the engineer's professional life.

I would like to spend several minutes discussing the preparation of a young person for an engineering career in this rapidly changing field and then, briefly, look at the challenges that must be met during that career in order to be successful. Since it is difficult, if not impossible, to predict the materials, methods, and techniques of engineering that we will experience in the year 2000, perhaps a much greater emphasis during college years should be placed on flexibility, management, communications, human relations, ethics, finance, marketing, and political awareness, along with the basic engineering knowledge.

This sounds like a 200 + hour curriculum doesn't it? Believe me, that too would be a major error. Somehow we must, within 4 to 6 years of college, filter out those who are not meant to be engineers and give those who remain the fundamentals upon which to build after college.

Perhaps the most important knowledge to be mastered in engineering school is not statistics, dynamics, or advanced mathematics, but how to think in a logical, methodical manner.

Since there will still always be a need for in-depth technical solutions, there may be a strong argument for two distinctly different courses to be taught in our engineering schools; one for professional engineers who will become the practitioners, and one for the technologist who will be dealing primarily with scientific facts. In medicine, this compares with the doctor and the specialist, or even the laboratory technician.

Today, there are schools that give a Bachelor of Technology (B.E.T.) as one route, and a Bachelor of Science (B.S.) as the other. However, the B.S. is generally so concentrated on technical courses that it is really technology. My degree was certainly this way. Please note that this change will not come easy since most teachers of engineering are highly skilled in technology and limited in recent experience with real world problems.

What can we do to prepare young people who wish to study engineering for a truly professional career?

Today, college technical training becomes outdated, or at least partly useless in five years or so. This cycle repeats itself again and again. I mentioned the rapid obsolescence of the tools used during my relatively short career.

My bookshelf is also still filled with college texts that are about as useful as the slide rule. The basic principles are there, but the methods of their use is entirely different today. The college classroom notes that I carefully saved were, likewise, of little or no use a few years into my career.

Another aspect of obsolescence, as a result of education and even on-the job training, is that of being too highly specialized. In 1970, there were major layoffs in the aerospace industry as there are again today. General Dynamics in Fort Worth has released 7,000 employees in the last 6 months, 1,000 of whom were engineers. Engineers who had spent their entire careers in wind tunnel testing, structural design of a radar antenna for fighter planes, etc., are suddenly jobless.

Many have left the field of engineering forever in order to support their families. Others went into a skills conversion program to learn a completely different area of engineering practice. Their college training had been so narrow and deep in one technical area, followed by a career that did the same, that they were not armed to cope with a changing job situation, any more than the Graham, Hudson or Studebaker automobile manufacturers were able to survive changing demands in the automobile field.

What can be done to prepare today's students as tomorrow's engineers?

1. Concentrate technical training on properties of materials, how they perform under environmental and structural stresses, why they react the way they do, and finally, how they should be used in problem solving.
2. Cover basic sciences, basic and advanced mathematics, and basic engineering principles as essential foundations for a career, <u>emphasizing the thought processes</u> that must be utilized in their practical applications.
3. Teach courses in how to think and reason logically <u>using real world problems as exercises</u>.
4. Include economics, some sociology, some basic psychology, management and leadership skills, ethics and within all of these, <u>require both oral and written work until communication becomes second nature</u>.

This describes a much broader and less specialized curriculum than is found in most colleges today. However, I feel that it, realistically, would prepare the young person for a highly successful career in the engineering field of their choosing.

What else does a young engineer need to think of in order to be the best prepared for tomorrow?

Without question, continuing education is essential. If what you learned in college is continually becoming obsolete, then you must learn its replacement.

Is there no end to school and studying? ... That's right!

However, it is much easier when your learning is directly related to the work that you do on the job. When you stop learning, you had better be prepared to retire. Perhaps the non-technical side of a career needs the same degree of continuing education effort.

Tomorrow's professional engineer must be prepared to keep himself or herself broad by working with many other people, both professionally and socially. In the profession, they should work and associate with engineers from other areas of employment in their particular field.

For instance, civil engineers should belong to and be active in ASCE. All professional engineers should support and work in their professional society, NSPE, in order to develop breadth and to see the viewpoint of others. (If you went to Texas A&M, for instance, consider being friendly with at least one non-Aggie, even if you don't trust him at first.) Join a service club, such as Rotary or Kiwanis.

Be active in a youth organization, such as Boy Scouts or Camp Fire. Serve on a public board or commission, help the United Way and your church, work on a political campaign (maybe even your own), and do all of these in enough depth to have fun and be of true service to others.

Baden Powell, who founded of the Boy Scouts over 80 years ago, said that true happiness only comes from helping others. You will find that in doing this, you will gain much more than you give. Do these things and you will continually broaden, and as result be alert to our changing society.

One final suggestion for achieving success is to assign a young engineer a mentor or big brother or sister, if you will.

Most of the valuable things that I know today, I've learned since graduation. Much of it was acquired in working with older, wiser heads that had already made their own myriad of mistakes. They also had the wisdom to let me make many of my own, but caught me before the fall was fatal.

One thing should be remembered. There will never be enough people in the world that have a foundation of broad-based education, that have applied themselves in depth to their job, no matter what it is, that have gone the extra mile to help their fellow man, that have been active in civil, professional, and technical societies, and that have always served their employer by doing more than asked. Jobs may come and go, but a person who has "gone the extra mile" will always be in demand. Give me 50 of them, and I will take on any engineering firm in the world.

In closing, let's again review the characteristics that tomorrow's engineer should have.

They are:

- ❋ Flexibility,
- ❋ Common sense,
- ❋ Education in fundamentals of engineering,
- ❋ The desire to always learn,
- ❋ People skills, and
- ❋ A willingness to be more than a technical person, be a "true professional".

Ethics in Emerging Technologies and the Role of Case Studies and Ethical Theories in Ethics Education

by
Jimmy H. Smith, Ph.D., P.E.[1]

What emerging topics are good candidates for on-line presentation?

The following are emerging topics in which professional ethics is considered vital:

1. Unethical Use of Internet Information (making it easy, quick, and cheap to cheat)
2. Personal National Identification System (privacy vs. national security)
3. The Changing Nature of the Engineering Profession
   (e.g., McDonaldization of Engineering – see following article by William Lawson)
4. Relationship of ABET requirements to ethics education
5. Micro-Ethics – (Related to the behavior of individual engineers)
6. Macro-Ethics – (Related to the behavior of the engineering profession)
7. The following engineering areas are becoming more than just very complex. In many instances, predicting the behavior of some engineering systems is not just "hard" – it is "impossible" to predict all of their behaviors. Thus, ethics must play an important role in the development and use of these complex technical systems. [2]
   a) Computer Software Development and Evaluation
   b) Bio-Engineering (Cloning)
   c) Bio-Medical Engineering (Technology-based medical tools)
   d) Pharmaceutical Engineering
   e) Communications (right to privacy)
   f) High-performance Materials
   g) Earth Systems Engineering

These are a few of the emerging ethical issues in engineering, where for some at least, the term "emerging" may relate to both ethics research and to ethics education.

[1] Director, National Institute for Engineering Ethics, Murdough Center for Engineering Professionalism, Texas Tech University, Lubbock, Texas

[2] "We can, and do, build systems not all of whose behaviors we can predict. We do, however, know that there will be some such unpredicted behaviors—we just don't know what they will be. The question then is: How do we *ethically* engineer when we know this—when we know that systems will have behaviors, some with negative or even catastrophic consequences—but we just don't know what those behaviors will be? Note that it wouldn't be an ethical question if we didn't anticipate that systems would have these negative properties. Ethicists and the courts alike have long held that if an engineer couldn't reasonably know the consequences of his or her actions, that's okay. But here we know! So how should we behave? How should we 'engineer'"? (See the article by Wm. Wulf, President, National Academy of Engineering, in this book.)

The McDonaldization of Engineering

by
William D. Lawson, P.E.[1]

The "McDonaldization of Engineering" is part of a broad social trend that has been identified and documented by contemporary sociologist George Ritzer in his book, *The McDonaldization of Society: An Investigation into the Changing Character of Contemporary Social Life* (1996). Building on the work of classic social theorist Max Weber, the McDonaldization thesis can be applied to trends familiar in engineering; for example, rampant commoditization and standardization of the engineering process. The point is that despite their many benefits, such activities also have a *downside*:

> McDonaldization … offers increased efficiency, predictability, calculability, and control... It also offers many more specific advantages in numerous settings. Despite these advantages, [there are] drawbacks to McDonaldization… Rational systems inevitably spawn a series of irrationalities that limit, eventually compromise, and perhaps even undermine their rationality (Ritzer 1996: 121).

McDonaldization exists in engineering in various ways, both external and internal to the profession. For example, on the external side, the concept of efficiency comes to mind. Engineers obviously take pride in efficiency and one of the core benefits engineers strive to deliver to society is the creation of more efficient systems – they may be highways, airports, heating and cooling systems, information networks, you name it – that is what engineers do.

But as applied to McDonaldization, efficiency assumes a different form: an efficiency that is externally imposed on engineers by society in the form of increased standardization, regulations, rules and procedures. This can get out of hand, or to use Ritzer's terms, it can "result in the irrationality of rationality."

Non-engineers who retain engineers to solve their engineering problems (societal leaders such as city councils, governmental agencies, asset managers, owners, etc.) can, in the name of increased efficiency, hamstring their engineering professionals to the extent that these professionals cannot do their work well. "One-size fits all" regulations, practice standards that discourage innovation, the focus on cost control of the engineering process at the expense of its intended goals – all of these hold strong potential to produce highly irrational results from a seemingly rational process.

---

[1]Lecturer, Senior Research Associate and Deputy Director, National Institute for Engineering Ethics, Murdough Center for Engineering Professionalism, Texas Tech University

A similar example, internal to the profession, is the control of engineering professionals by management through accounting models. A clear trend exists toward greater control of professional engineering work by outsiders. References in the economics and professionalism literature describe how accounting procedures increasingly govern professional engineering work, causing it to be *proletarianized*.

In absolute terms, no engineering happens that cannot be explained by the numbers – the financial model intended to monitor the work becomes the *de facto* and unyielding structure for all work activities. Even at face value, this is an extreme constraint to place on an entirely predictable task. It reaches the level of "incredible" when applied to professional work.

The Ethical Problem of Moonlighting
Questions and Answers

by
William D. Lawson, P.E.[1]

1. What are some things about moonlighting that might make it unethical for engineers?

Certainly one basic concern is that engineers who moonlight might not be giving the primary employer their best effort during normal work hours; that is, fatigue, distractions, and responsibilities from the second job might be detrimental to their regular work. Other ethical concerns raised by moonlighting have to do with issues of unfair competition, the potential for non-approved disclosure of business affairs and technical processes of the primary employer, and unauthorized use of the primary employer's equipment, supplies, and laboratory or office facilities to do outside work.

Further, and perhaps more significantly, outside employment holds the potential to create a conflict of interest between the employer and outside clients. It is for this reason that the NSPE Code of Ethics has, since its inception, contained language to the effect that "engineers shall not accept outside employment to the detriment of their regular work or interest. Before accepting any outside engineering employment, they will notify their employers." *NSPE Code of Ethics, Section III.1.c.*

2. What could a person do to make it ethical, or to avoid a conflict of interest when moonlighting?

The NSPE Code of Ethics section cited above offers sound advice: when an engineer is contemplating outside employment, he or she ought to notify the primary employer, fully and openly disclose the situation, and get explicit, preferably written, approval *before* doing the work. By all means the employee should comply with company policy.

Under many circumstances, such disclosure would be enough to allow moonlighting given the approval of both parties. However, not everything that is allowed is necessarily prudent. Engineers and their employers do well to carefully consider the risks and liabilities that can accrue to them (or their company) simply by virtue of association, especially if the outside employment involves engineering work.

For example, several years ago a friend of mine, who was employed by a national engineering firm in a large city, also did private, on-the-side consulting for residential clients. This he did with the explicit written approval of his primary employer, which they granted largely on the basis of policy which stated that the company did

---

[1]Lecturer, Senior Research Associate and Deputy Director, National Institute for Engineering Ethics, Murdough Center for Engineering Professionalism, Texas Tech University

not do any residential work – there would be no unfair competition. However, it so happened that one of my friend's on-the-side residential clients became dissatisfied and claimed damage, and this client not only sued my engineer friend but also his primary employer, the national engineering firm. Granted, naming the national engineering firm in the lawsuit was unfounded and an obvious grab for deep pockets, but it cost that firm a lot of effort and money to prove this in court. After the ordeal was over, that firm's policy about moonlighting became more restrictive.

All this to say, in addition to ethical concerns, moonlighting holds the potential for unfounded but nevertheless costly legal entanglements.

3. What is your opinion on doing work on the side in engineering? Do you think it is ethical, or do you think it is disloyal to one's company to do other work on the side?

The NSPE Board of Ethical Review (BER) has considered the ethics of moonlighting on various occasions. Cases on point are 62-19, 64-2, 97-1, 99-3, and 02-8. While the earlier cases get into fee competition issues which no longer are discussed in today's NSPE Code, the ethical reasoning of all these cases remains instructive for us. In sum, the BER does not forbid moonlighting, they simply require that it be done under full disclosure with prior approval, and with the proviso that no other ethical obligation is breached.

You raise the question of whether moonlighting is disloyal. This is a difficult question to answer, partly because loyalty in and of itself is such a complex subject. But if one assumes that engineers are free to have their own interests, develop their own relationships, and make their own commitments, all within appropriate boundaries, then I do not think moonlighting, on the face of it, would necessarily be an act of disloyalty.

4. Do you know of any state board laws or rules governing moonlighting practices?

The NSPE Code of Ethics section cited above is instructive here. Further, most state board laws include a section on "professional conduct and ethics" which addresses many of the same considerations as the NSPE Code, thus making compliance not only a matter of ethics but also a matter of law. For example, the Texas Engineering Practice Act includes explicit language to the effect that "The engineer shall disclose a possible conflict of interest to a potential or current client or employer…", and "Engineers shall act as faithful agents for their employers or clients", and "The engineer shall not use a confidence or private information regarding a client or employer to the disadvantage of such client or employer or for the advantage of a third party." Each of these legal provisions touches on ethical issues related to moonlighting.

Perspectives on Customs in Different Cultures

By
Jimmy H. Smith, Ph.D., P.E.
Professor and Director, National Institute for Engineering Ethics

While teaching engineering in Malaysia in the mid-1980s, I learned first hand that the customs and values which were instilled in me as a boy growing up in West Texas are not the same as those instilled in people 12,000 miles away. Such a revelation is hardly surprising, but what I was *not* expecting to learn was that the values of other cultures are not necessarily inferior to our own just because they are different, and in fact what one culture might view as unacceptable may have a justifiable basis to others.

An ancient and diverse culture, Malaysia has three principal religions – Islam, Buddhist, and Hindu – Islam being the predominant religion.

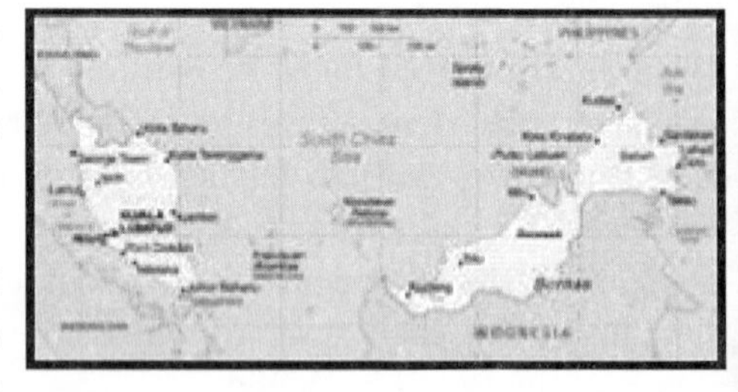

I had planned to live in Malaysia for at least one year and maybe several years and wanted to enjoy the independence of a personal automobile. My best friend in Malaysia, Hassan, tried to convince me to hire a car and driver rather than own and drive my own car, since traffic was hectic and people drove on the left side of the road (British influence). Being stubborn, I insisted on having my own automobile, so Hassan agreed to help me buy a small used car.

"...if I get stopped by the police... I should give the officer five ringgits"

Shortly after I had purchased the vehicle, Hassan and I had occasion to discuss local customs about driving in Malaysia. Hassan advised me that if I get stopped by the police, I should give the officer five ringgits (at that time, 5 Malaysian ringgits equated to 2 U.S. dollars).

Hassan's suggestion surprised me. I explained that in the US this would be considered a *bribe*, and I felt uncomfortable doing this. Out of curiosity, though, I asked Hassan what would happen if I did not give the police officer 5 ringgits. "*You'll get a ticket*!" was his response.

"How much will the ticket cost?" I asked, and Hassan replied, "Five ringgits." Hassan then explained to me that the difficulty comes in *paying* the ticket. "You'll spend most of a day at the Federal Building in Kuala Lumpur, standing in several lines, one after the other to pay the 5 ringgit ticket."

"Why do you have that custom, Hassan?" I inquired. His response actually made some sense to me. Hassan informed me that government servants such as police and customs agents like him, who were married and had a family, only received a salary of approximately half of what they would need for a modest living - at that time, 500 ringgits per month, equivalent to about 200 US dollars per month. Therefore, these public servants were allowed to accept additional "payments" up to about equal their salary. The combined funds, roughly 1,000 ringgits (US$400) per month, would be enough for a family to live on, although very modestly. "The way we look at it," Hassan continued, "the government pays for about half of the cost of police service, and the *guilty* pay for the other half."

The government pays half, and the *guilty* pay the other half...
now that had an appealing ring to it.

"*The government pays half, and the guilty pay half*" – now that had an appealing ring to it. I had not thought of looking at things in this way. In the US the guilty and innocent all pay the same in taxes for police protection. This piqued my curiosity.

Some time later I had occasion to further discuss this with Hassan. "How do you keep the police from abusing the privilege?" I asked. Hassan replied that the lifestyles of government servants are very closely controlled. For example, a government servant could not purchase a car without government approval, and if they did purchase a car, it had to be an older car with a very small engine. Further, they could only live in extremely modest housing, again, only with government approval. Such constraints kept government servants from abusing their privilege.

So regulation was the key. I gave Hassan's explanation of this issue serious thought. More importantly, our conversation made think through my own values and why I felt the way I did about this... as well as other issues in international settings.

In the end, I concluded that I could not accept this particular Malay practice as being a good one for me. Although paying money to avoid a ticket, get through customs quickly, etc., did directly benefit government servants, their lifestyles were so tightly controlled that the Malays lost what I viewed as very important: *personal freedom*. However, I also began to understand and appreciate how Hassan and others in his culture could feel okay with this.

Cultural differences do exist, and what I learned through this experience is that such practices afford an opportunity to not only gain understanding about other's values, but also to better understand and appreciate one's own values. But I also became more convinced that I should not engage in practices in which I was not comfortable, regardless of the local culture.

This is a very simple story that shows how our values and customs start in the sandbox, and those sandboxes exist throughout the whole world.

I have concluded that I have a personal and professional obligation to maintain my own standards and values, wherever I am, and should not "do in Rome as the Romans do."

Equally important, however, I have also concluded that a serious attempt should be made to understand different cultures and appreciate the fact that there can be legitimate cultural differences.

> "it is important that we judge other cultures with the same evenhandedness as we judge our own."

This causes me wonder about some aspects of our own "culture." For example, it seems rather common in our country to "pad" budgets simply because we know these budgets will likely be reduced. I confess... I have done this... moe than once.

Is padding the budget justified if our purpose is solely to make a sincere attempt to accomplish a job in the best possible fashion and with the interest of the public in mind? Or is this a form of *deception* that may, to some at least, appear as wrong as giving the police money to avoid receiving a ticket?

This brings to mind a few other questions. What is more important?

- Our Intentions?
- The Consequences?
- Basic Honesty?

Regardless of our answers to these questions, it is important that we judge the customs and practices of people in other cultures with the same evenhandedness as we judge our own actions, especially on issues which have an ethical component.

Not only is this the fair thing to do, but it is also the best way to understand and appreciate the views of others, while strengthening our own convictions of what is right and wrong for us.

Ethics for The Real World

Toward a Workable Definition of "Ethics" for Engineers

by

William D. Lawson, P.E.[1]

Have you ever tried to define *ethics*? I mean really tie it down in meaningful, understandable terms? As engineers, we recognize the importance of the topic. Yet when we survey the oceans of ink that have flowed concerning ethics, we are often left with the bewildered sense that defining *ethics* is sort of like nailing Jell-O to a tree – it is done neither easily, *nor cleanly*, if at all.

Lost At Sea

> "Defining *ethics* is sort of like nailing Jell-O to a tree."

Why should getting a handle on *ethics* be such a problem, especially since there seems to be so much "help" out there? Well, it can't be that no one has tried. Definitions for *ethics* abound, and their variety and range of content are impressive. As we wade in the shallow end of the ethics pool, immediately we encounter the one-liners, minnows if you will, such as *"live life well"* or *"do the right thing."* Notable primarily for their brevity, these otherwise catchy and appealing slogans regrettably provide little nourishment to the true seeker of ethical insight. Substance is what we want – so we must wade in farther. But to enter deeper ethical waters is to encounter ethics in its classic form, that is, *moral philosophy*. Quite suddenly we stumble over this subterranean precipice and realize that we are in *way* over our heads. The big fish are all there – the philosophers of the ages surround us – Socrates, Aristotle, Aquinas, Kant, Kierkegaard, Locke, Kohlberg, Rousseau, Sartre – to name a few. A class unto themselves, the considerable intellect of these philosophers is, well, considerable. But, unfortunately, we do not know how to appreciate their perspectives and this only leaves us disoriented and floundering.

> "When it comes to ethics as moral philosophy, it is common for us to not really get the big picture."

This suggests our first problem, *myopia.* When it comes to ethics as moral philosophy, it is common for us to not really get the big picture. Perhaps you have read the famous poem, "The Blind Men and the Elephant," by the nineteenth century poet John Godfrey Saxe. It tells of six *blind men* from Asia who go to "see" an elephant in order to determine what an elephant is like. The first blind man encounters the broad *side* of the elephant and immediately concludes the elephant "is very like a wall." The second happens to

[1]Lecturer, Senior Research Associate and Deputy Director, National Institute for Engineering Ethics, Murdough Center for Engineering Professionalism, Texas Tech University

feel a tusk and becomes clearly convinced that the elephant is like "a spear." The third blind man grabs hold of the elephant's trunk and boldly proclaims that the elephant is really just another type of snake. And on they go. You get the idea. Well, regarding ethics and moral philosophy, we engineers are not unlike those visually challenged men. As we grope around for insight, moral philosophers take us so close to the topic that at best we find it difficult to appreciate what they have to show us. From a distance it may be elephant hide but up this close, we only see mud and hair and flies.

This brings us to our second problem, which is *jargon*. We engineers use ethical jargon. I once heard a highly-educated electrical engineer describe the concrete lining of a drainage trench as "riffraff." Moral philosophers are no different from engineers when it comes to jargon (they drive on cement roads), yet we tend to subconsciously forget that our engineering jargon makes plenty of sense *to us*. Moral philosophers commonly use words we rarely if ever even think about. So when we swim in their pond we encounter the abstract and the unfamiliar: terms such as "metaphysics," "ontology," "epistemology," and "post-conventional morality." This might as well be Sanskrit. Understanding gets lost in the jargon.

Of course our third problem is *change*. We should have expected this: the elephant is *moving*. Our basic grasp of ethics is strong enough to tell us that ethics has something to do with moral standards. But when was the last time anyone agreed on *that topic* in this country? Even if we get past the other obstacles, what constitutes "good" and "right" is all relative, or so many would have us believe. This is incredibly demoralizing. If ethics is truly this slippery and subjective, why bother?

In view of these problems, many an engineer concludes that the better part of valor is retreat and we hastily turn and head for shore. Gasping and sputtering, thankful to be once again safe in our domain, we console ourselves that after all, we are engineers, not moral philosophers. Not enlightened, but alive. Perhaps we should try again… sometime.

Gaining a Footing

> "Originally and fundamentally, ethics was an *internalized* standard of conduct, behavior or self-regulation."

Maybe that is how you feel about *ethics* - in particular, engineering ethics. But there is help for us. Ethics need not be overly abstract or trivially simple. But to get past the myopia, the jargon and the change what we need is *context*, the big picture in terms we can understand.

Sociology (almost as abstract as moral philosophy, but not quite) provides this type of historical and social context for us. According to sociologists, ethics lies within the larger topic of *professionalism*. Simply stated, "a *profession* is a dignified occupation espousing an ethic of service, organized into an association, and practicing functional science," (Kimball 1992, p16). From this definition, we learn that ethics, also known as "the professional service ideal," "the service ethic," "the ethic of professional service" and by other similar terms, is a vital element of professionalism. This context is a benchmark for understanding ethics.

Taking it a step farther, the roots of the professional service ideal can be traced through the eighteenth and seventeenth centuries to the respected vocation of theology. Theology was one of the four classic professions of the time, the other three being law, medicine, and university education (Larson 1977, p4-5) (Hughes/Lynn 1965, p2). As an occupation, theology dominated colonial America during the period 1600 to 1760 (the word "profession" itself is religious in origin) and contributed developmentally to the meaning of profession. Dignified work began to be known as a "calling," and the spirit in which professionals did their work – putting other's interests before their own – became associated with professionalism as *the ethic of selfless service* (Kimball 1992, p105).

> "The professional...must not exploit the client for purposes of personal gratification."

In the mid-nineteenth century, the professional service ideal was commonly expressed as "to be professional [is] to be ethical" (Adams 1993). This highlights that originally and fundamentally, ethics was an *internalized* standard of conduct, behavior or self-regulation. The basis for this standard derives from the fact that professionals operate in areas that are generally immune to a client's oversight (Bachner 1991, p xii). Accordingly, professionals are in a position to exploit their clients, and clients must *trust* professionals to not take advantage of them. In the words of Greenwood (1957):

> The client's subordination to professional authority invests the professional with a monopoly of judgment. As the recipient of this privileged position, the professional in turn must not exploit the client for purposes of personal gratification...This accounts for the frequent synonymous use of the terms "professional" and "ethical" when applied to occupational behavior.

This relational characteristic has been said to constitute "the core of professionalism" (Kennedy 1986).

Higher Ground

If you have read this far, you already have my basic point – a clear, meaningful expression of ethics for the engineer is that, as professionals, *we must not exploit or take advantage of those who put their trust in us*. Stated positively, we professionals must treat others (society, clients, our colleagues) as we want to be treated. This is an excellent rule – *a golden rule* – to follow.

Of course, some of us want more input, more guidance. Here is where the codes of ethics come in. Most professions adopted written codes of ethics between 1904 and 1922, long after the professions organized into professional societies (Adams 1993). For example, the American Society of Civil Engineers was created in 1852, but ASCE's code of ethics was not added until circa 1914 (Wilenski 1964) (ASCE Website). Perhaps the codes were written in response to a perceived decline in morality brought on by the Industrial Age. In any event, codes of ethics attempt to represent the key values of the professional group. Ethical codes provide inspiration and guidance concerning the main obligations of the professional. Among other things they support those who seek to act ethically, they facilitate education and mutual understanding, and they contribute to the profession's public image (Martin and Schinzinger 1996, p106-107).

"The next time you read your code of ethics...imagine that you are in a room with the keenest, most accomplished, most respected colleagues in your entire profession."

Codes of ethics may be general, but unlike the writings of high-level moral philosophers, they are not overly abstract. The engineers who wrote them have distilled things for us. The next time you read your code of ethics, rather than thinking of the code as lifeless words on a page, imagine that you are in a room with the keenest, most accomplished, most respected colleagues in your entire profession. They are sharing their wisdom with us; it is worth listening to.

The Challenge to Engineers

Many would say that ethics has not fared well in this day and time. It is common to view motives with suspicion. But we are not powerless in this regard. We *can* do something.

We must first recognize that a fundamental fracture has appeared in the professional-client relationship. This critical flaw takes the form of *mistrust,* the perception that today's client can no longer totally trust professionals to "put their client's needs above their own." Fortunately, engineers know how to deal with this type of failure. We must re-enter the relationship at the point of departure; *we must build trust with clients and society.*

This is our challenge. If we build trust, if we do our work appropriately and wisely and *ethically* – we will fulfill our profession of service to society, today and into the twenty-first century.

References

Adams, Guy B., "Ethics and The Chimera of Professionalism: The Historical Context of an Oxymoronic Relationship," *American Review of Public Administration*, June 1993, V23N2, pp. 117-139.

ASCE Ethics, www.asce.org/aboutasce/ethics.html

Bachner, John Phillip, *Practice Management for Design Professionals, A Practical Guide to Avoiding Liability and Enhancing Profitability*, John Wiley & Sons, Inc., New York, NY, 1991, 371pp.

Greenwood, Ernest, "Attributes of a Profession," *Social Work*, July 1957, pp.45-55

Hughes, Everett C. "Professions," *The Professions in America*, edited by Kenneth S. Lynn, Houghton Mifflin Company, Boston, MA, 1965, 273pp.

Kennedy, R. Evan, "Professionalism. Is it Going or Coming?", *Journal of Professional Issues in Engineering Education and Practice*, Vol 112, No. 1, January 1986, pp. 49-52.

Kimball, Bruce A., *The "True Professional Ideal" in America, A History*, Blackwell Publishers, Cambridge MA, 1992, 429pp.

Larson, Magali Sarfatti, *The Rise of Professionalism, A Sociological Analysis*, University of California Press, Berkeley, CA, 1977, 309pp.

Martin, Mike W. and Schinzinger, Roland, *Ethics in Engineering, Third Edition*, The McGraw-Hill Companies, Inc., New York, NY, 1996, 439pp.

Saxe, John Godfrey, "The Blind Men and the Elephant," www.wordfocus.com/word-act-blindmen.html

Wilensky, Harold, "The Professionalization of Everyone?" *American Journal of Sociology*, Vol. 70, 1964, pp. 137-158.

- ASCE Ethics -
Edict, Enforcement and Education

By
Thomas W. Smith, III, Esq., M.ASCE[1]

The American Society of Civil Engineers ("ASCE" or the "Society") continuously strives to build a better quality of life for its membership, the engineering profession, and society at large. Implicit in every Society undertaking is the paramount importance of maintaining the highest standards of ethical conduct. Like other professional societies, an important means by which ASCE promotes high ethical standards is by maintaining and enforcing a Code of Ethics, and educating engineers and the public on ethics issues.

Edict:
Who Needs A Code Of Ethics?

Everyone has their own personal code of ethics, developed through education and experience. With the basic fundamentals identified in kindergarten and even earlier, we have each developed personal ethical codes, with input and guidance from numerous people, including family members, friends, teachers, mentors, co-workers, community and church leaders, coaches and role models. Recognizing the personal element of ethics, and expressly noting that ethics is a matter of an engineer's individual responsibility and honor, ASCE's Board of Direction resolved in the late 1800's not to adopt a code of ethics, explaining" that it is inexpedient for the Society to instruct its members as to their duties in private professional matters." So why do engineers need a professional code of ethics?

Ethical codes vary among individuals and also among corporations, governments and professions. History is filled with examples of the impact on society of varied ethical codes. From the corporate perspective, one can hardly pick up the newspaper anymore without reading about another corporate scandal. Through a long list of companies that most Americans can readily recite, including Enron, Arthur Andersen, Tyco, Worldcom and Health South, we have learned that ethical standards have a profound impact on corporate America. As Allen Greenspan noted testifying before Congress in 2002, "trust and reputation can vanish overnight," and we have seen repeated examples of such occurrences, demonstrating why good ethics is good business. Recognizing the importance of sound corporate ethics and governance, Congress adopted the Sarbanes-Oxley legislation in July 2002 in an effort to restore public confidence in corporate America with a legislated code of fiscal ethical responsibility. But Congress does not legislate every facet of corporate ethics.

[1]Managing Director, Corporate and General Counsel, American Society of Civil Engineers

Sound ethics is critical to the success of local, state and federal governments as well as to corporations,. Few people would have difficulty naming local and national government leaders whose power and credibility have been diminished or eradicated due to unethical conduct. Ethics issues can impede development and advance poverty in small communities and countries alike, examples of which include Bangladesh and Nigeria, which rank at the bottom of Transparency International's Corruption Perception Index list (http://www.transparency.org/). With new infrastructure construction likely to occur largely in developing countries, many with widely varying ethical standards, government ethics on a global scale remains a paramount concern.

Ethics also impacts professions. Many of the recent corporate scandals referenced above adversely impacted the credibility and public perception of the accounting profession. The legal profession has likewise suffered a longstanding struggle with credibility and public perception problems, and lawyer jokes remain a common favorite at most any gathering, ranging from parties to business meetings. While it is difficult to quantify the impacts of such public sentiment, I would note that the Virginia State Bar spends significant resources processing over 3,000 professional conduct complaints each year regarding Virginia's approximately 23,000 licensed attorneys. While many of these claims are proven to be unfounded, it causes one to wonder if the filing of unfounded claims is influenced by general public perception regarding the ethical standards and credibility of attorneys.

History demonstrates that trust, reputation and credibility remain important to the success of individuals, corporations, governments and professions. The engineering profession is no exception. Engineers are entrusted with the highest level of responsibility – protecting the public health, safety and welfare. Recognizing this responsibility, ASCE's Board of Direction defined the profession as a "calling in which special knowledge and skill are used in a distinctly intellectual plane in the service of humanity [in which] there is implied the application of the highest standards of excellence. . . in the ethical conduct of its members." Describing the engineering profession, President Herbert Hoover noted:

> *It is a great profession. There is the fascination of watching a figment of the imagination emerge through the aid of science to a plan on paper. Then it moves to realization in stone or metal or energy. Then it brings jobs and homes to men. Then it elevates the standards of living and adds to the comforts of life. That is the engineer's high privilege.*

With the engineers' high privilege comes a concomitant high responsibility - protecting the public health, safety and welfare. Entrusted with the highest level of responsibility, the engineers' credibility, trust, reputation, and high ethical standards remain paramount. But do engineers need a professional code of ethics?

I believe the answer is a resounding "yes." While surveys have shown that engineers enjoy a high level of trust among the public, the engineering profession has suffered from ethics breakdowns, notable examples of which include the Challenger space shuttle disaster, the Hyatt Regency walkway collapse, and the improprieties leading to the resignation of Spiro Agnew from the Vice Presidency of the United States. Engineers today work in a competitive business environment. Competition is global and sophisticated. Engineers are consistently pressured to reduce costs and increase productivity, to do more with less resources. Engineers must be well-rounded, with management expertise and familiarity with complex project requirements, responsibilities and risks. Sophisticated contracts, laws and regulations, and complex litigation are commonplace, and successful practice requires that engineers participate in these arenas. Engineers confront environmental and sustainable development concerns, tasked with ensuring that future generations can meet their needs. The electronic age and infrastructure security present new challenges and responsibilities. As a consequence, a written professional Code of Ethics proves a useful tool to assist the engineer in the frequently complex ethical decision-making process.

ASCE's Code of Ethics was adopted in 1914. Despite earlier concerns and delays, the Code of Ethics was approved by letter ballot of the membership. Although a significant first step, the original ASCE Code of Ethics contained six principles and addressed business issues, as opposed to addressing the personal professional ethics of the membership. The Society's Code of Ethics has been amended numerous times over the years, most notably to delete a provision which made it unethical "to invite or submit priced proposals under conditions that constitute price competition for professional services." This provision came under Department of Justice antitrust scrutiny in the 1970's. A subsequent Department of Justice investigation in 1992 led to voluntary revisions to the Code to eliminate language which prohibited "self-laudatory" advertising and to clarify prohibitions on unlawful consideration and contingency fees. In 1996, the Society amended the Code to incorporate the principles of sustainable development.

To address concerns regarding disparate international ethical standards, ASCE's Board of Direction voted in 1963 to add the following footnote to the Code of Ethics:

> On foreign engineering work, for which only United States engineering firms are to be considered, a member shall order his practice in accordance with the ASCE Code of Ethics. On other engineering works in a foreign country he may adapt his conduct according to the professional standards and customs of that country, but shall adhere as closely as practicable to the principles of this Code.

William Wisely wrote that this was a controversial footnote, described as the "When in Rome Clause." The footnote was no longer included in the Code of Ethics starting in 1977, which was the year Congress passed the Foreign Corrupt Practices Act to prohibit bribery of foreign officials by American corporations. The topic of global ethics remains of paramount concern today, and ASCE created this year a Task Committee on Global Standards of Professional Practice to work with engineering societies around the world to develop a standard of practice for engineers that will define their professional behavior in securing and performing engineering assignments.

ASCE's Code of Ethics remains an evolving document, benefiting from the input and experience of thousands of engineers over almost ninety years. While many ethics issues are gray and not black and white, the Code of Ethics provides a foundation for the engineer's ethical analysis and decision-making process. Likewise, discussing ethics issues with one's peers, calling ASCE's ethics hotline at 703-295-6061, and analyzing decisions under the assumption of public review and scrutiny, provide a good basis for sound ethical decision-making. While the Code of Ethics does not provide all the answers, it remains an effective foundation for ethical analysis and an important means by which ASCE advances the engineering profession.

Enforcement:
How Can ASCE Promote Compliance?

The Code of Ethics is not a stagnant document. To preserve the high ethical standards of the civil engineering profession, ASCE maintains and enforces its Code of Ethics. All Society members must subscribe to the Society's Code of Ethics, and it is the duty of every Society member to report promptly to ASCE's Committee on Professional Conduct ("CPC") any observed violation of the Code. Charges of unethical conduct may be brought by Society members and non-members and are referred to the CPC for investigation.

Established in 1923, CPC is charged with investigating charges of member misconduct. CPC comprises at least four past members of ASCE's Board of Direction. During the investigation phase, CPC acts like a grand jury. If CPC finds sufficient evidence to warrant disciplinary action, the case is scheduled for hearing before ASCE's Executive Committee. In conducting professional conduct investigations, CPC may solicit assistance from local Society members or Sections.

The Executive Committee considers proceedings for the discipline of a Society member upon the (a) recommendation of CPC, or (b) written request of ten or more Society members. Hearings are conducted in accordance with written procedures for professional conduct cases. Due Process is afforded to the member, including reasonable notice of the charges and the hearing, fair opportunity to hear the evidence, question witnesses and refute the evidence, and a hearing before an unbiased panel. The Executive Committee acts as a Judge or Jury. At this point, a CPC member serves a function similar to a prosecutor.

Upon finding a violation of the Code, the Executive Committee may take disciplinary action, other than expulsion, by a majority vote. Such action typically includes a letter of admonition or a suspension from membership. The most severe penalty is expulsion from the Society. The Executive Committee cannot expel a member, but can make a recommendation to the Board of Direction that the member be expelled. If the Executive Committee votes to recommend expulsion, the case is scheduled for hearing before the Board of Direction, with the same due process protections afforded at the Executive Committee hearing. A decision to expel the member requires a seventy-five percent vote of the Board. The Board may impose lesser disciplinary actions upon a majority vote. The Executive Committee and Board of Direction have discretionary authority to publish the action, with or without the name of the member. Such notice is typically published in *ASCE News*. The Executive Committee and Board of Direction also have discretionary authority to notify other professional organizations or registration boards of the action.

ASCE enforces the provisions of its Code of Ethics to preserve the high ethical standards of the Society and the profession. While taking care to ensure that such enforcement is not anticompetitive and that due process is afforded, with reasonable notice and a fair hearing, ASCE endeavors to advance the profession and the public interest by the fair enforcement of reasonable ethical practices.

Education:
An Ounce Of Prevention Is Worth A Pound Of Cure

As referenced above, the Code of Ethics is not a stagnant document, and all engineers should continue to familiarize themselves with the Code of Ethics and related responsibilities under applicable laws and licensing regulations. Recognizing the complexity of civil engineering projects and practice, and the critical responsibilities and wide-ranging considerations that civil engineers confront, ASCE's Board of Direction voted in 2001 to adopt Policy Statement 465, which "supports the concept of the master's degree or equivalent as a prerequisite for licensure and the practice of civil engineering at the professional level." Last year (2003), ASCE defined the Body of Knowledge (BOK) needed to enter the practice of engineering at the professional level to include "an understanding of professional and ethical responsibility." From personal experience, I can attest that civil engineering students appreciate and actively participate in ethics education and discussions, and such education and training is important both during the formal education process and throughout an engineer's professional career.

ASCE endeavors to educate engineers on ethics issues. The Society has case studies to provide guidance on ethical problems. The Society sponsors ethics seminars, and has adopted multiple policies on ethics issues, including Policy Statement Number 376 encouraging state boards of engineering registration to institute take-home examinations on professional ethics for professional registration; Policy Statement Number 130 supporting the establishment of rules of professional conduct for engineers and land surveyors consistent with the Society's Code of Ethics to guide licensees in their practice; Policy Statement Number 418 supporting implementation strategies to promote sustainable development; and Policy Statement Number 502, adopted in July 2003 to confirm the importance of the engineer's independence and duty to avoid conflicts of interest.

The Society makes available ethics videos, including a video tape of a mock Board of Direction hearing, the video and workbook entitled "*Testing Water...and Ethics*", and the National Institute for Engineering Ethics' video entitled "*Incident at Morales*." The Society publishes the Code of Ethics on the Society's web site and in the Official Register, and the Society has published Standards of Professional Conduct for Civil Engineers and Guidance for Civil Engineering Students. The Society also publishes papers on engineering ethics, including numerous articles in the Journal of Professional Issues in Engineering Education and Practice. The Society promotes the Order of the Engineer program and ring ceremonies, which focus attention on the obligation of the engineer to protect the public health, safety and welfare, and which help to "foster a spirit of pride and responsibility in the engineering profession, to bridge the gap between training and experience, and to present to the public a visible symbol identifying the engineer." The Society also awards annually the Daniel W. Mead prize for younger members and students on the basis of papers on professional ethics.

Recognizing the increasing complexity of the profession and the profound importance of engineering ethics, ethics education remains an important means by which ASCE and other professional societies can increase awareness of ethics issues, advance the science and profession of engineering to enhance the welfare of humanity, and enable engineers to be global leaders building a better quality of life.

Conclusion

Engineering ethics involves complex issues that are global in scale and critical to the Society, the profession and to the public at large. I believe engineers enjoy a positive public image, with a reputation for providing public value by honest and ethical means. By adopting and enforcing a Code of Ethics, and continuously educating engineers on the complex and evolving field of engineering ethics, ASCE and other professional societies can continue to advance the profession and improve the quality of life worldwide.

References

1. *The American Civil Engineer*, William H. Wisely, (1974).
2. Speech by Claudia Cositore, ASCE Professional Services Counsel, April 18, 1983.

Continuing Professional Competency: If at First You Don't Succeed...

by

William D. Lawson, P.E.[1]

Engineering licensure laws in 23 states now mandate continuing professional competency (CPC) as a requirement for professional engineer (PE) license renewal (NSPE @ 2002). This paper argues – from the *academic*, *technical*, *ethical*, and *practical* perspectives – that engineers should *advance*, and not step back from mandatory Continuing Professional Competency.

Academics: Practice What We Teach

The notion of competency is tightly linked to learning, and learning models continue to develop. In the 1990s, higher education practices began to apply continuous quality improvement, a.k.a. *Total Quality Management* – TQM, concepts to learner-centered educational processes focused on student outcomes (Halpern 1987: 6-7). This marriage of TQM and accreditation has yielded a model of higher education – similar to a production line – where an ordered sequence of defined operations results in a specified product or service, where assessment occurs at all levels, and where results are continuously monitored for conformance with desired outcomes; *viz.,* teaching and learning (Ewell 1993: 40). For engineers, the quintessential expression of this process-oriented model is the Accreditation Board for Engineering and Technology (ABET) *Engineering Criteria 2000* (ABET 1998: 14).

This educational model sees professional preparation as a *life-long* process (ABET Engineering Criteria 2000, Criterion 3*i*), no doubt in an attempt to address the difficulties associated with measuring and establishing competency in the first place (Curry and Wergin 1993: 348-9). The explicit requirement we place on engineering students and our academic programs today is that learning be *continuous*. Should not the same standard apply to practitioners?

Technical: How Deep is Your Pond?

Professional practice is tied directly to a system of knowledge that formalizes the skills on which the work proceeds (Abbott 1988: 35-58). Thus a professional, *by definition*, is knowledgeable, and on this view, the professional's body of knowledge is like a reservoir from which the professional draws as needed.

The body of knowledge is not static – scientific research continues to spawn unprecedented advancements in theory and technical specialization – nor is this specialized knowledge the sole property of the professional. Information technology

---

[1]Lecturer, Senior Research Associate and Deputy Director, National Institute for Engineering Ethics, Murdough Center for Engineering Professionalism, Texas Tech University

has opened up the knowledge domain so that it is no longer "...packed only in the professional's head or in a specialized library, where it is relatively inaccessible. It can be available not just to those who know, but also to those who *know how to get it*" (Haug 1975). CPC recognizes that engineering knowledge is growing and is increasingly accessible to the layman, and seeks to communicate to all stakeholders that the practitioner is keeping up and is accountable.

Ethics: Remind Me Again *Why* We're Doing This

As would be expected given the professions' inherent tie to expertise, competence is not only a technical but an *ethical* issue. Competence goes beyond the minimum standards for admission to a profession and incorporates the maintenance and improvement of skills and practices. "Despite the patent difficulty of doing so in the contemporary world, the professional is supposed to keep current with developments in his field, so that his clients do not seriously suffer relative harm from his failure to do so... Competence is for a purpose: *conscientious performance*" (Moore 1970: 13-14, emphasis added).

It is not hard to see why most engineering societies, including but not limited to the National Society of Professional Engineers and the flagship society for each of the three largest engineering disciplines: the Institute for Electrical and Electronics Engineers (IEEE), the American Society of Mechanical Engineers (ASME), and the American Society of Civil Engineers (ASCE), affirm in their codes of ethics that competence is an ethical requirement for professional practice.

Practically Speaking

While arguments exist both for and against mandatory continuing education for professionals (Kerka 1994) it is important not to confuse methodology with the desired outcome. The issue is *competency*, not education, and assurance of continuing professional competency remains an obligation of the engineering profession. Keeping in mind that "what gets measured is what gets valued" (Ruppert 1995: 17), CPC recognizes that the practice of engineering cannot flourish without careful attention. It is necessary to plan professional development, and constructing a framework within which competency can be demonstrated and analyzed offers a way of doing this.

Many concerns about mandatory CPC focus on implementation difficulties such as availability and diversity of programs, uniformity and reciprocity across states, and accessibility. The model CPC guidelines developed by the National Council of Examiners for Engineers and Surveyors (NCEES) in 1994 (NCEES @ 2001), as well as the rapid growth of distance learning via the Internet do much to alleviate these concerns. The practical barriers are becoming less formidable.

Conclusion

Academic, technical, ethical and practical voices show that engineers would do well to implement mandatory CPC. Our profession faced this type of issue in the early 1900s – the subject then was ethics – when those in search of moral high ground observed that "…no gentleman needed a code of ethics, and that no code of ethics would make a gentleman out of a crook" (Newell 1922: 133). The same thing has been said about CPC – the truly competent engineer will do the right thing and maintain competency without any external mandate. While the claim, "I'm from the Government and I'm here to help you," evokes skepticism, it is likewise naïve to persist in the view that competency can continue to be left unattended. Today's consumer demands excellence and continuous quality assurance from professionals, and mandatory CPC is a reasonable response to this challenge.

References

Abbott, Andrew (1988), *The System of Professions: An Essay on the Division of Expert Labor*, The University of Chicago Press, Chicago, IL, 435pp.

Accreditation Board for Engineering and Technology (1998), "Engineering Criteria 2000," in *How Do You Measure Success? Designing Effective Processes for Assessing Engineering Education*, American Society for Engineering Education, Washington, D.C., pp. 13-16.

Curry, Lynn and Wergin, Jon F. (1993), "Professional Education," *in Handbook of the Undergraduate Curriculum: A Comprehensive Guide to Purposes, Structures, Practices, and Change*, Jerry G. Gaff and James L. Ratcliff (*eds.*), Jossey-Bass Publishers, San Francisco, 1997, 747 pp.

Ewell, Peter T. (1993), "Total Quality and Academic Practice: The Idea We've Been Waiting For," *Change*, Vol. 25, No. 3, pp. 49-55, May/June 1993, reprinted as Article 24 in *CQI 101: A First Reader for Higher Education*, a publication of the AAHE Continuous Quality Improvement Project, American Association of Higher Education, Washington, D.C.

Haug, Marie R. (1975), "The Deprofessionalization of Everyone?" *Sociological Focus*, Vol. 8, No. 3., August 1975, pp. 197-213.

Halpern, Diane F. (1987), "Student Outcomes Assessment: Introduction and Overview," in *Student Outcomes Assessment: What Institutions Stand to Gain*, Diane Halpern (ed.), New Directions for Higher Education, no. 59, Vol. XV, number 3, San Francisco: Jossey-Bass, Inc.

Kerka, Sandra (1994), "Mandatory Continuing Education," *ERIC Digest No. 151*, Identifier No. ED376275, http://www.ed.gov/databases/ERIC_Digests/ed376275.html, accessed 08-30-01.

Moore, Wilbert E. (1970), *The Professions: Roles and Rules*, Russel Sage Foundation, 301pp.

NCEES (2001), "Continuing Competency, Engineers and Land Surveyors," http://www.ncees.org/engineers/cpc.html, accessed 08-30-01.

Newell, Frederick Haynes (1922), "Ethics of the Engineering Profession," article reprinted in *Professionalization*, Howard M. Vollmer and Donald L. Mills (eds.), Prentice Hall, Englewood Cliffs, NJ, 1966, pp. 133-137.

NSPE (2002), "Continuing Professional Competency: Status in the States," http://www.nspe.org/lc1-cpc.asp, accessed 03-30-2004.

Ruppert, Sandra S. (1995), "Roots and Realities of State-level Performance Indicator Systems," in *Assessing Performance in an Age of Accountability: Case Studies,* Gerald H. Gaither (*ed.*), New Directions for Higher Education, no. 91, Fall 1995, Jossey-Bass Publishers, San Francisco, 107 pp.

Law and Ethics

Heinz Dilemma - If It Is Legal, Is It Ethical? Vice-Versa?

Is there a direct relationship between the law and ethics? Let's first consider a well known ethical dilemma that has both a legal and ethical component known as the **Heinz Dilemma**[1] which was developed for a survey by a renowned philosopher, Lawrence Kohlberg. The story is an only illustration to examine how we think about laws and ethics, and is not based on a true story. The Heinz Dilemma:

> *A woman lives in Europe who will die from cancer unless she obtains an expensive drug that the doctors think will help her. Her husband, Heinz, cannot afford to purchase the drug. The local pharmacist invented the drug and is the sole source for obtaining it. He is charging ten times the cost of making the drug.*
>
> *The husband goes to everyone he knows seeking to borrow money, but he manages to raise only half the money needed to purchase the drug. When he asks the pharmacist to sell the drug at a cheaper price or to let him pay for it later, the pharmacist refuses. In desperation, Heinz breaks into the pharmacy and steals the drug.*

Was Heinz's action **legal**? Was Heinz's action **ethical**? Why or why not?

Is Having More Laws Better Than Having More Ethics?

Let's consider the question of whether we need to create more laws so that people will be more ethical. Recent cases in big business show that things can and do go wrong, even with the existence of excellent published standards of ethics. For example, Enron had an excellent and highly respected standard of ethics. However, the "culture" of at least some aspects of the company was not in keeping with their published standards.

A New York Times article[2] written by two attorneys contend that "toughening existing criminal laws and adding new ones might seem the best way to make sure that future Enrons and WorldComs won't happen." ***"But it won't work!"*** "The reason is both simple and all too easily ignored: Laws **lead people to focus on *what is legal* instead of *what is right***"."

They conclude: "Perhaps that's because we've turned what used to be moral questions into legal technicalities. In today's world, executives are more likely to ask **what they can get away with legally** rather than **what's fair and honest.**"

What is your opinion?

[1]Martin & Schinzinger, Ethics in Engineering, McGraw-Hill, 1989
[2]New York Times Article by David Skeel and William Stuntz, July 10, 2002

Ethics and Political Contributions[1]

by
Jimmy Smith, Ph.D., P.E.[2] and Dave Dorchester, P.E.[3]

Large political contributions to local political candidates made by engineers who later accept contracts from the elected officials they supported poses an ethical question. There are some engineers who say they feel they <u>must</u> contribute in order to be "eligible" for consideration to receive local government contracts. We hear the refrain from within the profession that "engineers are pressured into making contributions," and "it's a matter of survival."

In this matter, we are confronted with a clash between ethical principles of a profession which is also faced with the pressures of the business environment. We believe in the right of individual engineers to make campaign contributions to those running for office, and believe in the right of professional societies such as TSPE, through its' political action committee, to make contributions to individual political campaigns for the purpose of promoting better government in our state. However, it is naive to believe that very large political contributions made by individual engineers are "only in the interest of promoting better government," especially when the engineer or his/her firm later accepts contracts from the elected official(s) supported.

The National Society of Professional Engineers (NSPE) Code of Ethics Section II.5.b states: "Engineers shall not offer, give, solicit or receive, either directly or indirectly, any political contribution in an amount intended to influence the award of a contract by public authority, **or which may be reasonably construed by the public of having the effect or intent to influence the award of a contract...."** The NSPE Board of Ethical Review (BER) has considered this matter many times since its first case in 1962. In BER Case 88-2 (see following pages), the BER reconfirmed that: **"fundamental ethical principles stated in unequivocal terms cannot be bent or broken for economic expediency or gain."**

The writers believe that professional engineers who make *major* political contributions and subsequently accept work from the politicians they generously supported are presenting an image of the profession to the public that is not in keeping with the "professional conduct and ethics" aspects of being a licensed engineer in Texas.

[1]Originally published in the Spring 1996 Issue of the TexethicS Newsletter by the Murdough Center for Engineering Professionalism

[2] Professor and Director, Murdough Center for Engineering Professionalism, Texas Tech University

[3] Past President, Texas Society of Professional Engineers and past Board Chair, Texas Board of Professional Engineers

Political Contributions

NSPE Opinions of the Board of Ethical Review -- Case No. 88-2

Facts:

Engineer A is the principal in a small-sized consulting engineering firm. Approximately 50 percent of the work performed by Engineer A's firm is performed for the county in which the firm is located. The value of the work for the firm is estimated to be approximately $150,000 per year. Engineer A is requested to make a $5,000 political contribution, the maximum amount allowed by law, to help pay the cost of the media campaign of the county board chairman.

After subsequent thought, Engineer A makes a $2,000 contribution to the campaign of the chairman, a person Engineer A has known for many years through mutual public service activities as well as their activities on behalf of the same political party. The county board chairman serves in a part-time capacity and receives $9,000 per year for his services. Other members of the board receive $8,000 per year for their services.

As required under the laws of his state, Engineer A reports the campaign contributions to the state board of elections, and correctly certifies that the contributions do not exceed the limits set by the law of the state. These contributions and the contributions of other firms in the county are reported by members of the local media who appear to suggest that Engineer A and other firms have contributed to the campaign in anticipation of receiving work from the county. Engineer A continues to perform work for the county after making political contributions.

Question:

Is it unethical for Engineer A to continue to perform work for the county after making the $2,000 contribution to the campaign of the county board chairman?

References:

Code of Ethics Section II.3.a. - Engineers shall be objective and truthful in professional reports, statements or testimony. They shall include all relevant and pertinent information in such reports, statements or testimony.

Section II.5.b. - Engineers shall not offer, give, solicit or receive, either directly or indirectly, any political contribution in an amount intended to influence the award of a contract by public authority, or which may be reasonably construed by the public of having the effect or intent to influence the award of a contract. They shall not offer any gift, or other valuable consideration in order to secure work. They shall not pay a commission, percentage or brokerage fee in order to secure work except to a bona fide employee or bona fide established commercial or marketing agencies retained by them.

Section III.1.f. - Engineers shall avoid any act tending to promote their own interest at the expense of the dignity and integrity of the profession.

Discussion:
For many years, the engineering profession has been grappling with the ethical issues involved with political contributions by individuals to state and local candidates. Political contributions was the subject of a keynote address by the National Society of Professional Engineers at a recent national meeting and continues to be examined by a special task force charged with developing a political contributions policy.

Over the years, the Board of Ethical Review has had occasion to examine the question of political contributions. Case 62-12, the first case of its kind, involved engineers who were officers or partners of various organizations such as consulting firms, construction companies, or manufacturing companies who made it a practice to contribute to campaign funds on behalf of those seeking public office. The engineers also contributed as individuals to both major political parties and in some cases to rival candidates for the same office. The Board ruled that it was not unethical for an engineer to contribute to a political party or a candidate per se, but it is unethical to make contributions in the expectation of being awarded contracts on the basis of favoritism. The Board began its discussion by noting: "Here we must deal with motivation: what was in the mind of the contributor. It is beyond doubt that the engineer as a responsible citizen has and should have the same opportunity as others to hold political views and support the party or candidate of his choice for political office. Such interest and activity is to be encouraged." The Board noted however: "The implication of the facts, however, is that the political contributions were made to curry favor and place the engineer, and through him his firm, in a favorable position to secure contracts through the influence of the candidate elected to a public office which determines the award of such contracts." In concluding its discussion, the Board noted: "It is hardly possible to draw a precise line in dollar amounts for the purpose of defining when a political contribution becomes an improper incentive to secure contracts on the basis of favoritism. As in all ethical applications, the only sound rule is that when conduct may raise suspicion and doubt as to motive, it is the better part of wisdom to stay well within the line."

Thereafter, in Case 73-6, Engineers A, B, and C made political contributions in the sums of $150, $1,000, and $5,000, respectively, to a candidate for governor of the state in which the firms they are associated with as principals are located. The candidate they supported was victorious. Subsequently, the firms in which A, B, and C are principals, received several state contracts for engineering services with total fees ranging from $75,000 to $4 million over a two-year period. With two members dissenting, the Board found that in the absence of a showing of improper intent, Engineers A, B, and C were not acting unethically at the time they made their contributions, that Engineer A was not unethical for taking state contracts under the circumstances since the contribution was in a nominal amount, but that Engineers B and C were unethical for taking state contracts under the circumstances since their contributions were each over a nominal amount. The dissenting position criticized the majority conclusion that Engineers B and C were unethical noting: "As long as the present system of financing political campaigns is in effect, and in light of the conclusions reached in this case, any engineer who relies on governmental or public works type of engagements for a substantial portion of his practice would have to refrain from acting meaningfully and constructively in the political process. With candidates dependent on donations and contributions for financing campaigns, it is naive to assume that any elected legislator is going to heed the advice or requests for support of legislation and administration by persons who have not given him strong support of his campaign efforts including the financing of such efforts."

The Board also had the opportunity to discuss the issue of engineer political contributions to state and local candidates in other Cases (75-13 and 76-12). However, Cases 62-12 and 73-6 provide the greatest insight into the ethical dilemma faced by engineers who perform governmental or public works.

We do not wish to address the issue of contributions to political action committees which has been addressed in Case 75-13 and determined to be entirely proper. What we are faced with here is a fundamental clash between deeply rooted ethical principles and a profession faced by the pressures of the business environment. The language in the Code is clear; this Board has interpreted the language on more than two occasions and has been fairly consistent in its BER interpretation. Nevertheless, we continue to hear the refrain from many within the profession that "engineers are pressured into making contributions," and "it's a matter of survival."

We must respond, however, that fundamental ethical principles stated in unequivocal terms cannot be bent or broken for economic expediency or gain. It has been 26 years since the Board decided Case 62-12 and 15 years since its decision was affirmed in Case 73-6. The NSPE Code of Ethics language on the question of political contributions (II.5.b.) has not been modified in any substantive manner since that time.

While we recognize the difficulties encountered by many engineers who seek to perform public work and the pressures involved, we can find no justification for modifying our long-held view as enunciated in Case 62-12 and restated in Case 73-6 that direct contributions to candidates for political office in a nominal amount are permissible under the Code but that political contributions in excess of a nominal amount are violative of the Code.

Under the facts of this case, the requested political contribution of $5,000 was not a nominal contribution for the office of chairman of the county board and therefore was in violation of the Code of Ethics.

Nominal political contributions should be evaluated on a case-by-case basis depending upon the nature of the political office involved, the size of the jurisdiction which the public official serves, and other appropriate considerations based upon the unique nature of the office. But with most provisions of the Code, the greatest responsibility falls upon the shoulders of individual engineers who must make a decision based upon their own consciences as to what is appropriate.

In this particular case it is our judgment that a political contribution of $2,000 represents the upper limit of a nominal contribution and therefore is not in violation of the Code.

Conclusion:
It would not be unethical for Engineer A to perform work for the county after making a nominal political contribution of $2,000 to the reelection campaign of the county board chairman.

Board of Ethical Review

Eugene N. Bechamps, P.E.
Robert J. Haefeli, P.E.
Robert W. Jarvis, P.E.
Lindley Manning, P.E.
Paul E. Pritzker, P.E.
Harrison Streeter, P.E.
Herbert G. Koogle, P.E.-L.S., chairman

The Professional Engineer and Politics

A position statement from the Texas Society of Professional Engineers' Web Site www.tspe.org

Originally included in the Summer 2002 Issue of the TexethicS Newsletter, Published by the Murdough Center for Engineering Professionalism

In a free society, politics and government are inseparable. Politics shapes the character of government — and government affects Americans at the local, state and national level.

The engineering profession is directly affected by the course which government pursues regarding preservation of the free enterprise system and is particularly sensitive to actions of government which affect the professional climate in this country.

The importance of the relationship between the engineer and elected officials is exemplified by the remedy being sought for our state and nation's civil justice crisis.

The professional engineer must bring to the legislative process the practical, economic, legal and ethical viewpoint that the engineer is known to possess. Professionals and businessmen who neglect politics cannot lament unsound legislation. The engineer <u>must</u> participate in politics in order to influence legislation.

The Texas Society of Professional Engineers (TSPE) has developed an organization to affect government action at the state level while the National Society of Professional Engineers (NSPE) is providing the same service at the national level.

Through their legislative committees, executive boards, contact committees and staffs, TSPE and NSPE are working to ensure that the views of their members are voiced in legislative halls.

In addition, if you have any questions or concerns regarding the legislative or political process, please call the TSPE Legislative Information Service, 512/472-9286 or 800/580-8973. Members of TSPE may request copies of specific bills.

Texas Society of Professional Engineers Passes Resolution Related to Ethics and Political Contributions

On June 13, 2002, the Texas Society of Professional Engineers' Committee on Ethical Practices proposed the following resolution to the TSPE Board of Directors at its Annual Meeting in Arlington, Texas. The resolution passed by unanimous vote.

Whereas: The Texas Society of Professional Engineers (TSPE) strongly encourages engineers and engineering firms to be actively involved in the political process and to support political candidates of their choice with their time and financial contributions as desired and appropriate, and

Whereas: The Texas Society of Professional Engineers also strongly encourages engineers in Texas to be involved in providing or offering to provide engineering services to state, county, and city governments for the benefit of the citizens of Texas, and

Whereas: The Texas Society of Professional Engineers takes the position that political contributions should not be necessary or expected in order to be considered qualified, or to be awarded contracts, for work for public authorities, and

Whereas: The Texas Society of Professional Engineers endorses the National Society of Professional Engineers (NSPE) position on the matter of contributions as documented in the NSPE Code of Ethics Section II.5.b: *"Engineers shall not offer, give, solicit or receive, either directly or indirectly, any contribution to influence the award of a contract by public authority, or which may be reasonably construed by the public as having the effect of intent to influencing the awarding of a contract..."*

Therefore: It is resolved that the TSPE Executive Committee shall create a "Special Commission on Practices Related to Political Contributions" consisting of representatives from all TSPE regions and including invited participation by members of the Texas Engineering Alliance, ACEC-Texas, Texas Board of Professional Engineers and/or other appropriate representatives. This Commission will:

1) Examine the relationship between engineers/engineering firms receiving work from public authorities and the political contributions made to those authorities;
2) Determine the extent to which this issue is a professional or ethical problem
3) Identify ways that this issue ought to be professionally and ethically addressed
4) Recommend a position on the matter to the TSPE Executive Committee; and
5) Recommend a method of implementing the position that will be beneficial to, and in the best interest of, the citizens of Texas, the engineering profession, and our engineering society

Incident at Morales

An Engineering Ethics Story

Information and Study Guide

Developed and Distributed
by the

Murdough Center For Engineering Professionalism

The following pages contain:

Part A: Development of Incident at Morales

Part B: Suggestions for Use of the Video

Part C: Story, Cast of Characters, and Synopsis

Part D: Ethical Issues and Purpose of the Video

Part E: Questions Incident at Morales Raises

Part A: Development of Incident at Morales

Incident at Morales was developed by the National Institute for Engineering Ethics (NIEE), Murdough Center for Engineering Professionalism, Texas Tech University, with a grant from the National Science Foundation (Grant # SES-0138309) supplemented by significant donations from individuals, engineering societies, companies, and universities. Great Projects Film Company of New York City is the producer of the video. This study guide, the script and other information about the video may be obtained from the NIEE Internet site: www.niee.org.

Incident at Morales is a product of the combined efforts of a team with representation from several universities and individuals with experience in various engineering disciplines and philosophy.

Incident at Morales is not intended as a "quick fix" but as one tool that should be utilized to augment programs in engineering ethics. Typical programs in engineering ethics have several goals:

- ***Sensitivity***: to raise awareness of ethical aspects of professional work
- ***Knowledge***: to learn about professional standards
- ***Judgment***: to develop skills in moral reasoning
- ***Commitment***: to strengthen personal dedication to exemplary conduct

Part B: Suggestions for Use of the Video

The video is designed for interactive use with a discussion facilitator. The total running time of the video is thirty-six minutes; there are opportunities to pause for discussions after approximately twelve and twenty-four minutes.

At each break, the facilitator may engage viewers in a discussion of the ethical issues raised in the previous segment. At a university, the video may be used in three consecutive fifty-minute class sessions: the professor or facilitator might use one segment in each class session.

In a professional development workshop or seminar, two hours would be sufficient time for viewing and discussion.

The facilitator should view the video in advance and plan the discussion periods. The facilitator may decide to break a large audience into smaller groups - each consisting of three to six participants - for a more effective discussion period.

The facilitator should assign specific tasks to the participants. For instance, participants may be asked to generate questions for further discussion; suggestions for discussion questions appear later in this guide. Specific questions might require participants to:

- Identify ethical, technical, and economic issues and problems
- Identify affected parties (stakeholders) and their rights and responsibilities
- Identify social and political constraints on possible solutions
- Determine whether additional information is needed to make a good decision
- Suggest alternative courses of action for the principal characters
- Imagine possible consequences of those alternative actions
- Evaluate those alternatives according to basic ethical values

Actions can be evaluated by whether they honor basic ethical values such as:

- Honesty
- Fairness
- Civility
- Respect
- Kindness

Actions can also be evaluated by the following tests (cf. Davis, 1997):

- ***Harm test:*** Do the benefits outweigh the harms, short term and long term?
- ***Reversibility test:*** Would I think this choice was good if I traded places?
- ***Colleague test:*** What would professional colleagues say?
- ***Legality test:*** Would this choice violate a law or a policy of my employer?
- ***Publicity test***: How would this choice look on the front page of a newspaper?
- ***Common practice test:*** What if everyone behaved in this way?
- ***Wise relative test:*** What would my wise old aunt or uncle do?

Part C: Story, Cast of Characters, and Synopsis

Story

Incident at Morales involves a variety of ethical issues faced by a company that wants to quickly build a plant in order to manufacture a new chemical product to gain a competitive edge over the competition.

Potential technical and ethical issues arise from choices of designs, including valves, piping, chemicals, etc. The process to manufacture the product is designed to be automated and controlled by computer software. The process also involves high temperatures and pressures and requires the use of chemicals that need special handling.

Because of environmental considerations related to the chemicals used in the process, the company decides to construct their plant in Mexico. Technical, environmental, financial, and safety problems arise that involve ethical issues.

Cast of Characters

Fred .... Engineer hired by *Phaust* to design a new plant to manufacture a new paint remover

Wally .... Fred's supervisor at *Phaust*

Chuck .... Vice President of Engineering at *Phaust*

Dominique ..... Corporate liaison from *Chemistré* (parent company in France) to *Phaust*

Maria .... Fred's wife, a compliance litigator for U.S. Environmental Protection Agency

Hal .... Market Analyst at *Phaust*

Jen .... Research Chemist at *Phaust*

Peter .... Project Manager of the construction firm that builds the new plant in Morales

Jake .... Plant Manager for the *SuisseChem* plant in Big Spring, Texas

Manuel .... Plant manager for the new *Phaust* plant in Morales, Nuevo Leon, Mexico

Synopsis

Phaust Chemical manufactures "Old Stripper," a paint remover that dominates the market.

On learning that *Phaust's* competitor, *Chemitoil*, plans to introduce a new paint remover that may capture the market, executives at *Phaust* decide to develop a competing product. To save money in manufacturing the product, *Phaust* decides to construct a new chemical plant in Mexico and hires chemical engineer Fred Martinez, a former design engineer for the consulting company *Chemitoil*, to design the plant.

Problems arise when *Chemistré*, *Phaust*'s parent company in France, slashes budgets 20% across the board. In response, Chuck, the vice president of engineering at *Phaust*, strongly encourages Fred to reduce construction costs.

Fred confronts several engineering decisions in which ethical considerations play a major role:

- ❖ Whether to use expensive controls manufactured by Lutz and Lutz, which has an inside connection at Phaust

- Whether to line the evaporation ponds to prevent the seepage of hazardous substances in the effluents into the groundwater, although local regulations may not require this level of environmental protection
- Whether to purchase pipes and connectors made with stainless steel or a high pressure alloy

When samples of *Chemitoil's* new paint remover, "EasyStrip," become available, it is clear that to be competitive with "EasyStrip," *Phaust* must change the formulation of its new paint remover, which requires higher temperatures and pressures than originally anticipated.

Some unexpected problems arise:

- Leakage occurs in one of the connections
- The automatic control system fails so the plant manager offers to control the process manually

After the plant goes into full operation, an accident occurs, and the plant manager is killed while manually controlling the manufacturing process.

Part D: Ethical Issues and Purpose of the Video

Ethical Issues

A wide variety of ethical issues surface in *Incident at Morales*, including:

- Ethical responsibilities and obligations don't stop at the U.S. border
- Ethics is an integral (and explicit) component of ordinary technical and business decision-making in engineering practice. Engineers impact people and should be more concerned about people than objects
- Technically competent, ethically sensitive, reasonable people may have different perspectives and can disagree when faced with complex ethical issues
- Negotiations resolve some of the conflicts in the video, but some ethical conflicts remain unresolved. Ethical problems are sometimes resolved by rational methods and compromise
- Market stresses arise from competition with other companies and from pressures to advance a design and construction schedule
- It is sometimes necessary to make decisions under pressure with incomplete data, insufficient time, and insufficient information
- Guidance to help resolve ethical problems is available in the form of codes of ethics and actual case studies from professional and technical engineering societies and engineering licensing boards

Consideration of consequences of technical, financial, and ethical decisions is an important element of the video.

Purpose

The video is designed to help viewers become more aware that:

- ❖ Ethical considerations are an integral part of making engineering decisions
- ❖ A code of ethics will provide guidance in the decision-making process
- ❖ The obligations of a code of ethics do not stop at the United States border
- ❖ The obligations of engineers go beyond fulfilling a contract with a client or customer

Part E: Questions *Incident at Morales* Raises

Topics Considered

- ❖ Initial Ethical/Legal Issues
- ❖ More About Fred
- ❖ Wally's "One Rule"
- ❖ Company Slogan
- ❖ Effective Communications
- ❖ Marketing Decisions
- ❖ Budget Issues
- ❖ Regarding the L&L Controls
- ❖ Interaction Between Plant Designers and Plant Operators
- ❖ Safety Issues
- ❖ Personal Relationships
- ❖ Regarding International Cultural Issues
- ❖ Making Decisions
- ❖ Regarding the Software of the Cheaper Controls
- ❖ What if Automatic Controls Don't Work
- ❖ Margin of Error and Reasonable Care
- ❖ Trust and Candor
- ❖ If You Were in Charge

Initial Ethical/Legal Issues

1. How does a corporate culture affect how we practice engineering, and, in particular, how does it affect our dealings with ethical issues?
2. Was it ethical for *Phaust* to hire Fred, who recently did similar work for a competitor, *Chemitoil*?
3. What is Chuck's primary motivation for hiring a licensed Professional Engineer (P.E.)?
4. What issues are involved in hiring an engineer from a competitor?
5. How can Fred maintain a reputation for trustworthiness - being able to have insider information - while serving his new employer properly?
6. Although the lawyers note that Fred has no legal obligations to *Chemitoil* because he did not sign a non-disclosure agreement, does Fred have a moral

obligation to ensure the confidentiality of the information he may have learned at *Chemitoil*?

7. Does *Chemitoil* have an obligation to make sure that Fred is comfortable with what he should or should not disclose regarding his employment with their company?

More About Fred

8. Did Fred seem too young for the major responsibility of designing a new plant?
9. Is there anything that Fred should have asked about *Phaust* before accepting the job offer?
10. Should Fred have expressed his concerns more forcefully? If so, how?
11. Were all professional levels accountable for the oversights of engineering decisions?
12. How might professional or technical engineering societies be involved in this case?

Wally's "One Rule"

13. What is the impact of Wally's "One Rule" on Fred's ability to do his job? More importantly, does this interfere with Fred's ability to meet his professional ethical obligations in the course of conducting his job?
    a) How should Fred respond to Wally's "One Rule"?
    b) Is it possible for Fred to keep Wally, his supervisor, informed while fulfilling his obligations as an engineer to voice his concerns to the appropriate people at the appropriate time?
14. We recall Chuck saying "Fred you don't have to deal with these issues alone." Who should Fred be able to rely on to help him deal with "these issues"?
15. When Wally learned about the meeting with the environmental experts, was his reaction to Fred justified?
16. What was Wally's true reason in being so upset with Fred? He mentioned a very important element - employee bonuses. Was Wally referring to employee bonuses, or was he showing concern for his own bonus?
17. What do people really think they're being paid for? The employee's perception of what he/she is being paid for can influence his/her decisions and judgments with relationship to his/her position. Is Wally's perception of his position what influenced his reaction to Fred when Fred violated the "One Rule"?
    a) Is this part of normal or usual corporate culture?
    b) Should it be?

18. Should corporate leaders influence their employees' perceptions of their corporations with regard to its monetary compensation (i.e., salary, bonuses, benefits, etc.)?

Company Slogan

19. Corporate tagline: "We're fast at *Phaust*." Can "fast" mean both efficient and productive, or does expediency necessarily allude to irresponsibility?
20. Does "fast" mean "cheap"?
21. If you're going to put a project on a fast-track, how can you make sure you provide the resources to enable it to be done properly?

Effective Communications

22. The team meeting is marked by tension. Do ethical obligations suffer when the team is dysfunctional? Is some degree of disagreement appropriate? When does conflict become counterproductive?
23. Notwithstanding Wally's rule, if we are to develop a corporate culture, how do we encourage the kind of discussions that facilitate comfort in speaking freely of ethical, safety, and legal issues, as well as facilitate an *obligation* to speak freely about safety, ethical and legal issues?
24. What are some standards the leadership of an organization has to consider when creating an environment that creates this kind of a situation? Some questions we might ask ourselves:
    a) How do we set the appropriate standards?
    b) How do we make sure that we have communicated those standards effectively so that people not only understand them but also believe we're serious about them?

Regarding the L&L Controls

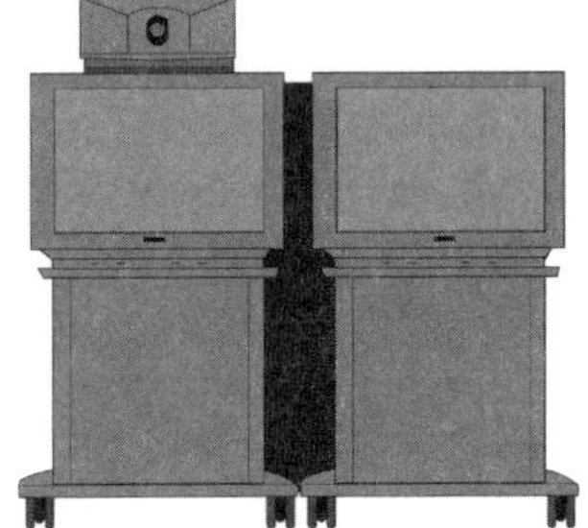

25. The sales representative for Lutz & Lutz Controls is Chuck's brother-in-law. Does this kind of a personal relationship tend to compromise not only the integrity but the effectiveness of the procurement process?
26. How does the relationship with L&L look from the perspective of other suppliers?
27. Should Fred take into account the possibility that there's actually a positive aspect in the long-term relationship between *Phaust* and L&L Controls?
28. How can a process be arranged so that we can assure that it is effective and that we are purchasing from the supplier that's going to provide the best

service and price for the organization?

29. Should we view that process from the perspective of other manufacturers of industrial controls?
30. What should Fred's role be in the procurement decision-making process?
31. What should be the role of others to make sure that the process is running effectively?
32. From whose perspectives should these decisions be made?

Marketing Decisions

33. Someone at *Phaust* suggested the name "Strip-Teasy" for the new paint-stripping product. Was this suggestion appropriate from an ethical viewpoint?
34. Should engineers have some input with regard to how the product is to be marketed?
35. Do marketing methods ever have engineering consequences?
36. Regarding Hal, from product development and market research, Wally whispered to Fred: "He's not one of us." What feeling does this statement convey? Is this likely to promote tension and/or dissention?
37. Should corporations have an obligation to have sensitivity training for their managers and their engineers?
    a) What has society taught people in this matter?
    b) How is that brought into the corporate boardroom?
    c) How is that transcended into the corporate culture?
38. What do you think you would do to promote trust and respect among your colleagues and other professionals in other departments?

Budget Issues

39. The French corporate headquarters mandated a 20% cut across the board. Is there a difference between cutting budgets across the board rather than giving a bottom line and allowing management to differentially cut in ways that impact less severely on the viability of the project?
40. Regarding the sudden cut in budget, Chuck says, "Sometimes you inflate budgets, and sometimes you build schedules with slack. That way, if something unexpected occurs, you're covered." Is this the same as covering for contingencies?
41. Did this attitude toward the budget promote "trust" in the company?

Interaction Between Plant Designers and Plant Operators

42. Should the engineers designing the project be in contact with people who have to maintain or operate it? Is this a serious consideration? How important is this relationship?

43. Is divorcing maintenance and operations from design a serious thing?
44. Is the separation of operations and design an ethical issue, or is it just a business issue?
45. When it becomes clear that the engineers are passing potential problems with valves and switches on to the future productivity of operations, does it seem that this is simply a way of sloughing off responsibility?
46. Is this action representative of the corporate culture of *Chemitoil*?
47. Does this attitude promote trust between designers and operations/ maintenance personnel?

Safety Issues

48. Were there any scenes that revealed a lack of proper protective equipment being worn?
49. Does not wearing proper protective equipment in a plant become a leading indicator of other corporate culture problems?
50. What about engineering college laboratories? Do faculty and students always wear appropriate safety gear?
51. Is this an ethical issue?
52. Do accidents just "happen," or are they "caused"?
53. When we think about engineers working with operations people, should there be a mutual respect for entering - what we might call - each other's "territory of responsible care"?
54. Whose responsibility is it to make sure that reasonable care and attention is given to safety?
    a) The plant operator?
    b) The manager of the lab?
    c) Anyone who observes the problem?
55. If we were to analyze the culture of *Phaust*, would the safety issues provide some good indicators about the entire culture of the company?

Regarding the Software of the Cheaper Controls

56. Fred decides to investigate the possibility of using the less expensive controls which, according to Peter, the project manager in Big Spring, has software that's "as buggy as a New York City basement." Why is it that software is so hard to get correct?
57. What is the responsibility of software engineers to take safety into consideration when they develop new software?
58. What responsibility do software engineers have for the quality of their products when the software will control a safety related process?

59. Do the ethical obligations of software engineers differ from those of other kinds of engineers (civil engineers, mechanical engineers, materials engineers)?
60. What are the differences between office software and process control software?

Personal Relationships

61. Is it appropriate for an engineer to discuss company matters with family members when he feels ethically bound to keep this information confidential?
62. How do employees and engineers decide whether to share some of their information from work when they go home?
63. Should anything be said at all?
64. The story shows a husband and wife (Fred and Maria) at home. Does this personal interaction have any effect on the story?
65. As an attorney for the EPA, did Maria's interest in how Fred was going to manage certain types of environmental issues that she had some specialized expertise in exemplify poor ethical standards?
66. Was it fair for Maria to use seductive techniques to win her way?
67. Is Maria's code of ethics an issue?
68. Was she flirting with some ethical issues that were outside the engineering area?
69. If Maria had been a schoolteacher or a nurse instead of an EPA compliance officer, would you have viewed this differently?

Regarding International Cultural Issues

70. What different challenges do international engineering practices offer?
71. What different opportunities do international engineering practices offer?
72. Do legal obligations change when we cross a state or national border?
73. Do our professional technical responsibilities and our ethical obligations change?
74. Do our obligations for the safety of our employees, the public, and environmental responsibilities change as we cross national borders?
75. Should we take the local culture into account? If so, how?
76. Do we have different ways of meeting our obligations?
77. Does it matter that *Phaust* is owned by a foreign firm who may not have the same cultural ideals as its North American management?

78. Should engineers act or react differently if they're owned by a foreign corporation?

Making Decisions

79. When Wally confronted Fred, Fred said that he was looking through some ethics manuals where he would have found a corporate or professional code of ethics. What would these codes say about his situation?
80. When making final decisions, was Fred trying to meet his ethical obligations within the constraints that he had?
81. We know that Fred decides to make a trade off in costs. Should Fred have decided to line the ponds rather than use higher quality controls?
82. Because lining the ponds was going to be more expensive, was Fred's decision to use cheaper controls and couplings a responsible one?
83. Are Fred's decisions the most effective ways to meet his obligations of ensuring the safety, health, and welfare of the operators and the public, as well as being a good employee and not exceeding his budget?
84. Is it ethically permissible to use a control system with sensors at the end of their normal operating range?
85. Should Fred have thought about future problems when specifying the cheaper controls and couplings?
86. Fred was unable to anticipate every deviation. Should he have been more prudent and tried to anticipate more of those deviations?
87. What options did Fred have?
88. At what point, if any, should the project have been stopped?
89. What would you have done?
90. Do you think that you would receive the same degree of criticism from your company if you violated an ethical standard compared to violating or missing a deadline or an objective?
91. Do you put ethical issues on the same level of importance as business objectives?

What if Automatic Controls Don't Work

Fred believed that the cheaper design was satisfactory until he learned that in order to produce a more competitive product, the pressure and temperature of the process had to be increased significantly. This exceeded the limits of the cheaper automatic controls.

The plant manager volunteers to control the process manually. Fred allows the plant manager to control the process.

92. Is it ethical to design an automated system, and then let your backup be a person?

93. Should there be an automated system where an individual supplements, rather than replaces, it?

Margin of Error and Reasonable Care

94. Did the design involve a margin of error?

95. What would you have done if you had been Fred?

96. Is it possible (or likely) that under pressure, we'll let liability insurance be our safety factor?

97. Do you feel that some companies may be willing to accept catastrophic failure as long as they are insured enough to cover the damages and liability?

98. Since nothing can be both totally safe and affordable, what is a reasonable amount of protection from failure?

Trust and Candor

99. What is the most critical element in effective relationships? Loyalty? Obedience? Money? Trust? Openness? Candor?

100. What roles should trust play in our professional and personal interactions?

101. What role should candor play in a professional or personal relationship?

102. Would candor imply effective communications?

103. Could you envision one definition of ethics as being "those activities and practices that tend to enhance trust"?

104. Although there are a lot of things that one will gain during the course of a professional career, there may be some things that can be taken away from you. Your job could be taken away (we see that happening all too often with downturns in the economy).
    a) Can you think of something that can never be taken away - unless you allow it to be?
    b) What about your "reputation for integrity"?

105. If someone says "I trust you," how does this make you feel about the relationship?

If You Were in Charge

When we are involved in our day-to-day work, we tend to look up to our leaders, supervisors, and/or bosses for guidance and inspiration in terms of how we conduct ourselves. It is important for us to put on another hat. Let's say to ourselves "Okay, we're in charge. We're the boss, we own the company, we have all the money we need to decide what we want to produce, where we're going to produce it, how we're going to produce it, and how we're going to set up the whole organization." We need to ask ourselves some questions:

106. What are the specific actions that we take as leaders, as the boss, as the owner, to make sure that everyone in our organization feels that they should conduct themselves to the highest standards of professional conduct and professional ethics?
107. What are the specific attributes that you would have in place in your company to make sure that happens?
108. What would you do if you were the leader and had all the controls at your command to set up your organization any way you wanted to in order to make sure that everyone in that organization would conduct themselves to the highest professional and ethical standards? Would some of the following actions come to mind?
    a) Clearly define your expectations of professional/ethical actions
    b) Communicate those expectations effectively and continuously
    c) Live the standards personally -- what people see in actions is what they're going to believe
    d) Create candor and open communication in the environment so that anyone within the organization feels free to bring up and discuss their thoughts, opinions, and ideas, but most of all, they feel free to bring up their concerns, problems, and news, be it good or bad, without fear of suffering some sort of retribution or reprisal

Acknowledgments

In addition to funding for the video provided by Grant SES-0138309 from the National Science Foundation, major donations came from:

Harry E. Bovay, Jr., P.E.
President, Mid-South Telecommunications;
Past President, National Society of Professional Engineers;

Victor O. Schinnerer and Company, Inc.

National Society of Professional Engineers

American Society of Mechanical Engineers

William J. Lhota, P.E.
Retired President, American Electric Power Energy Delivery

Steven P. Nichols, P.E.
Director, Clint W. Murchison Chair of Free Enterprise,
University of Texas at Austin

Robert L. Nichols, P.E.
Past President, National Society of Professional Engineers and the National Institute for Engineering Ethics

Donald L. Hiatte, P.E.
President, National Council of Examiners for Engineering and Surveying;
Past President, National Society of Professional Engineers

Jimmy H. Smith, P.E.
Past President, National Institute for Engineering Ethics

Leader of Private Fund Raising
E.D. "Dave" Dorchester, P.E.
Immediate Past President, National Institute for Engineering Ethics

Script Consultants/Advisors
Jose Guerra, P.E.; Jose Novoa, P.E.; Christopher Smith

Executive Producers, Producer, and Writer/Director

Jimmy H. Smith, Ph.D., P.E.
Project Director, Professor of Civil Engineering and Director
National Institute for Engineering Ethics, Texas Tech University

Steven P. Nichols, Ph.D. J.D., P.E.
Professor of Mechanical Engineering
Associate Vice President for Research and
Director, Clint W. Murchison Chair for Free Enterprise, University of Texas at Austin

Michael C. Loui, Ph.D.
Professor of Electrical & Computer Engineering
University of Illinois at Urbana-Champaign

Vivian Weil, Ph.D.
Professor of Philosophy, Director, Center for the Study of Ethics in the Professions
Illinois Institute of Technology

Philip E. Ulmer, P.E.
Past President, National Institute for Engineering Ethics
Safety Manager, General Communication, Inc., Anchorage, Alaska

Carl M. Skooglund
Retired Vice President and Ethics Director
Texas Instruments, Inc.

Frederick Suppe, Ph.D.
Professor of Philosophy, Texas Tech University

E. Walter LeFevre, Jr., Ph.D., P.E.
Past President, National Society of Professional Engineers
Professor of Civil Engineering, University of Arkansas

Patricia Harper
Program Coordinator, Assistant to the Director and Video Production Assistant
National Institute for Engineering Ethics, Texas Tech University

Producer : **Kenneth Mandel**
Great Projects Film Company, New York City

Writer-Director : **Paul Martin**
Great Projects Film Company, New York City

Information Regarding Purchases and Free Resources

Copies of *Incident at Morales* in VHS or DVD format
may be purchased by contacting

National Institute for Engineering Ethics
Box 41023, Lubbock, Texas 79409-1023
Phone: 806-742-6433 (NIEE) or 806-742-3525; Fax: 806-742-0444
Email: Ethics@coe.ttu.edu

For details and a short clip of the video, see
www.niee.org
Go to Products & Services

Free Resources Available

Study Guide and Power Point Presentation are available free at www.niee.org, Products & Services

The 24-page study guide contains suggestions for use of the video, the story line, list of characters, synopsis of the video, purpose of the video, over 100 questions about ethical issues that the story raises, and a suggested assignment for students and viewers. Twenty-six slides make up a power point presentation which may be downloaded free, modified to fit the presentation format and used as an introduction to *Incident at Morales*.

Information Regarding Prices and Conditions on Use of *Incident at Morales*

Prices and Discounts:

- **VHS Tape + 24 Page Study Guide**: $250 plus $10 S&H (36 minutes long with two pauses for discussion)
- **Interactive DVD + 24-Page Study Guide:** $500 plus $10 S&H (36 minutes plus 45 minutes of interactive discussions with executive producers and others; DVD also contains a Spanish subtitled version)

50 % Discount to Universities and Engineering for non-profit educational use.
60 % Discount to Universities and Engineering Societies for educational use on purchases of 5 or more copies in a single order.

Conditions on Use:

Incident at Morales is intended to be used in non-income-generating educational programs, unless NIEE is notified. If charges are imposed on viewers by societies, universities, companies, or individuals, a royalty fee of $100 for each such use is to be paid to NIEE.

Incident at Morales may be purchased for resale and used in marketing arrangements by organizations. Contact NIEE for details.

TEAM DISCUSSIONS ON *GILBANE GOLD*

Gilbane Gold Work Sheet

Assignment A: Being "personally" involved:

After viewing *Gilbane Gold*, list two or three major ethical issues in the video.

1. ______________________________

2. ______________________________

3. ______________________________

Assignment B: Being "responsible" for the future:

Assuming you are the President of Z CORP, list two or three actions you would take immediately (within the next 2 days)

1. ______________________________

2. ______________________________

3. ______________________________

"Gilbane Gold"

Produced by the National Institute for Engineering Ethics (NIEE) and the National Society of Professional Engineers (NSPE)

Synopsis

Gilbane Gold is the name given to dried sludge from the Gilbane wastewater treatment plant. It is sold to farmers as a commercial fertilizer. The annual revenue generated saves the average family about $300 per year in taxes. Several years ago the city of Gilbane established limits on the discharge of heavy metals to the sewers in order to protect Gilbane Gold from the build-up of toxic materials that could end up in the soil. The limits are more restrictive than federal limits but are based on the concentration of the discharge with no restriction on total weight of discharged material.

Z CORP, a computer components manufacturer, discharges wastewater containing small amounts of lead and arsenic into the city sewers. By the current city test standards, the discharge usually meets allowable levels. Z CORP people know of a newer test which shows that the discharge exceeds the limits. The ethical dilemma is: that acceptance of the new test might require additional investment in clean-up equipment by Z CORP. Word leaks and a TV investigation begins.

A complicating factor: Z CORP has just received a contract for five times as many computer modules as they presently make, but at a thin profit margin. Five times as much waste will be produced. Adding five times as much water will meet the city standards for discharge concentration, but Gilbane Gold will contain five times as much heavy metal as before. Z CORP's VP says changing the test standards would cause the company to lose money on the new contract. Her position: Z CORP should provide jobs; the city should worry about the environment.

Key Players

David Jackson ..... Young environmental engineer working for Z CORP

Tom Richards ...... Engineering consultant fired by Z CORP for espousing the new test standards

Phil Port .... Manager and head of Z CORP's environmental affairs department

Frank Seeders ..... Z CORP's engineering manager

Diane Collins .... VP in charge of Z CORP's Gilbane plant

Lloyd Bremen ..... Former state commissioner of environmental protection

Winslow Massin .. Professor emeritus, School of Engineering, Hanover University

Maria Renato .... TV reporter

Strategy for Using Cases to Study Ethics

There are many sources for case studies. We will discuss why we use case studies and then we will suggest a strategy for how case studies can be effectively used.

The most effective way to convince professionals that they will face ethical issues and that they have to make ethical choices is introduce them to making choices and decisions in a non-threatening environment. This is best accomplished by having them consider actual case studies of situations that other professionals have faced in the past.

Objectives: First, let's briefly discuss objectives for presenting ethics cases:

1. Encourage individual and group thinking and communication on ethical issues
2. Develop abilities to consider ethics issues from two standpoints:
   A. From a personal involvement viewpoint
   B. From a responsibility viewpoint

To accomplish these objectives, the workplace should exhibit an environment in which each individual is free to determine his/her own conclusions regarding ethical dilemmas. As in engineering design, there may be no single right answer. Unlike design, some ethical dilemmas may have no "right" answers, but a personal "best" answer does exist.

A suggested strategy:

In order to overcome our natural reluctance (at least initially) to discuss difficult ethical dilemmas in large groups, the group should break out into small group discussions after being presented with a specific ethics case.

1. Each small group should be asked to assume the role of one of the people in the case, discuss that person's situation in some detail, and develop a few questions that he or she might, or should be pondering. This encourages the group to think about the situation on a "**personal**" basis.

2. Next, to consider the various aspects of this ethical dilemma from a "**responsibility**" viewpoint, e.g., assume you are the President of the company or firm who is faced with resolving ethical issues.

Space Shuttle Challenger: Mission 51-L Launch

Prepared by:
Kurt Hoover, Graduate student, Department of Aerospace Engineering and Engineering Mechanics,
Wallace T. Fowler, Professor of Aerospace Engineering and Engineering Mechanics, and
Ronald O. Stearman, Professor of Aerospace Engineering and Engineering Mechanics.
The University of Texas at Austin

On January 28, 1986, the Space Shuttle Challenger was launched for the last time. The decision to launch the Challenger was not simple. Certainly no one dreamed that the Shuttle would explode less than two minutes after lift-off. Much has been said and written about the decision to launch. Was the decision to launch correct? How was the decision made? Could anyone have foreseen the subsequent explosion? Should the decision-making procedure have been modified? These questions are examined in this case study.

Background on The Space Shuttle:

The Space Shuttle is the most complicated vehicle ever constructed. Its complexity dwarfs any previous project ever attempted, including the Apollo project. The Apollo project possessed a very specific goal, to send men to the moon. The Space Shuttle program has a wide variety of goals, some of which conflict. The attempt to satisfy conflicting goals is one of the chief roots of difficulty with the design of the Space Shuttle. Originally, the design was to be only a part of NASA's overall manned space transportation system, but because of politics and budget cuts, it was transformed from an integral component of a system to the sole component of the manned space program.

The Space Shuttle was the first attempt to produce a truly reusable spacecraft. All previous spacecraft were designed to fly only a single mission. In the late 1960's, NASA envisioned a vehicle which could be used repeatedly, thus reducing both the engineering cost and hardware costs. However, the resulting vehicle was not as envisioned. It had severe design flaws, one of which caused the loss of the Challenger.

NASA Planning and Politics:

NASA's post-Apollo plans for the continued manned exploration of space rested on a three-legged triad. The first leg was a reusable space transportation system, the Space Shuttle, which could transport men and cargo to low earth orbit (LEO) and then land back on Earth to prepare for another mission. The second leg was a manned orbiting space station which would be resupplied by the Shuttle and would serve as both a transfer point for activities further from Earth and as a scientific and manufacturing platform. The final leg was the exploration of Mars, which would start from the Space Station. Unfortunately the politics and inflation of the early 70's forced NASA to retreat from its ambitious program. Both the Space Station and the Journey to Mars were delayed indefinitely and the United States manned space program was left standing on one leg, the space shuttle. Even worse, the Shuttle was constantly under attack by a Democratic congress and poorly defended by a Republican president.

To retain Shuttle funding, NASA was forced to make a series of major concessions. First, facing a highly constrained budget, NASA sacrificed the research and development necessary to produce a truly reusable shuttle, and instead accepted a design which was only partially reusable, eliminating one of the features which made the shuttle attractive in the first place. Solid Rocket Boosters **(SRBs)** were used instead of safer liquid fueled boosters because they required a much smaller research and development effort. Numerous other design changes were made to reduce the level of research and development required.

Second, to increase its political clout and to guarantee a steady customer base, NASA enlisted the support of the United States Air Force. The Air Force could provide the considerable political clout of the Defense Department and had many satellites which required launching. However, Air Force support did not come without a price. The Shuttle payload bay was required to meet Air Force size and shape requirements which placed key constraints on the ultimate design. Even more important was the Air Force requirement that the Shuttle be able to launch from Vandenburg Air Force Base in California. This constraint required a larger cross range than the Florida site, which in turn decreased the total allowable vehicle weight. The weight reduction required the elimination of the design's air breathing engines, resulting in a single-pass unpowered landing. This greatly limited the safety and landing versatility of the vehicle.

Factors Affecting the Launch Decision

Pressures to Fly:

As the year 1986 began, there was extreme pressure on NASA to "Fly out the Manifest". From its inception the Space Shuttle program had been plagued by exaggerated expectations, funding inconsistencies, and political pressures. The ultimate design was shaped almost as much by politics as physics. President Kennedy's declaration that the United States would land a man on the moon before the end of the decade had provided NASA's Apollo program with high visibility, a clear direction, and powerful political backing. The space shuttle program was not as fortunate; it had neither a clear direction nor consistent political backing.

System Status and Competition:

In spite of all its early difficulties, the Shuttle program looked quite good in 1985. A total of 19 flights had been launched and recovered, and although many had experienced minor problems, all but one of the flights could rightfully be categorized as successful. However, delays in the program as a whole had led the Air Force to request funds to develop an expendable launch vehicle. Worse still, the French launch organization Arianespace, had developed an independent capability to place satellites into orbit at prices the Shuttle could not hope to match without greatly increased federal subsidization (which was not likely to occur as Congress was becoming increasingly dissatisfied with the program). The shuttle was soon going to have to begin showing that it could pay for itself. There was only one way this could be done—increase the number of flights.

For the shuttle program, 1986 was to be the year of truth. NASA had to prove that it could launch a large number of flights on time to continue to attract customers and retain Congressional support.

Unfortunately, 1986 did not start out well for the shuttle program. Columbia, Flight 61-C, had experienced a record four on-pad aborts and had three other schedule slips. Finally, on mission 61-C, Columbia was forced to land at Edwards Air Force Base rather than at Kennedy Space Center as planned. The delays in Columbia's launch and touchdown threatened to upset the launch schedule for the rest of the year.

Not only did Columbia's landing at Edwards require it to be ferried back to the Cape, but several key shuttle parts had to be carried back by T-38 aircraft for use on the other vehicles. These parts included a temperature sensor for the propulsion system, the nose-wheel steering box, an air sensor for the crew cabin, and one of the five general purpose computers. At the time of the Challenger explosion, NASA supposedly had four complete shuttles. In reality there were only enough parts for two complete shuttles. Parts were passed around and reinstalled in the orbiters with the earliest launch dates. Each time a part was removed or inserted, the shuttles were exposed to a whole host of possible servicing-induced problems.

In addition to problems caused by the flight 61-C of Columbia, the next Columbia flight, 61-E, scheduled for March also put pressure on NASA to launch the Challenger on schedule. The March flight of Columbia was to carry the ASTRO spacecraft which had a very tight launch window because NASA wanted it to reach Halley's Comet before a Russian probe arrived at the comet. In order to launch Columbia 61-E on time, Challenger had to carry out its mission and return to Kennedy by January 31.

Politics:

NASA had much to gain from a successful Flight 51-L. The "Teacher in Space" mission had generated much more press interest than other recent shuttle flights. Publicity was and continues to be extremely important to the agency. It is a very important tool which NASA uses to help ensure its funding. The recent success of the Space Shuttle program had left NASA in a Catch 22 situation. Successful shuttle flights were no longer news because they were almost ordinary. However, launch aborts and delayed landings were more news worthy because they were much less common.

In addition to general publicity gained from flight 51-L, NASA undoubtedly was aware that a successful mission would play well in the White House. President Reagan shared NASA's love of publicity and was about to give a State of the Union speech. The value of an elementary teacher giving a lecture from orbit was obvious and was lost neither on NASA nor on President Reagan.

Dryden Flight Research Center EC82-21135 Phototographed 4JUL1982
Shuttle Challanger and NASA 747
NASA photo

Sequence of Events

Monday, January 27, 1986:
On Monday NASA had attempted to place Challenger in orbit only to be stymied by a stripped bolt and high winds. All preliminary procedures had been completed and the crew had just boarded when the first problem struck. A microsensor on the hatch indicated that it was not shut securely; it turned out that the hatch was shut securely and the sensor was malfunctioning, but valuable time was used determining that the sensor was the problem.

After closing the hatch the external hatch handle could not be removed. The threads on the connecting bolt were stripped and instead of cleanly disengaging when turned the handle simply spun around. Attempts to use a portable drill to remove the handle failed. Technicians on the scene asked Mission Control for permission to saw the bolt off. Fearing some form of structural stress to the hatch, engineers made numerous time consuming calculations before giving the go-ahead to cut off the bolt. The entire process consumed almost two hours before the countdown was resumed.

Misfortunes continued. During the attempts to verify the integrity of the hatch and remove the handle, the wind had been steadily rising. Chief Astronaut John Young flew a series of approaches in the shuttle training aircraft and confirmed the worst fears of Mission Control. The crosswinds at the Cape were in excess of the level allowed for the abort contingency. The opportunity had been missed and the flight would have to wait until the next possible launch window, the following morning. Everyone was quite discouraged especially since extremely cold weather was forecast for Tuesday which could further postpone the launch.

Tuesday, January 28, 1986:
After the canceled launch on Monday morning there was a great deal of concern about the possible effects of weather. The predicted low for Tuesday morning was

23° F, far below the nominal operating temperature for many of the Challenger's subsystems. Undoubtedly, as the sun came up and the launch time approached both air temperature and vehicle would warm up, but there was still concern. Would the ambient temperature become high enough to meet launch requirements?
NASA's Launch Commit Criteria stated that no launch should occur at temperatures below 31° F. There was also concern over any permanent effects on the shuttle due to the cold overnight temperatures.

All NASA centers and subcontractors involved with the Shuttle were asked to determine the possible effects of cold weather and present any concerns. In the meantime Kennedy Space Center went ahead with its freeze protection plan. This included the use of anti-freeze in the huge acoustic damping ponds, and allowing warm water to bleed through pipes, showers, and hoses to prevent freezing.

The weather for Tuesday morning was to be clear and cold. Because the overnight low was forecast at 23° F, there was doubt that Challenger would be much above freezing at launch time. The Launch Commit Criteria included very specific temperature limits for most systems on the shuttle. A special waiver would be required to launch if any of these criteria were not met. Although these criteria were supposedly legally binding, Marshall Space Flight Center administrator Larry Mulloy had been routinely writing waivers to cover the problems with the SRBs on the recent shuttle flights.

Engineers at Morton-Thiokol, the SRB manufacturer in Utah, were very concerned about the possible effects of the cold weather. The problems with the SRBs had been long known to engineers Roger Boisjoly and Allan McDonald, but both felt that their concerns were being ignored. They felt that the request by NASA to provide comment on the launch conditions was a golden opportunity to present their concerns. They were sure that Challenger should not be launched in such conditions as those expected for Tuesday morning. Using weather data provided by the Air Force, they calculated that at the 9:00 am launch time the temperature of the O-rings would be only 29° F. Even by 2:00 PM, the O-rings would have warmed only to 38° F.

The design validation tests originally done by Thiokol covered only a very narrow temperature range. The temperature data base did not include any temperatures below 53° F. The O-rings from Flight 51-C which had been launched under cold conditions the previous year showed very significant erosion. This was the only data available on the effects of cold, but all the Thiokol engineers agreed that the cold weather would decrease the elasticity of the synthetic rubber O-rings, which in turn might cause them to seal slowly and allow hot combustion gas to surge through the joint.

Based on the these results, the engineers at Thiokol recommended to NASA Marshall that Challenger not be launched until the O-rings reached a temperature of 53° F. The management of Marshall was flabbergasted, and demanded that Thiokol prove that launching was unsafe. This was a complete reversal of normal procedure. Normally, NASA required its subcontractors to prove that something was safe. Now they were requiring their subcontractors to prove that something was unsafe. Faced with this extreme pressure, Thiokol management asked its engineers to reconsider their position. When the engineers stuck to their original recommendations not to fly, Thiokol management overruled them and gave NASA its approval to launch.

Rockwell, the company which manufactured the Orbiter also had concerns about launching in cold and icy conditions. Their major concern was the possibility of ice from either the shuttle or the launch structure striking and damaging the vehicle. Like Thiokol, they recommended against the launch, and they too were pressed to explain their reasoning. Instead of sticking with their original strong recommendation against launch, the Rockwell team carefully worded their statement to say that they could not fully guarantee the safety of the shuttle.

In its desire to fly out its manifest, NASA was willing to accept this as a recommendation. The final decision to launch, however, belonged to Jesse Moore. He was informed of Rockwell's concerns, but was also told that they had approved the launch. The engineers and management from NASA Marshall chose not to even mention the original concerns of Thiokol. Somehow, as the warnings and concerns were communicated up each step of the ladder of responsibility, they became diminished.

Late Monday night the decision to push onward with the launch was made. Despite the very real concerns of some of the engineers familiar with the actual vehicle subsystems, the launch was approved. No one at NASA wanted to be responsible for further delaying an already delayed launch. Everyone was aware of the pressure on the agency to fly out the manifest, yet no one would have consciously risked the lives of the seven astronauts. Somehow, the potential rewards had come to outweigh the potential risks. Clearly, there were many reasons for launching Challenger on that cold Tuesday morning; in addition a great deal of frustration from the previous launch attempt remained.

Pre-Launch Events:
Although the decision to launch on Tuesday had been made late on Monday night, it was still possible that something might force NASA to postpone the launch. However, the decision to launch had been made, and nothing was going to stand in the way; the "press on" mentality was firmly established and even if all of Florida froze over, Challenger would launch.

The pre-launch inspection of Challenger and the launch pad by the ice-team was unusual to say the least. The ice-team's responsibility was to remove any frost or ice on the vehicle or launch structure. What they found during their inspection looked like something out of a science fiction movie. The freeze protection plan implemented by Kennedy personnel had gone very wrong. Hundreds of icicles, some up to 16 inches long, clung to the launch structure. The handrails and walkways near the shuttle entrance were covered in ice, making them extremely dangerous if the crew had to make an emergency evacuation. One solid sheet of ice stretched from the 195 foot level to the 235 foot level on the gantry. However, NASA continued to cling to its calculations that there would be no damage due to flying ice shaken lose during the launch.

The Launch:
As the SRBs ignited, the cold conditions did not allow the O-rings to properly seat. Within the first 300 milliseconds of ignition, both the primary and secondary O-rings on the lowest section of the right SRB were vaporized across 70° of arc by the hot combustion gases. Puffs of smoke with the same frequency as the vibrating booster are clearly present in pictures of the launch. However, soon after clearing the tower, a temporary seal of glassy aluminum-oxides from the propellant formed in place of the burned O-rings and Challenger continued skyward.

Unfortunately, at the time of greatest dynamic pressure, the shuttle encountered wind shear. As the Challenger's guidance control lurched the Shuttle to compensate for the wind shear, the fragile aluminum-oxide seal shattered. Flame arched out of the joint, struck the external tank and quickly burned through the insulation and the aluminum structure. Liquid Hydrogen fuel streamed out and was ignited. The Challenger exploded.

When the remains of the cabin were recovered, it became apparent that most of the crew survived the explosion and separation of the Shuttle from the rest of the vehicle. During the 2-minute 45-second fall to the ocean at least four of the personal egress packs were activated and at least three were functioning when the Challenger struck water. The high speed impact with the water produced a force of 200g and undoubtedly killed all the crew.

Post-Crash Events:
Since the crash of Challenger, NASA and external investigators have taken a look at both the shuttle and the sequence of events which allowed it to be launched. The SRBs have gone through significant redesign and now include a capture feature on the field joint. The three Marshall administrators most responsible for allowing the SRB problems to go uncorrected have all left NASA. Following the recommendations of the Rogers commission, NASA has attempted to streamline and clean-up its communication lines. A system for reporting suspected problems anonymously now exists within NASA. In addition, the astronauts themselves are now much more active in many decision making aspects of the program. The current NASA Administrator, Admiral Richard Truly, is a former shuttle astronaut.

<u>Safety and Ethics Issues</u>

There are many questions involving safety and/or ethics which are raised when we examine the decision to launch the Challenger. Obviously, the situation was unsafe. The ethics questions are more complex. If high standards of ethical conduct are to be maintained, then each person must differentiate between right and wrong, and must follow the course which is determined to be the right or ethical course. Frequently, the determination of right or wrong is not simple, and good arguments can be made on both sides of the question. Some of the issues raised by the Challenger launch decision are listed here:

1. Are solid rocket boosters inherently too dangerous to use on manned spacecraft? If so, why are they a part of the design?
2. Was safety traded for political acceptability in the design of the Space Shuttle?
3. Did the pressure to succeed cause too many things to be promised to too many people during the design of the Space Shuttle?
4. Did the need to maintain the launch schedule force decision makers to compromise safety in the launch decision?
5. Were responsibilities being ignored in the writing of routine launch waivers for Space Shuttle?
6. Were managers at Rockwell and Morton Thiokol wise (or justified) in ignoring the recommendations of their engineers?
7. Did the engineers at Rockwell and Morton Thiokol do all that they could to convince their own management and NASA of the dangers of launch?
8. When NASA pressed its contractors to launch, did it violate its responsibility to ensure crew safety?
9. When NASA discounted the effects of the weather, did it violate its responsibility to ensure crew safety?

References

1. *Actions to Implement the Recommendations of the Presidential Commission of the Space Shuttle Challenger Accident.* National Aeronautics and Space Administration. Washington, DC July 14, 1986.

2. *Challenger: A Major Malfunction.* Malcolm McConnell. Doubleday & Company, Inc. Garden City, NY. 1987

3. *Prescription for Disaster.* Joseph J. Trento. Crown Publishers Inc. New York, NY. 1987

4. *Report of the Presidential Commission of the Space Shuttle Challenger Accident.* The Presidential Commission of The Space Shuttle Challenger Accident. Washington, DC June 6, 1986.

Challenger Launch Decision Assignments

The problem faced by NASA managers on January 28, 1986, is simply stated - Given the existing weather conditions, the recommendations of the various engineering and operational groups, and the political pressures, should Challenger be launched?

Many conflicting factors were considered in reaching the decision to launch. Those responsible for high risk programs such as Challenger must attempt to identify and evaluate the risks. Specific questions which needed to be answered were:

(1) What level of risk was acceptable for launch? and

(2) Did the current conditions meet this standard?

Even properly identifying and evaluating all risks is not sufficient, because the potential benefits of taking each risk must be considered. Greater risks can sometimes be justified given the possibility of greater rewards. In the case of the Challenger, the people with the ultimate authority to launch came to the conclusion that the potential rewards justified what they believed to be relatively minor risks. The belief that the risks were minor, however, was not shared by many of the engineers further down the chain of responsibility.

Assignment A

Read the General Information provided on the Space Shuttle Challenger launch decision. Consider each of the following questions carefully in light of that information and write a complete and grammatically correct paragraph answering each.

1. Why did NASA decide to launch Challenger?
2. How safe is safe enough? How does one determine what is an acceptable risk?
3. Is it possible to develop a methodology for quantifying risks, or must each particular situation be addressed individually?
4. Were NASA administrators justified in writing Launch Commit Criteria Waivers for Challenger and previous shuttle flights?
5. At the time of the Challenger accident there was a general feeling among both NASA and the public that the space shuttle was no longer an experimental vehicle, but was now a fully operational vehicle, in the same sense as a commercial airliner. Was this a correct perception and why was it common?
6. Should someone have stopped the Challenger launch? If so how could an individual have accomplished this?
7. If you were on a jury attempting to place liability, whom would you say was responsible for the deaths of the astronauts? Are several individuals or groups liable?
8. How might the Morton-Thiokol engineers have convinced NASA and their own management to postpone the launch?
9. How might an engineer deal with pressure from above to follow a course of action he knows to be wrong?
10. How could the chains of communication and responsibility for the shuttle program have been made to function better?

Assignment B

Choose one of the following statements, research the topic, and write a two page paper in which you explore the impact of the topic on the Challenger explosion.

1. Following Apollo, the manned space program suffered from lack of funding and direction.
2. The design for the space shuttle is a series of compromises driven by poorly timed allocations of funds from congress.
3. To minimize R & D costs, only part of the shuttle system was made reusable and solid boosters were used instead of the safer liquid boosters.
4. NASA was under intense pressure at the time of the Challenger accident to prove that the shuttle was a viable launch vehicle.
5. A significant delay in launching Challenger would have upset the launch schedule for the rest of the year.
6. Flight 51-L (Challenger) was scrubbed the previous day leaving all involved frustrated and determined to launch as soon as possible.
7. No test data on any of the shuttle components existed for the low overnight or launch temperatures.
8. Problems with the seals on the SRBs had been known for several flights and waivers had been written for each flight.
9. Concerns about the O-rings were never revealed to the NASA administrators who had the final launch authority.
10. Morton-Thiokol initially recommended against launch, but when pressured by NASA reversed its decision.
11. The anti-freeze plan left large sheets of ice and icicles all over the launch structure. An analysis done at Houston showed no danger at lift-off due to falling ice.
12. Rockwell could not guarantee the shuttle's safety, but did not veto the launch. Their ice analysis showed some possibility of danger.
13. The ice-team recommended against launching, but was overruled by Mission Control.

Assignment C

Divide the class into small groups, no more than three to a group. Each group is to choose one of the four roles outlined below and develop a statement outlining the position represented by those in your role on January 28, 1986. Develop two statements: (1) what you think was the position of those in your role, and (2) the position that those in your role should have taken.

(1) NASA Management: You want to launch the Challenger as soon as possible. The delays are not only embarrassing, but threaten your funding and customer base. Challenger must launch on Tuesday to preserve the schedule. An analysis done by your engineers at Houston shows that the ice on the pad should not strike the Challenger when it lifts off.

(2) Thiokol Engineers: You believe it is not safe to launch, but have no hard data to back this up. Limited data from a previous cold weather flight indicates that temperature is important. Basic physics tells you that the O-rings will lose elasticity with decreasing temperature. You feel that both NASA and your own management are trying to solve the problem with a bureaucratic solution, when an engineering solution is called for.

(3) Thiokol Management: You must listen to your engineers, but at the same time you must please your primary customer. There is talk in Congress of awarding a second source contract. The last thing you want to do is admit that your product is defective. NASA is pressuring you to launch. If would be very damaging for your company if a delay is blamed on your SRBs.

(4) Rockwell Management: You are concerned about the amount of ice on the pad. Analysis by your engineers does not entirely agree with that done at Houston. Like Thiokol you must satisfy your customer. You would prefer not to launch, but are not sure that your reason to delay is good enough. Your objective is to try to convince NASA to delay without them pointing a finger at you as the cause.

Assignment D

Working in three person groups, develop a realistic procedure for making launch decisions which would have avoided the Challenger accident. Remember that the procedure must create a consensus among individuals and organizations with different objectives, backgrounds, and priorities. Part of your work will require that you develop a methodology to determine potential risks and benefits for launching the shuttle in less than ideal conditions. Remember that in the real world, personalities are often the dominant factor in a decision.

Assignment E

Working in three person groups, consider the problems of Allan McDonald and Roger Boisjoly. Develop a strategy to convince Thiokol management and NASA management that your safety concerns are valid. Consider the points of view of all of those who are pressing to launch. Remember that management often tends to view engineers as extremely competent in a specific area, but lacking a good understanding of the big picture.

The Developers of this publication express their sincere appreciation to Mr. Kurt Hoover, Dr. Wallace Fowler, Dr. Ronald Steadman, the Aerospace Department, and the College of Engineering at the University of Texas at Austin for providing this valuable source of educational information on engineering ethics.

For more information about the Challenger and other shuttles, see the NASA web site: NASA.gov or http://www.ksc.nasa.gov/shuttle/missions/51-l/docs/rogers-commission/table-of-contents.html

Backwards Math

The incorporation of ethics awareness in technical courses for engineering students and for consideration and discussion by practicing engineers is especially appropriate in today's environment. Engineering ethics is a topic which engineers simply cannot ignore given the nature of their work which, for example, may involve such things as designing and operating plants which may pollute and manufacturing products which may contain toxins and carcinogens. Not only do engineers have the potential to become involved in ethical issues which relate to the health, safety and welfare of society, but also they have recently borne the blame for many of the world's problems. Yet engineers are the ones who have provided most of the advances in technology which have greatly benefited mankind.

Because most ethical problems do not have clearly defined dimensions and right or wrong answers, answers which cannot be simply found in a Code of Ethics, instruction and/or discussion which generates an increased awareness of the professional and ethical responsibilities of engineers are important. For most dilemmas involving ethical or moral decisions, there is a single best personal answer even if there is no clear right or wrong answer. This is difficult for some to accept, as one of the trademarks of engineers is to see technical challenges as rights and wrongs, proven by equations.

There are a number of opportunities in engineering classes or in engineering offices in which the introduction of the case study in this module would be an appropriate entree to the discussion of ethics and professionalism. The case study presented here is reprinted from the May 5, 1980 issue of "Chemical Engineering" with the permission of the publisher, McGraw-Hill, Inc., New York, NY. The facts have been edited from a chemical engineering scenario (which discuss catalysts) to reflect the more typical experience of a transportation agency.

Situation

Jay's boss, a materials engineer for the State DOT, is an acknowledged expert in the field of pavement materials. Jay, also an engineer with the DOT, is the leader of an engineering/ research team that has been charged with developing a new pavement design, and the search for a suitable asphalt mix has been narrowed to two possibilities, Mix A and Mix B.

Jay's boss is certain that the best choice is Mix A, but he directs that tests be run on both, 'just for the record.' Owing to inexperienced help, the tests take longer than expected and the results show that Mix B is the preferred material. The engineers question the validity of the tests, but because of the project's timetable, there is no time to repeat the series. So the boss directs Jay to work the math backwards and come up with phony data to substantiate the choice of Mix A, a choice that all the engineers in the group, including Jay, fully agree with. Jay writes the report.

Questions: Consider the following questions and what would be your best personal answer.

1. What would you have done had you been Jay? Would you have written the report?

2. If you had refused to write the report, how could you have justified this refusal since you, along with all of the others on the team, felt that the test data were invalid and there was not time to duplicate the test? Is it not likely that experience is a better guide than one set of questionable test data? If yes, does this impact on the ethical nature of the dilemma?

3. Would it have been a good idea to write the report and also write a memo saying that you felt that you were being directed to do something which was unethical? If you did this, would it be done just to cover yourself if you are found out? Would that make what you did any more ethical than it would have been without the memo?

4. How about the alternative of writing the report, but not signing it? Would this be a satisfactory solution to an ethical dilemma?

5. Is this a case where you would become a whistle blower by going over your boss' head and reporting that you had been asked to write a false report?

6. If none of the above options seems to be a satisfactory action, what would you have done?

Murphy's Law: If ever an ethics case pointed up the pervasiveness of Murphy's Law, this one is it. If you have as much trouble making up your mind about the best course of action as did readers who responded to McGraw Hill's survey, there could be plenty of worthwhile discussion on this issue.

A useful suggestion was from another reader who suggested that there was no need to make a big deal of it. Report the "data as new data — which it is."

One reader commented that he didn't even want to think about this case because "It's bad enough being considered incompetent in your field, without being considered dishonest as well."

Clearly this module shows the personal nature of ethics. It also shows that lies will haunt you.

Deception in Business

Most professionals, not very far into their careers, will begin to find their careers progressing along one of two divergent paths. They will be moving into management or they will be in a technical ladder which will take them ever deeper into the technical operations of their organization.

Managers will all be involved to varying degrees in the hiring process and the business of evaluating and sometimes terminating employees. To a lesser extent, even the engineers on the technical ladder will have some of these responsibilities. The business of hiring, evaluating and firing directly affects people and decisions in these areas are heavily value laden while being filled with ethical dilemmas.

Federal law makes it illegal to discriminate on the basis of age in the hiring and firing of those who are more than 40 years old. Still, the opportunities for such discrimination are many and few would doubt that such discrimination is widely practiced.

Let's look at the following case study and see how you would handle it.

Situation

John was a 58-year-old professional who, in spite of a good record, had been laid off by a competitor company when it lost a government contract. The overall depressed economy in which his industry operated made his chances of getting another professional job highly unlikely. These chances were even further diminished by the fact that his was a type of work where youth is generally favored in hiring practices. John knew that age would work against him, even though he was vigorous, healthy and active. Only his gray hair gave away his age.

Before interviewing with you for a job with your company, he touched up his hair with dye so that the only gray was a bit which was confined to his temples. He knew that his real age might be learned in time, but he figured he would deal with that when and if it became a problem. Right now, his goal was to get a job similar to the one he had just lost. During the oral interview, John voluntarily told you that he was 45 years old. This was a lie.

After he has been on the job for a while and doing very satisfactory work, you learn that John lied about his age during the interview.

What should you do?

1. Given that John has been able to compete effectively against the younger professionals, was his deception important?
2. Was it inconsistent with the rules by which the game of business is played every day?

3. Is getting a job part of the game of business?
4. What, if anything, makes John's deception different from the bluffing and misrepresentation which commonly occurs in business?
5. What makes John's lie different from that which occurs every day when millions of people in the work world say "yes" to their bosses when they really believe "no" is the honest answer?
6. Was lying about his age less serious than would have been a lie about his G.P.A.?
7. If yes, is this true even though his G.P.A. was earned years ago when grade scales may have been different and since his work has proved his ability?
8. For the same reasons, would a lie about work history have been more important than the lie about age?
9. Why, particularly since his current work performance was highly satisfactory?

There are instances of unethical activities related to fundamental honesty in the corporate world. While perhaps not widespread, some executives practice some form of deception when negotiating with suppliers, customers, government officials, stock analysts or labor. For the good of the company, they conceal, tell less than complete truths, or actually make untrue statements and excuse their words—words which they define as bluffs—as being allowed, even required, by business dealings. This game may be played at all levels of corporate life in business and industry, but not in the better ones.

1. Is it less important for managers to be completely honest than for employees to be completely honest?
2. Does it matter if the employee is a professional or not?
3. Is there something which separates the rights of individuals from the rights of companies to practice the rights of business?
4. Do professionals have obligations to the public which a non-professional doesn't?

In the case of John, if you confront him about the lie and decide to continue his employment, does that confrontation terminate the concern, or is it valid to again confront him or consider the lie if it seems to have relevance at some time in the future?

1. If you decide to do nothing when you learn of John's lie, is it because his work has been highly satisfactory?
2. If John's work had been less good, would you be more apt to fire him for the lie?

The Tough Memo: Now He Saw It — Now He Didn't

Public Administration at the University of Arkansas - Revised 5/21/96

Used by permission. Contact Dr. William Miller, University of Arkansas

Context

A new political appointee, John Davis, has come to the State Department of Transportation. He was appointed to be Executive Director of the Agency by the newly elected governor. This particular governor's election campaign was based on "cleaning up the inefficiency and corruption in the state capitol."

John Davis begins his tenure as director of the Agency by sending out a tough memo. Anyone, he writes, caught taking any state property out of the office without written permission will be immediately fired. State property includes everything from a pencil to a computer. Further, he says in his memo, all staff are required to report any violation of this policy directly to him.

The Dilemma

Sam Miller is a career civil servant. Sam works as supervisor of the media department of the State Department of Transportation. One day, as he is talking to the department secretary, Jim Stevens enters the room and says hello. Jim Stevens is also a career civil servant and a very valuable member of the department's staff. Jim opens the storage cabinet and removes a piece of graph paper. He smiles at Sam and says, "My daughter has a school project and needs paper with lines both ways." Jim then leaves the room and Sam continues his conversation with the department secretary.

The next day there is a memo on Sam's desk from the director's office. The secretary has informed Director Davis of the "purloined graph paper". She rightly felt, the director Davis' memo states, that she was required to do so by the policy the director had made concerning taking state property. Now the Director wants Sam Miller to come to his office and be the "second witness" that is necessary for the director to fire Jim Stevens. Sam Miller sits as his desk and considers the situation. If he confirms the secretary's story, Jim Stevens will be fired and Sam will lose a very important part of his staff. Also a concern to Sam, Stevens has two kids in college and cannot afford to go without income now. Slowly Sam Miller climbs the stairs to the director's office. "Well," the director asks, "did you see the graph paper being taken?"

The Decision

After a brief pause, Sam says, "No. It may have happened that way, but I was involved in the conversation with the secretary and didn't really notice what Jim was doing." Though pretty frustrated, Director Davis accepts Sam's account and tells him to return to work.

What would you have done if you were Jim? Sam? Secretary? Director Davis?

Confidentiality Case: Useful Information

Public Administration at the University of Arkansas - Revised 5/21/96
Used by permission. Contact Dr. William Miller, Associate Professor of Political Science, University of Arkansas -- http://plsc.uark.edu/book/books/ethics/index.htm

Context

Charles Strait is a mid-level manager in the conservation department. As a part of the new emphasis on performance review for his agency, Charles is conducting a survey of commercial fisherman in the state. The purpose of the survey is to determine the effect that the programs of the conservation department have on the fishing industry.

Respondents to the survey are assured of both confidentiality and anonymity. The letter that accompanies the survey states that no specific responses will be connected with any survey respondents. To assure that, a stamped, self-addressed envelope with the department's own return address is sent out with each survey. Among the questions asked is the following question:

> My company often exceeds the fishing limits imposed by the conservation department.
> ____ Agree Strongly
> ____ Agree Mildly
> ____ Disagree Mildly
> ____ Disagree Strongly

The Dilemma

The problem comes when one company places its own return address on the reply envelope in addition to the conservation department's pre-printed address. Then, in reading the responses from this company, Charles notes that the respondent agreed strongly that his company often exceeded the fishing limits.

Unintentionally, Charles now knows about a company that should be more closely monitored. Though he has assured anonymity, it was the respondent's mistake that gave away the company's identity.

The Decision

Charles goes over the options in his mind. Commercial over-fishing can have disastrous consequences for sport fisherman and the ecology of a lake. Still, he is not supposed to know what he knows. Should he:

- Try his best to forget he knows what he knows?
- Assign extra inspectors to check for compliance at the respondent's company?

Would it change your view if the matter was more clearly related to the safety of the public?

Accepting Gifts And Amenities

One of the first ethical problems that many professionals face in their careers is the issue of what to do about accepting gifts and amenities. The issue is also faced by more experienced professionals. Consider the following actual experience of a chemical engineer.[1]

> **Case X**. Tom had been named the department manager of a large new chemical process unit which was in the design stage. Tom's responsibilities included forming the process unit staff, looking over the designers' shoulders to assure the plant was designed to be safe, operable and maintainable, and then starting the plant after construction. During his previous experience Tom had noticed that a new type of valve and valve operator could often be used in place of more common gate valves and their operators. In every case the new valve was less expensive and often gave a tighter shutoff than the gate valve. Tom convinced the project designer to add even more of these valves and operators to the design. This improved safety, because more flows could be shut off more quickly in an emergency.
>
> After a large number of valves had been specified and purchased, the salesman of the valves visited Tom and invited him on a very nice fishing trip to South America. Tom had not known the salesman, Jim, prior to the visit. He also had no direct purchasing responsibilities; he had just wanted the valves for increased safety in the new process unit.

Should Tom go on the trip? Engineering codes clearly state that accepting a bribe is unethical and unprofessional, but Tom's action cannot be fairly characterized as accepting a bribe. We might give a working definition of bribery as <u>remuneration for the performance of an act that is inconsistent with the nature of the work one has been hired to do</u>. If Tom took the trip, it would not be in remuneration for something inconsistent with his obligations to his employer. Furthermore, the offer was made after Tom's recommendations concerning the valves and (we shall assume) without any prior knowledge and expectation of the gift. We can, however, imagine a case that would involve bribery:

> **Case 1**. Tom had been named the department manager of a large new chemical process unit. His responsibilities included forming the process unit staff, looking over the designers' shoulders to assure the plant was designed to be safe, operable and maintainable, and then starting the plant after construction. During the design phase a salesman, Jim, approached

---

[1] *This case is a description of an actual experience of a chemical engineer. It is designated "Case X" because its place in the series of cases presented later is open for discussion. The other cases, which are hypothetical, are given numbers which roughly correspond to their place in a series, ranging from a clear case of bribery (Case 1) to a case which is clearly not bribery (Case 6)*

Tom and offered him a very nice fishing trip to South America in exchange for using his influence to get the valve sold by Jim's company specified by the designers. The valve was more costly and not as safe, but Tom recommended it anyhow. After the valves were purchased, the salesman invited Tom on the fishing trip to South America.

Even though the original scenario, unlike this one, is not a bribe, it does have some characteristics in common with this one. In order to see this, consider the following case, which has few, if any, analogies with a true bribe:

Case 6. Tom was named the department manager of a large new chemical process unit which was in the design stage. Tom's responsibilities included forming the process unit staff, looking over the designers' shoulders to assure the plant was designed to be safe, operable and maintainable, and then starting the plant after construction.

During his previous experience Tom had noticed that a new type of valve and valve operator could often be used in place of more common gate valves and their operators. In every case the new valve was less expensive and often gave a tighter shutoff than the gate valve. Tom convinced the project designer to add even more of these valves and operators to the design. This improved safety, because more flows would be shut off more quickly in an emergency.

After a large number of valves had been specified and purchased, the salesman came by and introduced himself, giving Tom a plastic pen worth about five dollars.

Few people would find objections with Tom's actions in this case, but consider the following cases, in which only the conclusion has been changed:

Case 2. After a large number of valves had been specified and purchased, Jim, the salesman, invited Tom to play golf with him at the local country club. Tom was an avid golfer and had wanted to play golf at the country club for some time, because it was the best course in town.

Case 3. After a large number of valves had been specified and purchased, Jim offered to sponsor Tom for membership in the local country club. Tom was an avid golfer and had wanted to be a member of the club for some time, but he had not been able to find a sponsor.

Case 4. After a large number of valves had been specified and purchased, Jim invited Tom to a seminar on valves to be held in South America. There would also be opportunities for fishing and other types of recreation. Tom's company would have to pay for transportation, but Jim's company would cover all of the expenses in South America. Tom was sure his manager would authorize the trip if asked, but some of the other managers in the firm believed such trips violate proper ethical and professional standards.

Case 5. After a large number of valves had been specified and purchased, Jim invited Tom on a very nice fishing trip to South America. Jim's company would cover all the expenses. Tom was sure that his manager would authorize the trip if asked, but some of the other managers in the firm believed such trips violate proper ethical and professional standards.

What are the relevant similarities and differences in these examples? Notice that there is not one single feature that distinguishes acceptable from unacceptable actions on Tom's part. The size of the gift, for example, cannot be used in any simple way to distinguish permissible from impermissible actions, for accepting an offer of a trip to a seminar in South America might be more ethically permissible than accepting sponsorship for membership in the local country club. Why?

Many companies set standards for accepting gifts and amenities. While this makes many decisions easier, such standards must still be applied to particular situations. Thus, the need for discrimination remains. Consider the following statements:

[General Dynamics Standards of Conduct.] It is a serious violation of our Standards for anyone to seek a competitive advantage through the use of gifts, gratuities, entertainment or other favors. Under no circumstances may we offer or give anything to a customer or a customer's representative in an effort to influence a contract award or other favorable customer action. It has been and will continue to be General Dynamics policy to compete solely on the merits of its products and services.

[Chase Manhattan Corp.] Staff members and their families may not solicit or accept any gifts of significant value, lavish entertainment or other valuable benefits intended to influence Chase's business. Staff members may not solicit or accept personal fees, commissions or other forms of remuneration because of any transactions or business involving Chase.

[R.J. Reynolds Industries] Employees will give no gifts to customers except items of nominal value which fit the legal, normal, and customary pattern of the Corporation's sales efforts for a particular market.

[Allied Chemical Corp.] With the exception of reasonable business entertainment and other activities permitted in accordance with the following paragraph, no employee of the Company shall give or transfer anything of value to or for the benefit, directly or indirectly, of the employee or agent of another person, including any customer, union representative or supplier. Reasonable business entertainment would cover, for example, a lunch, dinner, or occasional athletic or cultural event; gifts of nominal value ($25 or less); entertainment at Pleasantdale Farm or other Company facilities, or authorized transportation in

Company vehicles or aircraft. In addition, reasonable business entertainment covers traditional Company-sponsored promotional events.

[IBM] No IBM employee, or any member of his or her immediate family, can accept gratuities or gifts of money from a supplier, customer or anyone in a business relationship. Nor can they accept a gift or consideration that could be perceived as having been offered because of the business relationship. "Perceived" simply means this: If you read about it in your local newspaper, would you wonder whether the gift just might have something to do with a business relationship? No IBM employee can give money or a gift of significant value to a supplier if it could reasonably be viewed as being done to gain a business advantage. If you are offered money or a gift of some value by a supplier or if one arrives at your home or office, let your manager know about it immediately. If the gift is perishable, your manager will arrange to donate it to a local charitable organization. Otherwise, it would be returned to the supplier. Whatever the circumstances, you or your manager should write the supplier a letter, explaining IBM's guidelines on the subject of gifts and gratuities. Of course, it is an accepted practice to talk business over a meal. So it is perfectly all right to occasionally allow a supplier or customer to pick up the check. Similarly, it frequently is necessary for a supplier, including IBM, to provide education and executive briefings for customers. It's all right to accept or provide some services in connection with this kind of activity—services such as transportation, food or lodging. For instance, transportation in IBM or supplier planes to and from company locations, and lodging and food at company facilities are all right.

[Texas Instruments] It is TI policy that TIers may not give or accept any gift that might appear to improperly influence a business relationship or decision. If we receive any substantial gift or favor, it must be returned and our supervisor notified. This policy does not apply to items of small value commonly exchanged in business relationships, but even here, discretion and common sense should be our guide. In commercial business, the exchange of social amenities between suppliers, customers and TIers is acceptable when reasonably based on a clear business purpose and within the bounds of good taste. Excessive entertainment of any sort is not acceptable. Conferences accompanied by a meal with suppliers or customers are often necessary and desirable. Whenever appropriate, these meals should be on a reciprocal basis.

Questions for Discussion

1. Give as many characteristics of Case 1 as you can that make it a genuine instance of bribery. Compare this with the other cases. Where do you think the line should be drawn between permissible and impermissible action? Would you arrange the cases differently?

2. What arguments can you give for Tom's accepting Jim's offer in Case X? What arguments can you give for Tom's not accepting Jim's offer in Case X?

3. The corporate guidelines presented above are generally rather strict. Do you think they are too strict? Can you still imagine cases (such as some of those in Cases 1-6) where decisions would still be difficult by some of the guidelines?

4. Giving and accepting gifts in other countries presents a serious problem for some American companies. Some say, "When in Rome, do as the Romans do." Others say that American companies should abide by strict anti-bribery standards. Still others propose a middle way between these two options. An example of such an approach might be giving gifts to communities rather than individuals, thereby gaining business while not having to bribe individuals. What do you think is right?

5. There is an important distinction between paying a bribe and paying an extortion, a distinction that is recognized in principle even by the Foreign Corrupt Practices Act. For example, paying a government official a large sum of money to keep your warehouses from being set afire is probably a case of extortion rather than bribery. What are the morally important differences between bribery and extortion?

Suggestions For The Instructor

In order to begin thinking about bribery, it is helpful to list some of the characteristics of a "true" bribery situation, such as Case 1:

1. Tom had direct responsibility for specifying the valves.
2. The salesman approached Tom and made the offer before the valves were specified or purchased.
3. The valves specified were less safe and reliable and more expensive (or in some other way less desirable) than alternative valves.
4. There was in fact a causal relationship between the offer of the amenities and Tom's decision. In other words, Tom requested that Jim's valves be used because of Jim's offer.
5. Even though Case 1 involves bribery, the company would probably benefit from an ongoing cordial relationship with suppliers of the valves which Tom purchases. For example, obtaining service would probably be easier. [We shall assume that this is true.]
6. Tom rarely accepts amenities from suppliers with whom he does not do business. [We shall assume this.]
7. Knowledge of the gift may influence others to buy from Jim, even if Jim's product is not the best.
8. The gift was for a substantial amount of money.

9. Even if there had been no actual corruption, there was certainly the appearance of corruption. For example, consider IBM's test: "If you read about it in your local newspaper, would you wonder whether the gift just might have something to do with a business relationship?" By this test, there was the appearance of corruption. (In this case, of course, the appearance was not misleading.)

Here are some arguments against Tom' accepting Jim's offer in Case X:

1. Over the centuries, Western morality—and probably morality throughout the world—has tended to take an increasingly negative attitude toward bribery. This implies that the restrictions on actions closely related to bribery will probably also increase. The corporate codes cited in the student handout give evidence of the increasing restrictions, and by most of these standards Tom's accepting the offer would probably be out of bounds.
2. The size of the gift is morally troubling.
3. Knowledge of the gift could influence others to buy from Jim, even if Jim's products are not the most appropriate for them. This might operate as a kind of bribe-ahead-of-time for other people in Tom's plant or in other plants, even if Tom had no idea he would be offered a trip. They might say, "If we buy from Jim, we can expect a nice gift."
4. In morality, one of the important questions to ask is whether you would be willing for others to do the same thing you did. If every salesman offered gifts to people who bought—or recommended the purchase of—his products, and every purchaser accepted the gifts, the practice would of course become universal. Our first reaction is to say this would neutralize the influence of the gifts. You could expect a bribe from somebody, no matter whose product you recommended. Thus Tom might have been offered a nice trip to South America by whatever valve salesman made the sale. But this begins to look like extortion if not bribery: a salesman has to offer something to even have his product considered. Furthermore, smaller companies might not be able to offer the lavish gifts and so might not have their products considered. This would harm the competitive process. Finally, the gifts would probably tend to get larger and larger, as each salesman tried to top the other one. This would re-introduce the element of bribery. Thus the general acceptance would have undesirable consequences.

Here are some arguments in favor of Tom's accepting Jim's offer in Case X:

1. We have already pointed out that Tom's action cannot be an example of accepting a bribe in the true sense of the term. In order to be a true bribe taker, Tom would have had to make his decision because of Jim's offer. Since the trip was offered after Tom's decision and Tom did not know about the trip ahead of time, the trip could not be a bribe in the true sense.

2. Tom's company may stand to benefit from the personal relationship between Tom and Jim. It may make it easier to get replacement parts for the valves and to get other types of service from Jim's company.

3. Business life should have its "perks." Business and professional life involves a lot of hard work. Fishing trips and similar amenities add spice to life that is important in terms of job satisfaction and productivity.

Accepting the kinds of gifts that Tom took advantage of is quite common in Tom's industry. [We shall assume this.] It adds very little to the cost of the product. Any industry large enough to manufacture the valves in the first place would be able to afford such gifts as this without financial strain. It is true that, in taking the moral point of view, we must assume that everybody has a right to do what we do. But if every salesman offered trips and every person in Tom's position accepted them, would no harm result? Would things equalize? There might be a kind of "extortion" here, but this is just a word.

The Peter/Paul Dilemma

From Applied Ethics Case of the Month Program— Case 1005

The Situation

You are an engineer for a well-respected consulting firm. The firm has been very successful in the past, but recently business has slowed down due to a slump in the local and regional economy.

One of the projects being handled by the internal design group with which you work is for Excaliber Resorts, Inc., a reputable company which deals with the development of expensive, world class destination resort properties, including all of the roadway design, drainage and site civil work.

The estimated construction cost of the project your group is currently working on is in excess of $90 million, and this is only the first of several proposed phases. The project is going well. In fact, the work for this initial phase of the overall development is ahead of schedule and well within the budget.

Since the engineering work on the Excaliber project is slowing down, you and part of your group are assigned another project to work on at the same time. Unfortunately this new project for Wellfleet Corporation has been hanging around the office for several months, since the project manager went on vacation alone to Fiji, then was sued for divorce by his wife immediately after returning.

As a result, a lot of time has been charged to the job and billed to the client, but very little has been accomplished. While the deadline for the Wellfleet project is still well into the future, it is estimated that at least 50% more money is needed to complete the design properly than actually remains in the contract budget. Neither the Excaliber nor the Wellfleet project has a lump sum contract. Each is being billed using standard hourly multiplier rates.

You and the other members of your design group are told by the Vice President for Operations of your firm that since you are officially working on both projects, the majority of your time should be spent on the Wellfleet project, but your time should be charged to the Excaliber project, since there is more than enough left in that budget to finish the job. The Vice President also indicates that he will take the responsibility for these decisions.

What do you do?

NOTE: Digital copies of all Applied Ethics case of Month Cases are available at www.niee.org.

Alternate Approaches

1. Follow the Vice President's directions. After all, you are a subordinate, and are supposed to do what you are told by your superiors, especially the Vice President. If you don't, your loyalty to the company may be suspect and you may jeopardize your job. Besides, the Vice President assured you that he will take responsibility for this action.

2. Ask the Vice President for a written confirmation of his instructions and his assumption of the responsibility involved for such action. Also see if there is a policy covering the situation in the company's operations or personnel manual.

3. Do not do as the Vice President directed (i.e., do not post your time for one project to a different project). Say nothing and hope that the Vice President does not notice. However, keep copies of your signed time sheets for possible future reference. If the Vice President does notice and wants to change your time sheet, it is out of your hands.

4. Ask for the advice of one or more engineers you know personally who are working for other firms in the area to determine what the accepted standard of practice is in situations such as this.

5. Approach the Vice President and indicate that you cannot do as he directed, since it is a breach of honesty, fidelity to the client (which supersedes fidelity to the firm in this instance), responsibility (accountability) and professional integrity.

6. Approach the Vice President (as in '5' above), and suggest a way to rectify the situation, such as having the original project manager shoulder his rightful responsibility to get the project back on track, even if it means he has to work nights and weekends without pay.

7. Discuss the situation with the others in your team who are being told to do the same thing, so that you can approach the Vice President with the same reasoning as in '5' above as a unified group.

8. Discuss the situation with the others in your team, as in '7' above, but also offer the Vice President alternatives, such as each member of the team donating a few hours a week to the project without pay (after hours/weekends) to assist the original project manager in trying to rectify the situation, if all of you (or even some of you) agree to do so.

9. Spruce up your resume and start looking for another job.

Forum Comments From Respondents

The following comments synopsize the views of individuals responding with opinions differing in some regard to the alternatives indicated above. They are not necessarily indicative of the opinions of this program, its supporters, the program's Board of Review or the professional organizations associated with the program.

1. This happens more often than we like to admit. One effective approach is to point out to the Vice President that once you sign your time sheet, it is a legal document. Your signature is affirmation that the information shown is truthful. Should a close review of the projects occur and it became apparent that you had falsified your time sheets, you are personally responsible and liable, not the Vice President. At such a time it would be your reputation and your future swirling down the drain. It is highly unlikely the Vice President would admit requiring you to do what you did.

2. This situation is probably more common than we like to think. It happened early on in my employment with a state transportation agency. As I still work for that agency, you can see I followed the pack. Luckily, a change in management has allowed us to charge our time ethically and properly.

3. The most favored solution must include options to which the VP may respond. In addition, the VP should know that options are presented based on professional and business ethics, and not based on the unwillingness of the employee to share in the burden. Finally, the burden of accountability should be shared by the previous project manager to assure that history doesn't repeat itself with that individual or set a precedent in the organization.

4. Clients are not a dime a dozen. They are people; they deserve respect; but its gotta start somewhere. When an action occurs, someone surely should be held accountable for the consequences. Good or bad, it's time for the vacationer to pay for the vacation. If teamwork is valued, everyone should help out, but project managers have to know prior to taking an assignment that they are responsible for the end result.

5. I think you need to give the VP some alternatives if you can, but I'm not sure I agree that all of the team members should "donate" their time.

6. I agree that the core values of honesty and fidelity are paramount in this situation. I don't agree that it's a question of fidelity to the client over fidelity to the firm. Often we forget that fidelity to our employer means to the entity, not our supervisor. The firm is not well served by dishonest, possibly criminal, behavior which could severely damage the firm's reputation and risk losing a major client. If the VP doesn't respond, then this is a case where it's perfectly appropriate to go over his head.

7. Go over your Vice President's head and discuss the situation and your discomfort with it with his superior.

8. Try to discuss the situation with the VP's boss and see if he has been instructed to do this by his boss. If this is the case, start looking for another job.

9. It is the consultant's requirement to the client to report all charges fairly and accurately. One of the projects may be funded by a different agency. During a closing audit the data may reveal errors in charging which will negatively affect the reputability of both consultant and client.

10. Do as you're told, but keep documentation regarding the actual accounting of your time in case you need it in your defense later.

11. Make a personal documentation of the discussion with the Vice President. Also, if the Vice President presses the issue, refuse to sign your timecard for the reason of not wanting your signature on a document that is falsely representing the number of hours worked and leave the billing to the discretion of the supervisor. Although this is still a gray area, you have presented and documented your disagreement with the decision and cleared your name from this decision. (College senior)

12. Approach the Vice President and indicate that you cannot do as s/he directed, since it is a breach of honesty, fidelity to the client (which supercedes fidelity to the firm in this instance), responsibility (accountability) and professional integrity. Suggest ways to solve the situation which don't place the burden on people who were not responsible. The original project manager should be held responsible but placing the whole thing on him may not be practical or possible. The next level of responsibility rests on the managers of that project manager, who should have been aware of what was happening. Ultimately, the burden needs to rest on the people who were responsible and the company as a whole. Since the size of these projects are so large, and the money involved so much, it may not be feasible for individuals to take much of the burden. Such being the case, it appears that the company you work for should use some of its own earnings to cover the losses that occur due to its bad management practices. If the Vice President is not willing to act ethnically [sic], then you may have to go to higher authorities, such as the President of your company, national ASCE, or even other companies who are being cheated. (College student)

13. Since the VP was not able to think of [an ethical way to handle the situation], he should not be a VP to begin with, so please have him step down and find a replacement. (College senior)

14. Fill out your time record in accordance with the time you actually spent and keep a record of your charges. If the VP wants to cheat, let him change the records. Of course. If I find out he changed my time sheet, I might be compelled to write him a letter with my resignation and send a carbon copy to his client!

The Fetid Favor Fiasco

From Applied Ethics Case of the Month Program - Case 1007

The Situation

You are an architect with a 10-person firm in Placidville, in the central part of the state. Placidville is a small city with a population of 83,576, based on a 1996 census. It also is the location of the state's largest university (about 23,000 students). Due to the business and recreational opportunities in the state, the population has been steadily increasing over the past 15 years, and an aggressive building program was initiated at the university five years ago to construct and equip approximately 16 new classroom and laboratory buildings over the following 20 years.

Last weekend your wife invited mutual friends, Ted and Alice Hammer, to spend the weekend with you at your cottage on a lake in the northern part of the state. Ted is a senior project manager with Quality Construction Co., a local commercial and institutional building construction firm. His wife, Alice, is an activist in the Placidville community, and generally known for her on-going campaign for integrity in government.

During the weekend, Alice mentioned that at a recent zoning board meeting, Gus Olson, one of the other board members, told her that the construction company he works for as a carpenter (Shreud Contractors) has assigned him as the construction foreman for a garage being built for Ray Vandergrafft. Alice recognized the name and knew that Ray is one of several project managers for the Capital Construction Projects Office at the university. Gus laughed and said that something as small as a wood-framed garage was an unusual project for Shreud Construction, since they normally were involved in heavy steel erection projects for large structures in the region.

Ted interjected that he had heard Shreud Contractors recently was awarded a contract with an estimated budget of $500,000 to make remedial repairs to the steel superstructure of the university's aging football stadium. He understood the contract had been awarded directly to Shreud on a time-and-materials basis to avoid the expense to the university of preparing extensive bid documents and going through a competitive selection process. He said it was interesting that the garage construction for Vandergrafft was going on at the same time as the remedial work on the stadium.

In fact, Ted said, he had learned a few years ago that his own firm had previously been involved in a similar situation when they were the contractor for a new engineering test facility at the university and at the same time had built a large addition to Ray Vandergrafft's kitchen because his wife, Olga, was a gourmet chef. Ted said he had decided not to make an issue of it, since the project had been completed for some time, and he had been told that the construction materials and appliances for the kitchen had come from surplus materials from the new lab site and incentive gifts from the laboratory equipment suppliers. He also had learned that the construction crew worked on the kitchen when there was not enough to do at the new engineering test facility site.

At that point your wife said she had driven by Fred Facade's house a few days ago, saw a Shreud pickup truck parked in the driveway and noted that there was remodeling going on at the house, apparently to raise the roof and add more space on the second floor. Fred is the University Architect.

You have done architectural design work for the university over the years and have contributed to their alumni giving campaign. In the past, you suspected that some contractors and architects received favored treatment from the university, especially the Office of the University Architect, but this is the first time you have heard anyone detail a specific situation. You know the University President on a first-name basis, as well as the Vice President for Finance, to whom the University Architect reports. You have also known Fred Facade on a casual basis for more than 18 years.

What, if anything, do you do?

<u>Alternate Approaches</u>

1. Do nothing. You do not know that there is anything illegal or underhanded going on based on what you have heard so far. Just because Ted Hammer's firm participated in the construction of Vandergrafft's kitchen a number of years ago, things have changed in the industry and it is highly unlikely anyone would do that kind of thing in this day and age.

2. Leave it alone. This sounds like a case of sour grapes on the part of Gus Olson for having been put on such a small project while the really interesting remedial work on the stadium is being carried out by others in Shreud Contractors. Also, Ted Hammer is not about to say anything supportive of Shreud Contractors anyway, since Shreud got the stadium remedial repairs project handed to them directly without bidding. After all, Quality Construction and Shreud Contractors are competing contractors in the same town and often go after the same construction projects.

3. Do nothing. It is commonly recognized that this sort of thing goes on all the time, and making an issue about it is not going to put you in a positive position for more architectural design work with the university.

4. You are outraged! Call Vandergrafft at his house and tell him what you have heard. Also tell him that if you ever hear of his doing such a thing again, you'll raise such a stink that he will be forced to resign from his position as a project manager with the Capital Construction Projects Office at the University and will have difficulty finding another job within 2,000 miles, if then.

5. Do nothing until you can get verification from someone else that Shreud Contractors are working on Vandegrafft's garage without charging for the labor

and/or materials. It may be that Shreud has a contract with Vandergrafft to build the garage and Gus Olson, not being part of the Shreud management group, may be assuming things that are unwarranted or unsubstantiated.

6. Arrange with Ted Hammer to have one of his people ask around the local carpenters, teamsters and laborers union halls to gather as much information as possible about Shreud doing both the stadium project and Vandergrafft's garage at the same time so that you and Ted can put a coherent case together before talking with anyone else.

7. Make some discreet inquiries around town, particularly among lumber suppliers, to see if either Vandergrafft or Shreud Contractors has recently purchased lumber and had it delivered to the Vandergrafft home.

8. Arrange for a quiet lunch away from the campus with the Vice President for Finance at the University to discuss what you have heard, and to express your concerns about what appears to be kickback incentives in the form of labor and materials for the Vandergrafft garage. Indicate that you are concerned because you have contributed to the University's fund raising campaigns on a regular basis for years and feel that your firm has not been getting their fair share of the architectural design work at the university.

9. Arrange for a quiet lunch away from the campus with the Vice President for Finance at the University to discuss what you have heard, and to express your concerns about what appears to be kickback incentives in the form of labor and materials for the Vandergrafft garage. Indicate that you are concerned because you have contributed to the University's fund raising campaigns on a regular basis for years and do not like to see the money spent on personal projects for selected faculty or members of the university's administration.

10. Arrange for lunch with the University President to discuss what you have heard, and to express your concerns about what appears to be kickback incentives in the form of labor and materials for the Vandergrafft garage. Indicate that you are concerned because you have contributed to the University's fund raising campaigns on a regular basis for years and feel that your firm has not been getting their fair share of the architectural design work at the university.

11. Arrange for lunch with the University President to discuss what you have heard, and to express your concerns about what appears to be kickback incentives in the form of labor and materials for the Vandergrafft garage. Indicate that you are concerned because you have contributed to the University's fund raising campaigns on a regular basis for years and do not like to see the money spent on personal projects for selected faculty or members of the university's administration.

12. Fred Facade is a fellow professional and respected in the community. He undoubtedly has not realized the appearance he has created, nor the

harm it could do to him personally and professionally. You owe him the courtesy of letting him know that you know, and giving him time to clean up his act. Take Fred to lunch and have a heart-to-heart chat. Let him know that it is wrong to accept kickbacks and suggest he find a quiet way of terminating his relationship with the contractors.

13. You have absolutely no respect for Fred Facade. He has tarnished the stature of architects in the community and deserves severe sanctions. You vow to collect as much incriminating evidence as you can over the next couple of weeks, then send it off to the state Board of Registration for Architects, as well as the American Institute of Architects, and accuse Fred of unethical practices. You will also demand that his license be suspended pending a full and thorough investigation.

14. You know that the administration at the university would probably not admit the situation, especially if it is a case of kickbacks and doing work under the table for selected university personnel without payment from these individuals. Therefore, you should call the local newspaper publisher and confidentially transmit the information you obtained last weekend at your cottage about Shreud Contractors, without giving the names of your sources, but indicating that these individuals are "usually reliable sources of information."

15. Write a letter to the local newspaper for publication on the Letters to the Editor page recounting the information you received last weekend (but not naming your sources), and registering a heartfelt concern about the way the university shows favoritism and does business in the local community, at the expense of the fund donors and taxpayers.

Forum Comments From Respondents

The following comments synopsize the views of individuals responding with opinions differing in some regard to the alternatives indicated above. They are not necessarily indicative of the opinions of this program, its supporters, the program's Board of Review or the professional organizations associated with the program.

1. Get the facts. Cocktail party trials and convictions are often reversed by a higher court.

2. Substantiate the rumors, then notify those who have the authority to perform a criminal investigation of your suspicions.

3. Suggest to Ted's wife that she find out what means is necessary for the city to justify an investigation of the business being carried out by the university. If the facts are sufficient, let the city investigate.

4. Remember the source of the information. Ted is a competing contractor and likely is carrying some hard feelings about losing a big contract. Alice is an activist who is looking for this sort of thing, so she may be jumping to conclusions because it supports her cause. Either one should not be enlisted to further research

the issue. Make inquiries yourself or through others who are detached from the situation. If independent research indicates something is going on, report it to the president of the university.

5. I have a hard time believing kickbacks are legal, so another alternative, which might actually be more ethical than giving the university a chance to cover things up, might be to report the allegations to legal authorities, and let them decide about investigating.

6. Staying anonymous will allow me to separate business from friendship. Fred Façade may be a friend of mine, but his practice is keeping food off of my table. I'd rather not go to the president or vice president because they could be getting the same kickback and it's just not as well known.

7. In a vaguely similar case, several of my colleagues simply refused to act on allegations of misbehavior without 'proof'. They also felt that attempting to gather more facts about the case ran too large a risk of unfairly harming the reputation of the individual in question. This seemed to represent the mainstream ethical thinking in this particular group of colleagues. In this case, however, I do not think that it would be ethical to ignore the allegations.

8. You are obligated to make your questions known, without making accusations and without raising the topic of your previous donations. Find the lowest ranking person in the university hierarchy who is sufficiently independent, and simply tell that person what you have heard and observed. Make no more direct accusations, but suggest that the issue needs to be explored. To ignore this is to shirk your responsibility as a citizen; to get further involved or to try to use undue influence is meddling.

9. Invite both the university president and vice president of finance to a quiet lunch. During this lunch explain the rumors that have been heard and that the individuals may (or may not) have good intentions, but that the outward appearance is very bad. It is up to the university to address this issue as it sees appropriate. I would not discuss my contributions to the university fund nor that I also pursue work at the university (these are not relevant to the situation). I would request that one of the two get back to me to let me know what they had found.

10. The appropriate line of action is to report what you have heard to either the Campus Ethics Committee in written form, or to the President's Office in written form, or both. A copy of the material should be retained for future reference. It's not the job of the person reporting alleged incidents to investigate/judiciate [sic] cases.

11. A confidential complaint to the state architectural licensing board should be enough to start an investigation. If there is impropriety, it will be clearly shown. The individuals involved deserve to be investigated as there is an obligation to avoid the appearance of impropriety.

It's Just the Nearness of Who?

From Applied Ethics Case of the Month Program — Case 1019

The Situation

Tim Mover, a young professional engineer with about 10 years of experience, is employed by Global Engineering and Construction, Inc., a large, international design and construction firm. He is proceeding quickly up the corporate ladder and was recently assigned as regional manager in a branch office several hours by airplane away from the main office. Although the office is small, he has a staff consisting of an assistant manager and a secretary. Tim is aware that the firm is negotiating a multi-million dollar contract that will require someone with a background such as his to be the overall project manager. Although Tim is relatively young for such a position, he suspects that his current job as branch manager was arranged in order to evaluate his potential and ability to handle the multi-million dollar project. If he gets the new assignment, he will move to the project site several states away within the next two years.

As for Tim's current assignment, his office is responsible for project development and preliminary designs, as well as supervision of on-going contracts. Tim's assistant, Roy Stalward, is a seasoned, field-oriented engineer who is 15 years older than Tim. Roy is highly respected in the home office and his opinion in construction matters is rarely questioned. Roy has had 10 years of experience in this same branch office. His area of responsibility is managing the on-going projects, thereby relieving Tim to concentrate on developing new projects, project planning and client relations in the area.

Tim's initial impression of Roy confirms Roy's reputation. However, within a month of Tim starting work in the branch office, he observed Roy in the company of a particularly attractive younger woman. As Roy is not married, Tim thought nothing unusual about the occurrence. The next week, however, Tim attended a meeting with one of the firm's major subcontractors, Hotspark Electrical Co., and the same young lady was present. It turned out that she is the daughter of Hotspark's owner, as well as the general manager of the electrical subcontracting company. Over the next six months Tim twice observed Roy and the young woman at unusual times and in locations which suggested something other than a strictly business relation between them. Finally Tim confronted Roy directly and asked if he had a personal relationship with the young lady. Roy said that he did not.

Tim now faces a dilemma. Tim is Roy's supervisor, but because of Roy's seniority, Tim is not in a position to reassign or fire him. Tim suspects that Hotspark Electrical is receiving preferential treatment at the expense of his firm, Global Engineering and Construction, but he has been unable to document anything specific. Regardless of documentation, Tim is concerned that compromises in the quality of the work performed by Hotspark will inevitably occur in the future on some on-going projects. Global does not have an explicit policy preventing personal employee relationships with subcontractors. Nonetheless, Tim was once told that the president of Global takes a personal interest in "conflict of interest" situations, particularly if they cost the firm money.

Should Tim do something, and if so what? (Roy is still held in highest regard by everyone else in the firm, and Tim probably only has a year or so left in this branch office before going on to manage the multi-million dollar project.)

Alternate Approaches

1. Tim should do nothing, since it would be a case of Tim's word against Roy's, and Tim has no documented evidence of any specific "favors" having been given to Hotspark.
2. Tim should go to Global's main office, have a meeting with the company president and apprise him/her fully of the situation, emphasizing Roy's value to the firm. He should then ask for advice on how to handle the situation from that point on, in the absence of specific written company policy.
3. Despite the fact that Tim is being groomed for increased managerial responsibilities in a different location, he should discuss the situation with his immediate supervisor on a confidential basis. He needs to explain that he has found nothing that indicates improper activities are taking place, but there is a perception. He should seek advice and direction from this immediate supervisor.
4. Tim should ask the president's office for a specific written policy covering situations such as this as an addition to Global's company employee policy manual, without indicating that there is an immediate problem, or who might be involved. Then he should use the new written policy to take appropriate action with Roy.
5. Tim should check the bidding procedures and history of Hotspark's involvement with Global, including an informal audit of the quality of Hotspark's work. If there are irregularities in that regard, Tim should confront Roy for an explanation. If he is not satisfied by Roy's response, Tim should then speak with the president of Global. However, if there are no irregularities with Hotspark's work, Tim should do nothing more than caution Roy on "potential conflicts of interest".
6. Tim should do nothing at the present time, but should keep a log of any future observed "meetings" between Roy and the subcontractor's representative, as well as keep close documentation of any unusual costs for any projects where Hotspark is used as a subcontractor. Tim can then confront Roy with this evidence before either firing or transferring him to another location.
7. Tim should call Roy into his office and tell him that the perception is that Roy has a "special" relationship with one of the subcontractors, which, while not against explicit company policy, nonetheless could attract the attention of

the president of Global. Tim should also mention that the president takes a personal interest in conflict of interest situations, especially where preferential treatment of a subcontractor may result in loss of profit to Global. Tim should also tell Roy that he is a valued employee of Global and a situation like this, if not handled properly, could impact his future with Global if it became public knowledge.

8. Tim should talk with Roy and explain that while perhaps nothing unusual is taking place, there is a perception of impropriety. This perception will have a damaging effect on Global's reputation as a general contractor. He should also suggest that Roy spend an equal amount of time with representatives of other electrical subcontractors who work with Global. Hopefully Roy will appreciate the concern and take appropriate action.

9. Tim should get to know the young woman's personality and character by taking her out to lunch. In this way, he can more accurately assess her motives and capabilities in her interactions with Roy.

10. Tim should seek out his own personal relationship with the young woman from Hotspark.

Forum Comments from Respondents

1. Additional possibilities for Tim to consider:
    - ❖ What has been the degree of ethical conduct by Roy in the past? Would his ethical principles override lust?
    - ❖ The relationship could be harmless despite appearances..but considering human nature it could be real.
    - ❖ Can Tim act on suspicions alone? There is a risk that his hunch may be wrong.
    - ❖ If Tim reassigns Roy or fires him, what would be the consequences to Tim's career? What would be the consequences to the company?
    - ❖ What are the motives of the young woman?

2. I once had a somewhat similar case in my office and it was nearly a no-win situation. Denial is always the first thing you hear. Next the defenders go on the offense and try to attack your own credibility. It's best to proceed very cautiously and try to maintain a good relationship with both parties, yet be firm in your position that this sort of behavior just can't continue.

3. Tim should find an explicit company policy on the issue. If there is none, then Tim should caution Roy about a perceived preferential treatment relationship. Because Tim is seeking higher managerial positions, he should handle the issue himself to prove his managerial capabilities. He should then consult his supervisor to be sure he has acted appropriately.

4. Tim needs to sit down with Roy. He must keep in mind that Roy is probably aware of his ethical responsibilities to the company. Tim must explain that he is knowledgeable about Roy's abilities and they are well earned. But no one is without fault. He should explain to Roy that his personal life is just that – his personal life – until it becomes unethical. He should also let Roy know that additional attention will be about the activities of the company the young woman works for to insure that no irregularities are occurring.
5. Tim should follow Roy and the young woman, take pictures of their encounters, and then suggest that Roy end the relationship, or things could get ugly. Tim should let Roy view the pictures so he knows that Tim means business. [editor's comment: hmmmmmm?!]
6. A point to consider: In the corporate world, especially when viewed through the public eye, the appearance of impropriety is just as damaging to the reputation of that company or employee as the act itself. So even if no preferential treatment is occurring as a result of the relationship, it is nonetheless damaging to the company and its reputation with other subcontractors and the public. Based on that, the managerial staff and president have a right to be informed of the situation.

Santa in the Summer

From Applied Ethics Case of the Month Program — Case 1024

The Situation

Rod Traverse is a civil engineering student at a well-known university in the mid-west. Because he did well in his surveying course during his junior year, he is working for the summer before his senior year for the state Department of Transportation (DOT) on a road construction project 140 miles from his hometown. His duties include working closely with the state's on-site resident engineer, Jim Upwright and several other state highway construction engineers for the project. Ethel Hicks (known to her friends as "Eth") is Upwright's supervisor at the DOT headquarters and visits the site every couple of weeks to see how the project is progressing.

Every Friday afternoon about 4:00 p.m., Rod and the DOT engineers get into their cars or trucks to drive home for the weekend. Since he works a good bit of the time reducing survey data and keeping records in the state's construction trailer, Rod has noticed individual foremen for the three separate bridge contractors working on the project putting a box or other article in the back of the resident engineer's pickup truck about 3:30 p.m. on most Fridays. These boxes and articles have included a new set of tires, a mountain bike, a case of Duggan's Dew o' Kirkintilloch Scotch whiskey, and a shotgun.

There are several more bridge structures to be designed and built under another contract for the project. Upwright will be asked to make comments and give recommendations regarding the three bridge contractors presently on the project, if they show interest in obtaining the additional work.

Since Rod's work is part of a summer credit course program at the university, Upwright will also be required to communicate with Rod's advisor at school (Dr. R. E. Serchur) and recommend an appropriate grade for Rod's summer work course.

Under the circumstances, is Rod obligated to say anything about the gifts to anyone, and if so, to whom and when?

Alternate Approaches

1. Rod is only a summer hire and should keep his nose out of things that are none of his business. He has heard that things like this happen on some construction projects. The last thing he wants is to jeopardize his chances for a good grade in the summer work course, which could impact his job opportunities the following spring when he graduates.

2. Rod should contact his faculty advisor, R. E. Serchur, over the weekend and ask for his advice about the most appropriate course of action, if any, he should take.

3. Rod should approach one of the bridge contractor foremen to try to clarify the facts of the situation through them, before confronting Upwright (perhaps the items in the pickup truck were intended for charity or they may be purchasing items that Upwright wants at a discount and Upwright is reimbursing them for the cost of each item).

4. Rod should send an anonymous letter to "Eth" Hicks in guise of a taxpayer who frequently observes things on the project, and has noted the apparent transfer of gifts from contractor personnel to a pickup truck with a license plate number that coincides with that of Upwright's pickup. The letter should suggest that these actions appear improper and should be looked into.

5. Rod should discretely inquire of someone in the DOT who is not connected with the project what the policy is with respect to DOT personnel accepting gifts from contractors. If it is against DOT policy, then Rod should blow the whistle to "Eth" Hicks.

6. Rod should ask Upwright directly about what has transpired with the contractors' foremen and explain that while nothing wrong may be occurring, the appearance of impropriety exists to the casual observer.

7. Rod should approach one of the contractor's foremen and mention how nice it would be to have a Bose Lifestyle 20 sound system to put in his fraternity room during the ensuing college year.

8. Rod should photograph the Friday gifts for a few weeks with dates, times and the license plate on Upwright's pickup clearly visible. He should show Upwright copies of these photos, emphasizing how practical Rod's summer job is, then suggest that his efforts surely deserve an A grade and a glowing recommendation for his university file.

9. Rod has no control over the situation. He should leave the construction project as soon as possible and take an incomplete in the summer work course, so that he won't be included as part of the situation should anything happen (someone blows the whistle) before the summer has ended.

Forum Comments from Respondents

Comments from students:

1. Something needs to be done, but Rob should first discuss the incident with his faculty advisor so as to place his grade in a better position if things turn ugly. Then he should put out a 'feeler' as described in #5, and evaluate his final decision based on the response received. If it is against policy, he should report it to "Eth"; if it is not, he would only be exposing himself to crossfire while accomplishing nothing, and should let the issue drop (though I still disagree with such a practice occurring).

2. Rod should talk with Upwright about what he has seen and indicates that it at least appears to be improper. If he then determines that it is wrong, he should express his disappointment as an engineering student at the bad example being set for him, although I don't think I would blow the whistle.

3. Because Rob is only summer help, he needs to be careful about how he handles the situation. I have found that it is often helpful to seek council and guidance from someone with a lot of experience. That is why he should talk with his faculty adviser. He will help steer him in the right direction, and give him the courage to do what is right.

4. If his faulty advisor is a professional engineer, he should have some good advice for him.

Comments from practicing engineers

1. Rod needs to consult with one or two of the other assistant engineers, preferably those who are candid and will tell the truth. Presuming they all agree that the situation does not look good, they need to approach the Resident Engineer as a team and lay out the problem to him as soon as possible. Rod should not try to solve the problem on his own because this sort of thing is not easily solved one on one.

   If Rod finds the whole system is corrupt, it is still beyond his control to solve it and he needs to go to the ethics ombudsman in the state government and seek advice from them. For good measure, he needs to discuss his perception of the problem with his faculty advisor.

   (I was in a very similar situation when I started working as a recent graduate for a state DOT. I wish now I had followed my own suggestions then.)

2. I can envision a situation where this young man might be somewhat (or perhaps terribly) disillusioned about the profession he has selected. If he's a sensitive soul, he might even seek out a personal meeting with his advisor and ask if all engineers are expected to sell their souls for a "mess of pottage."

3. One wonders what the Resident Engineer is doing in return for the gifts. Perhaps he has the concrete testing technician pulling all his sets of cylinders from the first truck of the day and ignoring all the others (by leaving the site early?). Or allowing the contractor to place the subgrade in excessively thick lifts. Or allowing the subgrade material to contain excessive amounts of organics. Or allowing a substantial number of the required rock bolts in a rock cut to be shorter than the length called for in the project specifications. Or making "corrections" to the calculated yardage so carefully calculated by the student engineer.

4. Rod has found himself in a pickle. State employees are in a particularly bright public light and most cannot accept any gift. Some cannot even accept lunch. Is Rod "obligated" to do something? Is a witness to a bank robbery "obligated" to report what s/he has seen? If Rod ignores the rather obvious graft, he jeopardizes his reputation and his ability to discern right from wrong while facing himself in the mirror every morning. His course of action is clearly to confide in his resident engineer's supervisor, Eth Hicks. He should clearly document what he has seen and then set a meeting with her to discuss what he has seen. The worst that can happen to Rod is that he fails to get credit for the course; with any luck, though, the faculty advisor will understand the situation and speak to Ms Hicks instead of the Resident Engineer for a recommendation regarding Rod's grade in the course. Otherwise, if the DOT fails to back up Rod, he has learned a very valuable lesson. He doesn't want to work there anyway!

   Should that actually happen, he should document what he has seen and send it to the State's governmental oversight agency and let them investigate. After all, it's tax dollars that are being put at risk, not to mention the integrity of the system, and the livelihoods of many other firms unfairly cut out of competition for future work.

Now You Have It, Now You Don't

From Applied Ethics Case of the Month Program - Case 1016

The Situation

The setting is Windsore County in northern California. During the El Nino winter, heavy weather triggered a large debris flow landslide in the mountain area of the county. The landslide partially buried ten summer homes situated along the bottom of a narrow canyon. During the emergency, the Windsore County Public Works Department hired Nearby Engineers, a local geotechnical engineering firm, as a sole source emergency selection to evaluate the landslide. Nearby Engineers determined that a significant additional debris flow hazard still remained. As a result, the County condemned the ten summer home properties and developed a schematic plan to protect other summer homes downstream by construction of an earth "dam" and debris collection basin, plus a large storm drain.

The Windsore County Public Works Department then issues a Request for Proposal (RFP) for engineering services to design the dam, debris flow basin and storm drain. The RFP stated that a firm would be selected on the basis of qualifications and experience (as they typically selected consultants in the past). They would then negotiate with the selected firm to arrive at an appropriate fee and contract (standard Qualifications Based Selection - QBS procedure).

Trueheart Engineers, Nearby Engineers, and two other engineering firms responded to the RFP, each submitting qualifications statements and a proposed scope of services. The County Public Works Department formed an interview and selection committee comprised of Public Works Department professional staff members, who reviewed the submittals, interviewed the four firms and ended up selecting Trueheart Engineers for the project. Trueheart was notified of their selection. They met with the Public Works staff, refined the scope of work, and negotiated an acceptable fee estimate and contract provisions. The contract was then to be approved by the Windsore County Board of Supervisors.

Shortly thereafter, Trueheart Engineers was notified by the County Public Works staff that the contract approval was delayed and that they were also requesting a formal proposal from Nearby Engineers, which was a deviation from their previously announced QBS selection process. After the Public Works Department received Nearby Engineers' proposal, the Director of Public Works overrode the staff's previous recommendations to hire Trueheart Engineers, and he recommended to the Windsore County Board of Supervisors that Nearby Engineers be hired instead. Notably, Nearby Engineers' scope of work and contract language was the same as that negotiated by Trueheart and the fee was the same as Trueheart's. Nearby Engineers was hired for the project.

Nearby Engineers, Trueheart Engineers, members of the County Public Works interview/selection committee and the Director of Public Works are all licensed professional engineers and belong to various professional organizations concerned with ethics.

Questions for Review and Consideration

Question 1:
You are the President of Nearby Engineers. How do you feel about submitting an additional proposal when you know that Trueheart Engineers have already been selected for the project by the Public Works staff?

Question 2:
You are a professional engineer on the County Public Works Department staff and a member of the interview/selection team for this project. The Director of Public Works has overturned the announced QBS selection process as well as your professional evaluation of the proposed consultants. What, if anything, should you do?

Question 3:
You are the Director of Public Works, and although it is not known to your staff, you have been instructed by a member of the Board of Supervisors to hire Nearby Engineers regardless of the announced QBS selection procedure. What, if anything, do you do?

Question 4:
You are the President of Trueheart Engineers. You and your staff spent considerable professional time after you were notified that you were selected for the project, refining the scope of work and contract - time which otherwise could have been spent on billable work for other projects under contract. What, if anything, do you do?

Alternate Approaches (four parts)
The alternate solutions proposed with the case history are shown below, with the percentage of votes received tabulated for each proposed solution. Not all of those replying cast votes for the solutions proposed with the case history, preferring to recommend their own. These alternate solutions and additional comments received from the web site visitors and members of the Board of Review are shown in the Forum section, below.

Solutions Proposed for Question 1.

1. Your firm did all the emergency engineering work for the project, so this job was really supposed to be yours. Therefore there is nothing wrong with submitting the additional proposal for the project assignment.
2. Some people may think that submitting a proposal for the project assignment is contrary to the Engineering Society ethics code, but the code is old and inflexible, and really doesn't address this type of situation.

3. You were hesitant to submit a proposal and deviate from the "normal" QBS process, because you are concerned about setting a precedent that might work against your firm in a future RFP with the county.

4. All is fair in business. If the county is willing to bear any additional scrutiny by requesting a proposal from your firm, who were you to turn down such an opportunity?

5. There is nothing wrong with submitting a proposal for the final engineering work as requested by the county, even though the committee had recommended Trueheart Engineering for the work. After all, the Director of Public Works does have the final say in these matters. It is not as though you went behind Trueheart's back to get the work.

6. Submitting an additional proposal for the final engineering design as requested is not unethical, because there must have been something about the Trueheart proposal the Director of Public Works didn't like. If you didn't submit this additional proposal, the Director would feel that you are not responsive to his requests, and that would be a black mark on your record the next time a project comes up for design.

7. You should have inquired if the County Public Works Department had terminated negotiations with Trueheart Engineers. If the negotiations had not been terminated, Nearby Engineers should not have submitted the additional proposal.

8. You should have notified Trueheart Engineers that your firm had been requested to submit an additional proposal. That would then have allowed you to pursue the project without any problem.

9. You were relieved that your firm continues its close relationship with the Public Works Department Director and with the Board of Supervisors. In the future, you must remember to downscale your work scopes to the level of Trueheart Engineers.

10. There is a fine line between unethical behavior and good business. In this case you have worked hard for a number of years to develop a close relationship with the Director of Public Works and the members of the Board of Supervisors. It is no surprise that they came back to you for another proposal.

11. If the shoe was on the other foot, Trueheart Engineers would have done the same thing. Additionally, you needed the work to meet your payroll commitments to your employees.

12. You are pleased that things worked out the way they did. After all, if your firm was good enough to be selected to do the preliminary engineering, why shouldn't you be good enough to do the final design work?

Solutions Proposed for Question 2

13. You should quit your job in disgust.

14. You should ask the Public Works Director what the basis was for overturning the department's selection committee recommendation, which was based on the QBS (qualifications based selection) procedure, as advertised. If you are satisfied with the reasons given, you should express your objection to the way the matter was handled.

15. You should ask the Public Works Director what the basis was for overturning the department's selection committee recommendation, which was based on the QBS (qualifications based selection) procedure, as advertised. If there were reasons that would truly benefit the project, then fine. If not, you should start looking for another job.

16. Accept the decision of the Public Works Director, since it was never your decision to make any way.

17. You realize that even though you thought the QBS process was the right way to select the design firm and understood it to be one of the advertised criteria for selection, your boss overrode your recommendations. All you can do is be quiet and not jeopardize your job. After all, your job is in reality fairly political, and you'll just have to accept that.

18. You should get together with the other staff members on the selection committee and try to reach a consensus about how to deal with the situation. Then proceed according to whatever consensus is reached.

19. You should call the State Board of Registration for Professional Engineers, get the name and telephone number of the chair of the Ethics Committee, call that person, explain the dilemma as follows, "On a hypothetical basis, what would you recommend to an engineer who finds themselves in the following situation..?"

20. This type of thing should not go unnoticed. You should leak the Director of Public Works' refusal to accept the recommendation of the staff selection committee to the local press without divulging your name.

21. You should get together with the other staff members on the selection committee and send letters of protest to the local section of the American Society of Civil Engineers (ASCE), the state chapter of the American Public Works Association (APWA), the National Society for Professional engineers (NSPE) and as many others as come to mind, signed by each of you on the committee, and as many others in the department as are willing to do so.

Solutions Proposed for Question 3

22. You should have quit your position in disgust.

23. You should accept the decision of the Board of Supervisors, since it was never yours to make anyway.

24. You should call the members of the department's selection committee into your office and tell them that you had no choice in the issue, since you were directed by a member of the Board of Supervisors (who are your bosses) to award the contract to Nearby Engineers. You should also tell them that this information is confidential and not to be mentioned outside of the office.

25. You should have publicly declared the QBS selection process null and void because of changes in the selection process as dictated by the Board of Supervisors. In that way the contract could then be awarded to Nearby Engineers without repercussion, despite any ethical implications in the situation.

26. You should have advised the member of the Board of Supervisors that such a procedure is unethical and declined to carry out the instructions, even though it may jeopardize your continued employment as Public Works Director.

27. You should have tried to determine if the member of the Board of Supervisors who directed you to hire Nearby Engineers is speaking for the entire Board, or just herself, then made a decision whether to follow those directions or canvas the rest of the Board on a one-on-one basis.

28. You should have appealed to the full Board of Supervisors and explained to them that the procedure demanded by one of the Board members was unethical and not in keeping with accepted engineering procedures. If the Board had refused to rescind the instructions, you should have told them that you declined to participate in the revision of the design award and asked that someone else negotiate the contract with Nearby Engineers.

29. You should have terminated negotiations with Trueheart Engineers and advertised a new request for proposals (RFP) to be reviewed by a new selection committee, which should have included the member of the Board of Supervisors.

30. This is just another instance of this type of heavy-handed political maneuvering by some members of the Board of Supervisors. You should have leaked the directive from the member of the Board to the local press.

31. You should have done exactly what you did do, and keep quiet about who demanded the change in engineering firms. After all, your job is very political, and you must keep peace with the Board of Supervisors if you expect to get things done without a lot of interference. There will be other opportunities for Trueheart Engineers to submit proposals for new projects in the future.

32. This is the way things are done in the public sector most of the time. The decision was not yours, even though you stand to take the heat for it, if there is any. Trueheart Engineers are big folk and they should understand that the way the game is played is not necessarily the way it is advertised or appears to the public. You have to go along to get along in this type of public works position.

Solutions Proposed for Question 4

33. You are outraged! You should make an objection to the way your firm was treated in an open meeting of the Board of Supervisors. It is unlikely they will override the Public Works Director's "recommendation", but you'll feel a lot better in any event.

34. You should call the President of Nearby Engineers, indicating that you are prepared to show her a draft of a formal ethics complaint and lawsuit, unless they are willing to decline the contract (in which case you will not proceed with the complaint and lawsuit).

35. You should contact each of elected representatives in the County and expose the deviousness of the Public Works Director.

36. You should file a charge of professional misconduct against the Public Works Director with the State Board of Registration for Professional Engineers, since you understand that Nearby Engineer's scope of work, contract language and fee are identical to yours.

37. You should file a claim with the County for the cost of the time spent in preparing the proposal and in negotiations, claiming bad faith on their part, realizing that the claim will most likely be rejected and you have spent even more lost time in on a poor situation.

38. You should contact your corporate lawyer and file a damage suit against the County and its Board of Supervisors (individually and jointly) for treble damages based on all of Trueheart's costs in responding to the County's fallacious Request For Proposals and Trueheart's legal costs, on the basis of racketeering and blatant violation of public policy and law.

39. There is nothing you can do to help your firm in such a situation, except try to figure out how bearing the brunt of this unfair and unethical treatment by the County can be turned into some sort of advantage for you in the future.

40. You should grit your teeth and do nothing! You've been had! No matter what you do, it will reflect badly on the Public Works Director, and thereby greatly decrease the likelihood of your firm obtaining work from the County in the future.

41. You have learned a valuable lesson by this experience. That is, not to begin work on a project without a signed letter of authorization to proceed or a signed contract so that if the project is cancelled or you are taken off the project you can still bill the client for the time spent to date.

42. If your firm is largely dependent on work from Windsore County in order to make a reasonable profit and provide sufficient work for your staff, you should consider converting it to an internet shopping and delivery service for housewives in the surrounding area.

Between a Buck and a Hired Place

From Applied Ethics Case of the Month Program - Case 1029

The Situation

Clayton is the Director of Public Works in Springfield, a medium-sized rural town with a stable population. A licensed engineer, he has held this position for fifteen years and is highly regarded by his peers and well-liked by those he supervises.

Besides Clayton, the other full time employees in the Public Works department include the City Engineer/ Deputy Director for Public Works, plus two assistant engineers, a special projects coordinator, a survey crew, three inspectors, and an executive secretary, Eleanor. All employees are on the city merit system. Eleanor is 55 years old and has worked in the Public Works department in Springfield for 35 years. Earlier directors have seen her as indispensable, and she has worked for many years above and beyond her assigned duties.

In the last five years, however, things have not gone well for Eleanor. A series of family tragedies has left her feeling beaten down. Her husband had a stroke one year ago and is now an invalid. She and her husband have no children, and it is only her salary and health insurance that keep him from a lower quality nursing care home. Eleanor herself suffers from poor health, and now has only a fraction of the energy she used to have, this due to the strain of looking after her husband every day, as well as keeping up their house by herself.

Clayton is sympathetic to the problems that Eleanor is having. He is aware of the contributions she made to the Public Works department in the past. For years, he believes, she was underpaid. Her actual worth to the department was much greater than any rating she could have received in the old sex-biased system. Because of her poor health and other problems, however, it is now questionable whether she achieves what is minimally required in her position as an executive secretary (her present merit position).

Considering these things, Clayton has kept Eleanor on the payroll but six months ago hired another secretary, Pauline, part-time for 30 hours per week, to compensate. Eleanor now spends more of her time as a receptionist than as an executive secretary. While Pauline has expressed no resentment to anyone, she does notice she is doing all the work while Eleanor is getting executive secretary pay for doing essentially nothing.

To further complicate matters, one of the largest employers in the region recently shut its doors, going off the tax rolls and cutting about 8 percent of the local jobs in Springfield. The ripple effect is projected to reduce next year's tax revenue by 15 percent and the Springfield City Council has demanded that the City Manager make the budget balance.

The ambitious new City Engineer/ Deputy Director for Public Works, Allan, who just completed an executive MBA program and has his sights set on a high-profile public works career in the nearby state capitol, meets monthly with Clayton to discuss project issues and department business. At this month's meeting, Allan informs Clayton that he thinks Eleanor is rated too high for the work she presently does and that the Public Works department is not getting its money's worth out of her. In light of budgetary needs that the department has, Allan proposes letting Eleanor go. There are measurable skills and reasonable expectations that she no longer demonstrates competence in.

Clayton agrees that logically, the merit system, and the good of the Public Works department, seem to dictate such a decision. He feels, however, that the department owes Eleanor some help in her more troubled times, at least until she is eligible to collect her pension. The civil service regulations don't give Clayton much flexibility. He can keep Eleanor on in her present position, or let her go with minimal benefits. What should he do?

This case is based on "Compassion Case: For Loyal Service—A Humane Resource" by the Public Administration at the University of Arkansas, revised 7/19/96, and is used with permission.

Alternate Approaches

1. Let Eleanor go, immediately. Clayton's course of action is clear and he must not be squeamish about it. He has a fiduciary responsibility to the citizens of Springfield that requires him to terminate Eleanor if her performance is substandard and counseling is ineffective. Harsh as it may seem, the rules clearly direct Clayton to concern himself with Eleanor's performance. Her personal problems are not within his purview as director to act on.
2. Let Eleanor go, soon. Clayton should take Allan's advice and let Eleanor go with her earned benefits, being sure that he first takes time to carefully document her case so it cannot be construed as age discrimination. He should delegate to Allan and Pauline the task of compiling the necessary documentation. Harsh as it may seem, this is not the welfare department.
3. Let Eleanor go, unless... Clayton should talk with Eleanor, explain the negative impact her performance is having on the Department, and give her a six-month probationary period in which to improve her performance. The cold facts are that everyone must pull his/her own weight or leave. If Eleanor does not improve enough to warrant her current position and salary, Clayton must let Eleanor go.
4. Keep Eleanor on, as is. Eleanor's loyalty must be rewarded, and Clayton must not be stingy about it. Eleanor is so well-liked and has performed so well in the past that Clayton can justifiably overlook her present shortcomings, for years if necessary. There are more than enough people in the office, Allan and Pauline in particular, to help Eleanor in her time of need.
5. Keep Eleanor on, but seek help from the City. Clayton should retain Eleanor in the department at her current position and salary and arrange for appropriate

counseling. Most likely Springfield has a qualified psychological consultant available through their Human Resources Department who can seek out available sources of external support for Eleanor in caring for her husband and the costs associated with that care.

6. Keep Eleanor on, but seek help from Pauline. Eleanor should remain on the payroll in her current position and salary, but Clayton should continue to incrementally transfer Eleanor's responsibilities to Pauline. He should let Pauline know that she will assume Eleanor's job (and salary) upon Eleanor's retirement.
7. Demote Eleanor. Clayton should keep Eleanor on the payroll till age 62, but explain that she will have to take a step-down in position to receptionist and cannot remain as an executive secretary. While this will mean a reduction in pay, it is better than nothing and will supplement her health insurance coverage until she becomes eligible for Medicare.
8. Transfer Eleanor. Clayton should ask the civil service group to find a job for Eleanor in another department that is less demanding and is more compatible with her age and work abilities, albeit at lower pay. This will take some of the pressure off Eleanor, and relieve the Public Works Department of an unnecessary burden.
9. Reprimand Allan. Clayton should inform Allan that the Public Works Department would likely not exist to provide the opportunities Allan now enjoys were it not for many past years of selfless service by persons such as Eleanor. Clayton should remind Allan of the importance of loyalty, and require that he take sensitivity training. Further evidence of not being a team player on Allan's part would be grounds for his dismissal.
10. Hide and Watch. Clayton is partly, maybe mostly, responsible for much of Eleanor's predicament since for the past 15 years, as Director, if he believed Eleanor was underpaid, he had methods of advancing her salary that he chose not to use. In the name of cutting "waste" out of government, for which he happily took full credit, Clayton watched Eleanor give 110% at menial pay, in effect mortgaging her health, motivation, spirit and effectiveness for the sake of the Department. Clayton's true motive, then and now, has been self-promotion; Clayton realizes that to comfortably reach his own retirement he must keep the City Manager happy. Citing his favorite line from Casablanca, "This is the beginning of a beautiful friendship", Clayton should quietly direct Allan to deal with the situation (Eleanor) as he (Allan) has proposed.

Forum Comments from Respondents

1. Demote Eleanor. Clayton should keep Eleanor on the payroll till age 62, but explain that she will have to take a step-down in position to receptionist and cannot remain as an executive secretary. While this will mean a reduction in pay, it is better than nothing and will supplement her health insurance coverage

until she becomes eligible for Medicare. Most likely Springfield has a qualified psychological consultant available through their Human Resources Department who can seek out available sources of external support for Eleanor in caring for her husband and the costs associated with that care.

2. Promote Eleanor, give her a raise, and have her coach Pauline to learn the extensive roles of their position.

3. Besides the other two chosen paths, there are some possibilities that Clayton seems to neglect. While it is true that Eleanor may be overpaid for her current job performance, is it possible that the budget may be cut in other ways, maybe not necessarily in its employees but in it expenditures? Also, if Clayton does come to the conclusion that he must fix the Eleanor situation, one very key thing he will have to do is talk to Eleanor about her current job performance and the financial situation of the department. Clayton needs to inform her of his options as her employer and communicate why this is a difficult decision for him. Pauline may eventually take Eleanor's position and salary when Eleanor retires or maybe even makes her own request for demotion, leaving, or otherwise. But clearly, out of respect for Eleanor, she needs to be informed of the situation and possible outcomes.

4. Eleanor should be given an early retirement package which saves the city from a great deal of legal grievances. She will have the time she needs for her husband, and her position can easily be taken up by a more energetic and lower-paid person. She would probably prefer to retire now than continue working. Logic is no longer what society operates by. Our society is emotionally and irrationally charged. Even with a good lawyer for the city, the juries of today would inevitably harshly penalize the city with an enormous lawsuit worth more than Eleanor's present wages for the next 2000 years. Early retirement plans are the best compromises in today's increasingly aging work force population.

5. Eleanor should be talked to, and put on probation for a 6 month period. During that time, Clayton should fully assess her performance. If she is unable to fulfill her job, she should be demoted, possibly to receptionist. She will still keep an income, and will be eligible for retirement.

6. Give Eleanor a probationary period in which to prove that she can earn her current pay. If after 4-6 months she proves that she is unable to earn the pay that she receives, she should be given an appropriate job (and pay) with the company for her skills and workload ability. She has been loyal to the company and should at least be allowed to be on the payroll until her retirement and pension kicks in if it is her choice to do so.

7. Keep Eleanor on but give her 6 months probation to improve or be demoted to Pauline's position. Explain to Eleanor that because of her past value she will not be let go but that at her current performance, she cannot retain her position and pay.

8. Dismiss Eleanor but give her a severance pay package that would roughly equal the difference between her actual salary and the salary she should have been receiving over the past 15 years.
9. Clayton should talk to Eleanor, and when he does, he should tell her that if her work does not improve in the next six months, that she will be demoted, but not fired.
10. Clayton should talk with Eleanor directly about the problem of her performance and what solutions might be available to help her personal situation. Eleanor is still a bright woman albeit burned out; and she probably knows she is not performing at a high level and that her work is being given to another person. Enlist Eleanor to become part of the solution. If she cannot improve within a reasonable time, provide a position meeting her current performance and at a reduced salary.
11. There is no evidence given to indicate that Eleanor has ever been counseled regarding her decline in performance and the impact her attendance is having on the department. Therefore, she must be counseled and given a chance to improve, Clayton should begin documentation and progressive discipline up to and including termination.
12. Eleanor should be informed of her negative effect on the department, given time to improve, and given the resources to improve. Given her long-term commitment to the Department, a substantial probationary period should be given, perhaps one month for every year worked, or 35 months. If you were to view this as a kind of severance pay (in the event that no improvement occurred), it certainly would not be out of line with the golden parachutes that executives routinely get. However to ensure a successful outcome, the social service support should be augmented with significant feedback every six months against a set of benchmarks, agreed on by both Clayton and Eleanor.

Comments from Board of Review Members

All managers/supervisors are confronted with this type of issue from time to time. Clayton, Director of Public Works, must be fair to Eleanor, but at the same time, he must meet his obligations to his employer, the City of Springfield. It is unfortunate that Eleanor was not properly compensated in the earlier days, if such be the case. But, that is history, and it is not appropriate to try to compensate after the fact. It may well be a bad reflection on the performance review procedures, and Clayton needs to reevaluate the process. Clayton has an ethical obligation to be fair with Eleanor. Her personal situation, however, should not be a factor in the decision reached. On the other hand, her prior service and contributions to the Public Works Department should be considered. Clayton also must be fair to the Department and its staff. It is not fair to the other staff members to compensate Eleanor for services not performed. It is assumed that the Department, or the City, has an established policy on a termination package. If the policy allows flexibility, then Clayton should be as liberal as the policy will permit. Eleanor should not receive any benefits that would not be given to some other person being terminated.

Don't Ask, Don't Tell

From Applied Ethics Case of the Month Program — Case 1030

The Situation

Reilly M. Karful, a wastewater process engineer with Slud, Gefl, Owsdown & Hill (SGOH), has spent the last few weeks designing an equalization tank for the wastewater treatment plant for the town of Whiteside. It so happens that the existing extended aeration treatment plant is located inside one room of a pre-engineered metal building, except for the headworks, which consist of a screen. The design that Reilly is proposing includes installing a new tank inside the treatment room and relocating the screen on top of the tank.

During a design review, the reviewer, Yvonne Moore-Karful (distant relation to Reilly) of the architectural firm Plansem & Drawsem, notices that sewage plant headworks are covered by National Fire Protection Association (NFPA) Code Section 820, which requires certain areas to be Class I, Division 1 or Class I, Division 2; i.e., "explosion proof." Concerned about the potential problem, Yvonne calls to let Reilly know that moving the screen into the treatment room would require the room to be classified as explosion proof, and she follows up this call with a memo and her notations on the design drawings. Reilly decides to double-check this information and he reviews NFPA 820 himself. The code indeed appears to require the headworks and the existing treatment room to be explosion proof because the plant does not have primary settling tanks.

Reilly informs the project manager at SGOH, Dante McWaves, about the matter, and gives him a copy of Yvonne's memo. With Reilly in his office, Dante phones the client, Bull Parker, the town commissioner, to make him aware of Reilly's concerns. Bull happens to be cleaning his shotgun when he gets Dante's call. News of cost increases and delays does not please Bull, and he pointedly reminds Dante that the project is behind-schedule, that Whiteside needs this addition to the wastewater treatment plant to avoid non-compliance with State regulations, and that Whiteside is not a wealthy town. As he trips the action on the shotgun, as if to place a shell in the chamber (Dante hears this over the phone), Bull tells Dante that he'll have the County Fire Code Official, Bobby Burns, get back to Dante.

Reilly notices the color drain from Dante's face and that his hands visibly shake as he hangs up the phone. Within minutes, the fire code official calls and informs Dante that the existing room is already explosion proof. What luck! The problem is solved and Dante, still pale yet with obvious relief, informs Reilly that he doesn't want to bother Bull with any more problems. Dante then clearly and unequivocally directs Reilly to finish the project.

But Reilly is concerned. While Bobby Burns' statement satisfies Dante, Reilly suspects the fire code official could be wrong. Though Reilly has limited experience with fire codes, he has noticed that the electrical sockets, light switches, lighting fixtures, and junction boxes appear normal and don't give the appearance of being explosion proof. While part of him wants to pursue the matter, he also realizes that exploring his doubts about the existing room being explosion proof will not only further stress out Dante, but could result in additional project delays, in Whiteside having to fund an expensive retrofit to bring their existing facility up to code, and in further annoying Bull.

What, if anything, should Reilly do?

Alternate Approaches

1. Get on with it. Reilly should accept the fire code official's statement, as well as Dante's directive, and get on with the project without delay.
2. Cover your backside. Reilly should accept the fire code official's statement, as well as Dante's directive, and get on with the project, after first taking the time to carefully document his concerns about the entire situation in a detailed memo to file.
3. Use the opportunity to learn. Since Reilly is admittedly inexperienced with explosion-proof switches and fixtures, he should "quietly" do some homework on the internet and with electrical contractors or suppliers to determine how to identify whether the fixtures in the treatment plant room are explosion-proof, and he should go out to the site and make the appropriate inspection. If something is indeed amiss, Reilly should estimate the cost for explosion-proofing the treatment room, and compare this with the costs of alternative designs which will not require explosion-proofing. He should not charge this effort to the project, but do it on his own time, and once he feels he has the solution, he should outline a recommended course of action and notify Dante of his findings.
4. This is too big to handle alone. Realizing that Bull has effectively intimidated Dante, Reilly should, without Dante's knowledge, approach SGOH's managing principal, Mr. Owsdown, and share his concerns about the fire code official's seemingly "convenient" statement, and about Dante's acquiescence in the matter. Reilly should await further instruction while Mr. Owsdown deals with Dante.
5. Comply with the code of ethics. In the interest of the public health and safety, Reilly should pursue his concerns until he feels the matter is satisfactorily addressed. At the least, this would consist of doing some homework and having a qualified independent third party knowledgeable of the NFPA requirements – perhaps Yvonne – inspect the treatment room with him. If something is indeed amiss, Reilly should engineer appropriate solutions. Further, Reilly should bill all this work to the project, even though it will surely result in a budget overrun for Dante, but that is a small matter compared to the public safety and welfare.
6. There are many ways to solve a problem. Reilly should call a friend at the local newspaper and – with the agreement that he be identified only as a

"reliable source" – share his concerns about the fire code official's seemingly "convenient" statement, and about Bull Parker's intimidation tactics. He knows that his newspaper friend will make appropriate inquiry, and this will pressure Bull and Bobby to act responsibly without Reilly having to get wrapped up in SGOH's company politics.

7. Implement company loss prevention training. After giving Dante time to pull himself together, Reilly should approach Dante and share his concerns about the fire code official's seemingly "convenient" statement. Further, he should suggest a game plan that shifts the liability off of their firm, such as requesting that the fire code official provide written confirmation of the explosion-proof nature of the treatment room. This gets Reilly's and Dante's firm legally off the hook if an explosion occurs.

8. Run toward the roar. After giving Dante time to pull himself together, Reilly should approach Dante and share his concerns about the fire code official's seemingly "convenient" statement. Further, he should suggest that both he and Dante visit Bull Parker to explain their concerns, as well as the ramifications of negligence and the fact that an engineer may not ignore codes in a design. While risky, this puts the issue back in the client's court.

9. Go to the source. After giving Dante time to pull himself together, Reilly should approach Dante and share his concerns about the fire code official's seemingly "convenient" statement. Further, he should suggest that both he and Dante visit the fire code official and arrange for a meeting and inspection of the existing treatment plant room to look into their concerns. Perhaps the fire code official has never been in the room, or has mistaken it for another, properly explosion-proofed room.

10. Check it to them. Reilly should inform Dante that he refuses to continue with the project, and resign from the firm if necessary. While this will be costly to him, it is surely better than being involved when the sludge hits the fan!

Forum Comments from Respondents

1. I suggest solution #3 (use the opportunity to learn) with the following changes. If something is obviously amiss, a little homework will reveal that to Reilly. Then he has some real facts to take a stand on. If Dante, Bull, or the fire official still refuse to accept the new plans to bring the facility up to code, Reilly can bring the issue to Mr. Owsdown or other officials and insist on an independent inspection. If after doing his homework, it is still not clear to Reilly whether the facility is explosion proof, it should be acceptable, probably mandatory for a routine inspection by the fire official or other qualified inspector to OK the renovation in writing. At that point, there should be no need to doubt the integrity of the fire official, or to insist on additional independent inspections.

2. Although there is intimidation involved, ensuring that the room is secure in the event of a fire is more important than coming in on budget and on time. Even if it costs Reilly his job, he must take action whether he likes it or not to verify the safety of those that will be in that area.

3. Doing your homework and double-checking is the way to go. This is an opportunity for Reilly to do his homework on the explosion proof aspects of the code and this will give him experience for the future. He ought to also double check with the fire official after explaining his concern with Dante. This, I feel, is not disloyal to the client, employer or the public. Since they do indeed need to comply with code, Reilly will be looking out for the best interests of both the company and client, while also following the code of ethics.

4. If going to the source doesn't prove successful in fixing the problem, then Reilly should continue up the chain of command until the problem gets due attention. If this is all to no avail, then a little attention from the local media would surely do the trick. Although Reilly puts his job in serious jeopardy if the media route is chosen, he can rest assured that he did everything within his power to maintain the safety of plant employees.

5. Before visiting Bull, Reilly needs to do some research as to whether the room meets the "explosion-proofed room" standards. Further, he should explore some possible solutions to the problem so that he can suggest a viable approach.

6. Reilly should take the time to place this situation into the five ethical standards and think it through: The Golden Rule Standard, The Professional Ethic, Immanuel Kant's Categorical Imperative, The Utilitarian Rule, and The "60 Minutes" Test. From his conclusions to these five ethical standards, Reilly should then make his decision about what should and needs to be done.

7. Go to the source. If Dante refuses to go to the fire code official, then Reilly should inform Dante that he refuses to continue with the project, and resign from the firm if necessary.

8. After choosing solution #2 (cover your backside), I would also take the approach on solution #3 (use the opportunity to learn) and go to the internet and learn about fire-proof rooms and do further inspection to see whether the room complies with the information gathered by the research. If it does not, I would document the condition and take necessary actions.

9. I don't feel like I am ready to handle these types of things yet (Ed. note: this response is from a graduate engineer with 3.5 years experience).

10. To comply with the code of ethics, means must be made to secure safety. Research and professional insight should be introduced while simultaneously asking for a letter of proof from the fire code official to cover themselves!

When the information is gathered and the situation is clearly understood, then throw the ball into Bull's court!

11. Going to the source is a great approach, but I would include Yvonne in the meeting. She was the design reviewer who made the original comment about the explosion proof room. If the local inspector is able to convince her, then the problem is solved and the project can proceed. If she is not convinced, then Reilly is obligated to pursue a solution that the reviewer will accept.

12. If I were Reilly, I would pursue solution #8 (run toward the roar). This option calls for a meeting with Reilly, Dante and Bull (i.e. engineer, project manager and client). The purpose of the meeting is to allow input from all parties and inform the client (Bull) of the ramifications to ignoring the code. In my opinion this is the best option suggested because it addresses the problem in accordance to the Code of Ethics.

13. The course of action should be solution #5 (comply with the code of ethics). This option is the best choice for several reasons. First, public safety is to be recognized as the top priority of any project. Fire codes are written as a direct result of tragedy. They are lessons learned the hard way. There are both legal and humanitarian motivations to building "up to code". The safety of the public is best considered by meeting with Bull and addressing the issue head on.

    Secondly, although Bull may be an intimidator he is still the client. Reilly has a duty to the client to keep him informed of what is going on and to guide him towards the right engineering solution. The option of running toward the roar is the only option that takes the situation back to the client in a professional manner. This option is supported by canons 1 & 4 of the ASCE Code of Ethics.

    Further, Canon 1 of the Code of Ethics instructs engineers to view the safety, health and welfare of the public as their primary goal. By meeting with Bull to "clear the air," Reilly would in fact be serving the public by educating Bull of the problem and the need to handle it correctly. By addressing the issue directly with Bull, Reilly would assure that whether Bull continues to use SGOH, he is at least aware of the consequences of his decisions.

    Canon 4 of the ASCE Code of Ethics states that engineers should act as faithful agents or trustees for their employer or client. In this case Reilly and Dante need to take the roles of both engineers and lawyers. As engineers they need to inform their client of the technical aspects of the undertaking and to advise their client of the technically correct course of action. They can accomplish this by informing Bull of their findings and/or concerns and suggesting a workable solution. They can give

Bull examples of similar situations, and their outcomes, where the decisions were not made correctly. Reilly and Dante have a responsibility to convince Bull that although the decision may be hard and will certainly lead to cost overruns, it is the best decision for safety and financial reasons (doing it right once is always cheaper than doing it twice).

As "lawyers," Reilly and Dante need to protect their client by strongly advising him to choose the legally correct course of action. They need to educate Bull of the legal reasons why he should heed the precautions of re-inspecting the room and re-designing it in accordance to fire code. Reilly and Dante need to convince Bull that not only is he making the right choice for the town but he is also protecting his own reputation.

In conclusion, although I feel that the "run towards the roar" option is the best choice for this situation, I do feel that it falls a little short of perfect. For example, there is nothing said about what will be done if Bull doesn't come around and continues to insist on proceeding with the project as is. There is also no mention of when to inform the principal (Mr. Owsdown) of the situation. Dante and Reilly work for Mr. Owsdown and he should be informed as soon as possible and prior to the meeting with Bull as he may have valuable suggestions based on his experience for how to deal with the situation. I believe that if Reilly and Dante do a competent job of explaining the situation, Bull will make the right choice. But, if Bull does not accept the situation, Reilly and his firm should pull out of the program and document why in a letter sent to both Bull and an engineering firm that would take over the work.

14. While Reilly should educate himself, once he discovers that the room is not explosion proof, all cost to redesign / re-engineer a solution should be billable.

15. I think these actions are a little premature but a combination of aspects from solutions #8 (run towards the roar) and #9 (go to the source) would be the most efficient and ethical. Reilly should call the fire code official back and ask the exact requirements for explosion proof rooms. Then after personal inspection, if there is any question that the room is not satisfactory Reilly should arrange a meeting with the fire code official. Further, Reilly should ask the fire official to sign off that the room is indeed explosion proof. This way, Reilly has covered himself and his firm.

16. Reilly should begin by documenting everything as it occurred. Then, he should approach Dante and visit the fire code official with him. Reilly should also enlist a third party to inspect the room, such as Yvonne. Once these things were done, Reilly should accompany Dante to visit Mr. Owsdown to make him aware of the situation. At this point, Reilly should ask for Mr. Owdown's support and suggestions. Finally, Reilly should

approach Bull with all the facts gathered and offer a solution that meets code. If Bull still ignores the facts, Reilly might consider "leaking" the story to the media to put some pressure on Bull. If it comes down to Reilly being forced to complete a faulty design, he should resign. Regardless of the documentation, Reilly's name would be associated with the structure, and he should not wish to have any part in the disaster that might ensue from an improperly designed building.

Comments from Board of Review Members

1. Reilly recognizes that he is not the official who certifies fire code issues, but remains concerned about the situation nonetheless. As the engineer of record for the project, he might send a letter to the county fire official explaining his concerns and formally requesting a signed letter certifying that the building meets the Class I, Division I requirements. Reilly would keep this letter for his files and forward a copy to the owner.

2. Reilly rightly guards his professional reputation and realizes that this project must succeed both in terms of scheduled delivery and quality of design. Consequently, Reilly might contact another fire inspector and request an independent evaluation of the building's fire protection. If the independent appraisal finds the building not satisfactory, Reilly would need to inform the city that repairs are required.

3. Reilly should make a memorandum for record of the telephone call and provide copies by certified mail to the fire official, city and all other parties involved in the project. If no one disputes the memo, Reilly is legally covered in the event of a future problem.

4. Reilly should put his head together with Dante and settle on a game plan that shifts the liability back to the County (after all, it's the County's project). Dante needs to have the fire code official, Bobby Burns, provide a written confirmation of the explosion-proof nature of the treatment room. If Bobby Burns is not willing to provide such a written confirmation, then Dante should write him a letter, certified, return receipt requested, summarizing the conversation and concluding with a statement to the effect of, "If any of the foregoing is not in accord with your recollection of our conversation, it is important that you provide me with a written clarification of our discussion." This helps manage the liability aspects of the problem.

5. Reilly should make an appointment with Yvonne to inspect the treatment room. If Yvonne is an expert on NFPA codes, she should be able to tell if additional concern is warranted. Reilly is admittedly not an expert on NFPA codes and he should be able to obtain guidance from Yvonne.

6. Reilly should prepare a brief internal memo to Dante stating his concerns regarding the potential code violation and the associated safety issues as they apply to the people currently and potentially working in the reconfigured

treatment plant. He should ask for an opportunity to discuss his concerns in detail with Dante. Additionally, if the matter of who is the responsible, licensed, Professional Engineer for this design has not been made clear, Reilly should ask Dante to clarify the firm's policy regarding professional responsibility for this project. On the assumption that Reilly will be asked to sign and seal the documents, then he should state his need to check the actual condition of the facility with regard to code compliance and to assure that the reconfigured facility will meet the fire code. If the firm decides to ignore the issue and proceed without addressing the apparent present and future code violations, Reilly should "polish-up" his resume and look for a more professionally oriented employer.

7. Many, if not most, codes of ethics provide that the health, safety and welfare of the public is paramount. This, therefore, requires the engineer to take all reasonable steps to protect the health, safety and welfare of the public. The facilities (equipment, switches, etc.) not meeting the building code requirements for explosion proof certainly places the public's health and safety in question. Reilly has an ethical obligation to take action. He should first advise the fire code official that he has observed that the electrical system does not meet the fire code. If the fire official declines to take any action, then Reilly should advise his supervisor, Dante McWaves, about his concern. If Dante McWaves declines to take any action, then Reilly should write a memorandum to Dante McWaves, with a copy to Bull Parker (the town commissioner) and the fire code official, advising about the violation of the building code and requesting that he be authorized to make the appropriate revisions even though additional cost and time will be involved. If Dante McWaves declines to give the authorization, or declines to respond, Reilly should request being removed from the project. This may very well cost Reilly his employment, but the health and safety of the public is paramount, even at the expense of employment.

Was That "Piracy" Or "Privacy"?

From Applied Ethics Case of the Month Program — Case 1031

The Situation

Lawrence, the managing principal of NorthLink Consultants, is pleased at his firms' new information technology (IT) capabilities. Knowing that effective use of IT offers a strategic competitive advantage in the marketplace, Lawrence observes an increase in cooperation on projects and the office is using much less paper for memos and policy directives. The company web site is growing as staff, engineers and clients contribute to the site.

Lawrence, however, worries that several employees are spending an excessive amount of time on email. He suspects that much of this email activity is directed at family and friends on the Internet and outside the firm. He had reminded the employees of NorthLink's policy which states that email is for company business and emails are considered part of the firm's property.

Still, Lawrence feels that there is way too much time when employees are emailing in inappropriate ways. He approaches his systems engineer, Gwen, with a question. Since all the computers are connected on the computer network, could she access the employees' email files on their PCs? Gwen replies that such an examination of the files on the PC workstations is possible. Her own feelings, however, are that such an attempt to "reach out and touch" the users' PCs would be a breach of trust. In fact, some employees might be so offended with this intrusion of privacy that they would leave the firm.

Lawrence responds that company policy clearly informs employees that the email files are the property of the firm. They should understand that it is part of his supervisory responsibility to see that they use the email properly. Gwen argues that employees might well use the email to talk about issues that they do not want management to see. These may be legitimate company issues, but are not meant to be shared with the management. Adamant in his resolve, Lawrence states (as he walks out of Gwen's office) that, by tomorrow evening, he expects to be able to access all of the email files on each of the PCs.

Gwen is very disturbed. This policy will open up communications she feels should be regarded as private except when some formal legal decision requires them to be opened. However, since it is her job she knows she cannot refuse to perform a technical change in the system, and she feels she must allow Lawrence access by tomorrow evening.

Question: What should Gwen do?

NOTE:

This case is based on "Reach Out and Touch Someone" by the Public Administration at the University of Arkansas, and is used with permission.

1. Comply, willingly. Gwen should do what she is told. NorthLink employees are using company equipment to make personal email transmittals and doing so on company time, despite having been informed that doing so is against company policy. They have no reason to complain, nor should Gwen complain or feel uncomfortable about following her supervisor's orders.

2. Comply, reluctantly. Gwen should stifle her conscience and abide by Lawrence's request. She does not have the responsibility for strategic vision, running the company, or any other business decision, and should not insert herself into that process. Lawrence is in charge of this section and responsible for implementing company policy as it pertains to IT activities. Times are tough and she needs the job. After all, Gwen is not the one spending company time on private business.

3. Refuse, flatly. Gwen should refuse to access the email files for Lawrence in the manner he has requested on the basis that his action is unethical. Further, she should inform Lawrence that she is prepared to resign if he forces the issue.

4. Refuse, conditionally. Gwen should refuse to access the email files for Lawrence in the manner he has requested on the basis that his action may not be legal. Further, she should inform Lawrence that she will not access the files without express written approval from the human resources department, the employee union, and the firm's legal department.

5. Analyze, carefully. Since she has access to the employee email files, Gwen should offer to do a confidential analysis for Lawrence of the email files to determine the apparent volume of personal messages, as well as which employees seem to be using the system the most for personal emails, but she will not review the nature or content of any of the messages.

6. Document, clearly. Gwen should help Lawrence put together a brief agreement form for each employee to read and sign that reiterates the company's policy regarding the use of its equipment and time for personal email communications, and which clearly states that the employee agrees that the company has the right to review employee email communications on a random, unannounced basis, for compliance with the policy.

7. Monitor, quietly. Gwen should propose an option to Lawrence that instead of trying to read the emails, she can install a clandestine tracking system that keeps a daily log of internal and external email volume (sent and received) by individuals. This system will provide weekly reports of email activity to Lawrence, which he can use to manage IT resources and activities.

8. Monitor, openly. Gwen, as administrator, should suggest that she monitor the email of those employees who are putting the most strain on the email system and, if the email is not related to company business, counsel them privately to cut it out or risk the loss of their job. Prior to setting her on this

course of action, Lawrence should announce to the staff that, in accordance with the Company Policy, email will be read and individuals engaging in email correspondence unrelated to Company business will be subject to the written corrective action policy (counseling, warning, formal reprimand, suspension, termination). All discipline above counseling would be performed by Lawrence and/or his partners.

9. Inform, quickly. Gwen should quietly inform all employees of Lawrence's abrupt decision, immediately, and suggest that individuals clean out their email files so as to not face Lawrence's ire. She should note the date and time of Lawrence's remarks to her, and, for her own file only, reasons for her opposition.

10. Download, surreptitiously. Gwen should realize that this is the perfect opportunity to see how the firm's partners, including Lawrence, spend their email times. After making sure that her own email box is clean as a whistle, she should make a copy of each manager's inbox for perusal later, since you never know when this type of information might come in handy.

Forum Comments from Respondents

1. Comply, but it will not work and Lawrence will need to come up with an alternative. Gwen will do her job, and ultimately, Lawrence will NOT be satisfied. People will use other accounts like Yahoo, HotMail, MSN, and so on. People will only use the company account for related business; however, they will continue to use the network for personal business.

2. Similar to monitor openly, Gwen should do what Lawrence asks her to but only after informing all employees about the change in structure of the system. The goal is to make sure that employees aren't comitting "theft of payroll funds" by using time inappropriately. In response to Gwen's privacy concerns, a system should be set up in which the employees can communicate to one another via emails specially marked as ones not to be viewed by management. A non-management employee (perhaps Gwen) should be given the authority to monitor these emails.

3. Combine Approach 6 and 8. First, the managing director should be counseled regarding the ethical and legal aspects of his decision. Specifically, he should understand that the review and disciplinary process must ensure the equal treatment of similarly positioned employees - if a star performer and a weak performer are both found to have abused the email system, the managing director must be prepared to treat the employees in a similar fashion to avoid a lawsuit by the poorly performing employee. (Of course, that information might diminish his desire to review the emails.) Second, if employees have not already signed an acknowledgement that they have read and agreed to the employee policies (including a clear statement of policy, including the company's right to review employee email communications on a random, unannounced basis, and the consequences of breaching the policy), then a signed acknowledgement should be obtained

in order to help avoid misunderstanding or miscommunications. If employees have already signed such an acknowledgement (often done at the time of hire) then that step appears to be unnecessary. However, there should be an announcement that the review process has begun, who is involved in the process, how confidentiality is being protected, and that the email review will be prospective only. As for the process of review, an employee in a position of trust (perhaps the director of HR or legal counsel) should guide the process. Given Gwen's concerns with the directive, she might not be in the best position to adequately or properly manage the process. All discipline above counseling would be performed by Lawrence and/or his partners; however, accurate records should be kept of the disciplinary process and the reasons supporting the actions taken against an employee.

Comments from Board of Review Members

1. Whether or not the company has a "right" to review the email, doing so will generate a huge moral and morale problem within the firm, regardless of what information would be revealed. Employees have a right to some privacy, and this is a change in company policy that supercedes this right. Until now, the company has not performed email monitoring, nor stated it would. To institute this policy without warning is not appropriate. Gwen should try explaining this to Lawrence again (and other Partners), with an admonishment to notify the employees in advance of instituting this spying. She should let him know that he is opening himself, and the company, up to lawsuits and loss of good employees who will not tolerate this kind of behavior.

2. One key consideration is the sensitivity of the content of the emails even if they are all relevant company business. Inappropriate disclosure (even to Lawrence) of personnel decisions could cause serious (and unnecessary) personal and/or professional embarrassment. For example, I am aware that the Department of Defense (DoD) does routinely screen emails sent through DoD email systems. They also have monitoring programs that automatically compile and report visits to "inappropriate" internet web sites. The most common inappropriate type of site is pornography, but that is not the only type of site prohibited. Service members have been tried and convicted of computer crimes based on their emails and website visitation histories.

3. A very interesting question; we are addressing it at our firm right now but on a slightly larger scale. Our firm is networked with T1 lines between all offices and high-speed internet access from every desktop. Recently our MIS Manager noticed that the server hard drives were becoming filled up, which was surprising given their size and previous company history of hard drive capacity utilization. He checked and, lo and behold, there were gigabytes of totally inappropriate files that had been downloaded and saved to the company's servers. Virtually all of them were of a nature that would cause legal difficulties of one type or another (either with owners of

intellectual property or, in some cases, with law enforcement personnel). After conferring with our Chief Operating Officer, he did a file wipe with no prior notification to those who had saved those files. At the next "all hands" company conference call, the announcement was made that our MIS Manager would be using his Network Administrator technical capabilities to, 1) identify who was embezzling company property (paid "on-the-job" time, PLUS bandwidth, PLUS file space) in this fashion, and 2) forward a copy of the offending file (including date, time, duration of time on line, and identity of the employee who did the download) to our CEO, who will not tolerate such activity.

How does this story relate to your case history? In several ways. First, diversion of company property to employee personal use is theft. This includes the time the employee spends conducting personal business when they are at work. Consulting firms like mine sell their time to their clients; clients would not pay for time used by a consultant's employees who were taking a nap when they were supposedly conducting a field investigation. This situation is equivalent.

Second, the problem is larger than simple inappropriate email use. Chat rooms and internet browsing are highly addictive activities. Employees who are not being properly supervised can literally spend hours in this fashion.

Third, downloaded files offer the added risk of exposing the firm to significant adverse legal liability ("If you did not know that our intellectual property had been pirated and was resident on your server, you should have known and I'm sure the court will agree.").

4. Gwen is obviously sensitive to the employee's side of the issue. Lawrence should recognize that sensitivity and suggest a compromise. Gwen, as administrator, should monitor the email of those employees who are putting the most strain on the email system and, if the email is NOT related to company business, counsel them privately to cut it out or risk the loss of their job.

   Prior to setting her on this course of action, Lawrence should announce to the staff that, in accordance with the Company Policy, email will be read and individuals engaging in email correspondence unrelated to Company business will be subject to the written corrective action policy (counseling, warning, formal reprimand, suspension, termination). All discipline above counseling would be performed by Lawrence and/or his partners.

5. When I worked as a consultant, my company, I believe, had a similar policy to NorthLink's. However, I don't ever recall being explicitly informed about it; the policy was just something you assumed was "out there," and I don't recall it ever being talked about. To my knowledge, the issue of invasion of employee privacy was never raised, probably because we had reason to believe that management was monitoring email, even though they never talked about it.

6. An interesting case, indeed. My company's policy is pretty much like NorthLink's. We have the capability to monitor email but we rarely do it. That is, we only monitor email if there is some evidence that it is being abused. We have a number of firewalls that prevent employees from visiting inappropriate sites, and we have not had any complaints about intruding into people's personal lives.

7. Gwen has two choices:
   (1) do what she is instructed to do, or
   (2) resign her position.

   Little would be gained by resigning since the next system engineer will be required to provide the information so the privacy of the employees would still be breached. Gwen might request a 30-day delay in carrying out the instructions so she can check with the firm's legal counsel to verify that no statues were being violated and also investigate what other firms are doing.

   She might also suggest that she review the emails (instead of Lawrence reviewing), and she would talk with those who are violating the firm's policy. She might also suggest that it would be better not to review past emails, but rather publish a notice that all emails henceforth are subject to review by management.

   The best procedure would have all employees sign an acknowledgement of the policy. Very likely Gwen's suggestions will fall on deaf ears. In that case she should proceed carrying out Lawrence's instructions.

Ye Olde Water Main

From Applied Ethics Case of the Month Program — Case 1033

The Situation

Shadyvale, a picturesque town in upstate New England, is having water problems. Three years ago an engineering consultant issued a report which stated, among other things, that the existing water main in Shadyvale was generally in good condition but was extremely old. Further, the water main no longer is large enough for all of the properties served. Since that time, the Town Selectmen have been trying to secure the funds needed to replace the old main with a new main of larger diameter, but they still cannot afford to do so.

However, the State Department of Transportation (DOT) is planning a highway reconstruction project in Shadyvale. Warren, a senior planning and programming engineer, is the senior DOT engineer responsible for this project. Although he now lives in the Capitol City, Warren was born and raised in a small village not far from Shadyvale, next door to his boyhood friend, Earnest "Red" Anderson. With a twinge of nostalgia and fond memories of a more peaceful time, Warren confidently delegates the project to one of his subordinates, Dianne, a young engineer intern about to sit for the PE exam.

Dianne is well liked, highly competent, and by all indications, will go far in the Department. As per standard practice, Dianne initiates the design layout for the Shadyvale project to avoid conflicts with the existing utilities, including the old water main. She understands that State DOT policy unambiguously requires that only unavoidable utility conflicts will be paid for as part of highway projects and that other utility work is to be considered as a betterment that must be paid for by the local municipality. This is quite fresh on her mind, since due to a downturn in the State's economy, several layoffs have occurred in the State government, and recent memos have repeatedly emphasized fiscal belt-tightening.

With the design at about 30 percent completion, Dianne submits a set of drawings to Warren for his review and comment. She also recommends a site visit to observe the route, tie down several details, and resolve various design issues. On the appointed morning, Warren and Dianne secure a State vehicle and make the very pleasant drive to Shadyvale. The stress of the city fades with each mile as they drive through some of the most picturesque scenery in the State.

"I'll definitely have to bring the wife back in September to enjoy the foliage," muses Warren to himself as they drive into town. Shops, restaurants, and even the local garage still have that old-town feel. As expected, it is a wonderful day in the field, and Dianne busily gathers the information she needs to take the drawings to the next level. In the meantime, Warren takes the opportunity to explore the town. To his pleasant surprise, he learns that his friend, Red Anderson, still lives in Shadyvale, and in fact is the senior Town Selectman.

Upon learning of Warren's being in Shadyvale, Red invites Warren and Dianne to lunch at one of the local cafes. In addition to catching up on old times, Red recounts Shadyvale's water main problem and how the town just sees no way to get together the kind of money needed for such a project. Dianne is well aware of the location of this water main – she knows Warren is too – she made a conscious decision to design around it. But neither of them says anything, not wanting to spoil such a friendly visit. After a good meal, good conversation, and plans for Warren and Red to get together in the near future, Warren and Dianne depart from Shadyvale.

On the way back to Department headquarters, Dianne comments to Warren about how, in view of their conversation with Red, she feels disappointed that she must align the new closed drainage system for this project on the opposite side of the road from the old water main. The facts are, were the proposed drainage system to impact the existing water main, it could be dealt with simply as an added share to the Engineer's Estimate for the Town's cost, and Shadyvale would only be responsible for the difference in price between the size of the now-impacted existing water main and the proposed larger size (about $25,000), rather than for all of the water main work (over $350,000). "It's just too bad," she sighed.

Warren had been silent, listening to Dianne's comments and line of reasoning. After she finished her thought, just as they were pulling into DOT headquarters, Warren looked directly at Dianne and stated in an odd sort of way, "Yes, the citizens of Shadyvale are really fine people, and they deserve the best the State DOT can offer. I want you to do a really thorough job on the layout, and I'll back you 100 percent." Then, as they were getting out of the car, Warren added, "By the way, Dianne, you had mentioned that you were going to send me a reference form for your PE license. Please get that to me and I'll complete it right away."

It is now the next morning. Dianne is at her desk, looking at the Shadyvale drawings. It seems obvious to her based on Warren's clandestine remark that he wants her to change the design so that Shadyvale's old water main is impacted, thus requiring the State DOT to bear most of the cost of replacing it. This could be justified easily enough, and no subsequent reviewer would question the issue if Warren signs off on this as he says he will. After all, it is not like Warren is asking her to embezzle funds for personal use or anything. But Dianne knows this action is not consistent with DOT policy. A voice inside screams, "What about your ethics?" Immediately another voice replies, "What about your PE license recommendation from Warren?" Yet another pipes up, "What about the deserving citizens of Shadyvale?" And on it goes.

How should Dianne proceed with this matter?

Alternate Approaches

1. Go along to get along, compliantly. Dianne should do what Warren wants – not rock the boat. She is getting her marching orders from Warren, she is under his direct supervision, and she should embrace this opportunity to learn from him how to handle these types of situations.

2. Go along to get along, cautiously. Dianne should follow Warren's apparent hint and design the reconstruction project so that the water line is replaced. If she feels that Warren cannot be trusted, she should keep records of conversations, meetings and review comments by him.

3. Share the wealth, generously. Adding the water main to the highway project is a "win-win" for everyone involved. The increase in the State project costs are negligible while building considerable community good will. Shadyvale will obtain plenty of clean water at an affordable cost, Warren will fulfill his sense of hometown obligation, and Dianne will receive the favorable PE recommendation she seeks. It is after all the responsibility of a professional engineer to look beyond policies and seek the best solution to every problem.

4. Take refuge in ambiguity, innocently. Dianne should proceed to develop the plans as per her understanding of DOT policy, pretending to have not picked up on Warren's subtle "hint" to move the water line. In her transmittal memo to Warren, Dianne should take pains to point out how she has complied with his request to be very thorough in her design work on this project. This puts the issue squarely in Warren's court.

5. Straddle the fence, technically. Dianne is not a decision maker, but she does have the opportunity to prepare and present two plans. One plan should follow the letter of the DOT directive and the second should follow a "cost-share" approach to include replacement of the water main. Since Warren seems very interested in influencing this project, offer him the opportunity to recommend the project to be sent forward.

6. Straddle the fence, politically. Don't let policy stand in the way of common sense. Since Shadyvale will benefit if the design is prepared in a certain way, there may be someone from the Selectmen who has influence in certain circles and could convince someone on the DOT board to grant a variance to the policy in this case. Maybe the Selectmen could ante up funds that would increase the town's share of the water line replacement cost which would justify the policy variance. Dianne should consult with Warren on this approach to negotiate a win-win solution and efficient expenditure of public funds.

7. Face the facts, squarely. Dianne should consult with Warren to clarify his intentions. His statement, made as they were getting out of the car, could be

interpreted to mean either design in accordance with DOT regulations and policy, or prepare the design to cause replacement of the old water main. If he is indeed asking her to ignore DOT policy, Dianne should talk with Warren about her concerns regarding moving the alignment and explain the reasons that this is a bad idea and try to talk him out of doing this.

8. Agree to disagree, cordially. Dianne should ask to be removed from the project – she should make it clear to Warren that she would not be able in good conscience to do what he has asked, and point out that it is within Warren's prerogative to engineer the alignment change himself, or direct someone else to make the alignment change.

9. Adhere to policy, strictly. Dianne has clear guidance from the DOT to minimize all costs to the State. This guidance is fair, causes no real harm to the people of Shadyvale, and does not violate the ASCE Code of Ethics. There is no reason, other than her concern about possibly losing a favorable PE recommendation, for Dianne to not follow the DOT directive.

10. Blow the whistle, loudly. Dianne should flatly refuse to change the alignment – Warren is asking her to do something unethical and something that will cost the DOT more than it should. She should simply say she cannot do this and put the alignment in the best location. Further, she should go over Warren's head to his boss and point out that he is encouraging her to violate DOT policy.

<u>Forum Comments from Respondents</u>

1. Warren's comments could be taken either way. If Warren wants a violation of DOT policy, Dianne could take option 5 and force Warren to take responsibility for the change in alignment. If he does so, Diane can go along with this, recognizing that Warren is the senior engineer and has the ultimate responsibility for the design. Or, Dianne could blow the whistle. If she blows the whistle she could do it loudly and publicly, or she could do it quietly by leaking a copy of the decision memo to someone more senior than Warren or outside the DOT. Either way, Dianne should look into getting a different person for her reference for her PE license.

2. Dianne is aware that, given the State's tight financial situation, the decision to alter the plans to please Warren could result in someone else in the DOT getting laid off. There are plenty of other reasons to adhere to DOT policy, but I feel that this is the major one.

3. Instead of keeping this a big secret, Dianne should ask Warren to discuss this situation openly with DOT and Shadyvale to see if something can be worked out. Dianne should take a proactive approach to the problem, without trying to draw attention to certain individuals. If the DOT is willing to work with Shadyvale, great. If the situation is treated as a secret then it is dishonest.

4. I would not mind if my gas tax helped a few people get some needed clean water.

5. In this case, I believe Dianne should ask Warren to clarify his intentions. By doing what he wants, she will save her PE recommendation. This is against policy, but sometimes you have to stray from what is mandated. No rule or law is absolute; there are and will always be exceptions and under varying circumstances what is "right" may not agree with regulations. As for the money, the residents of Shadyvale are going to pay about the same amount either directly or indirectly. The funds necessary for carrying out such projects come from the people and it will all even out in the end.

6. Option 3 is preferred because everyone is winning here. Shadyvale will benefit from the project and the DOT will not lose much. The most important thing is that the project will be done.

7. It seems that if Warren really wants Dianne to bend the rules for the project he should be willing to do the same. She gives him two plans, one of them is what he wants, the other one is more reasonable. Leave it up to Warren whether or not coffee with an old friend justifies $350,000 worth of added costs for a starved state budget. If he wants to bend the rules, Dianne can do her job but leave the responsibility of the choice where it belongs, with Warren.

8. Dianne should face the facts squarely, analyze the pro's and con's of the alignment options and let Warren decide. Warren is the responsible engineer and he owes Dianne clear guidance. Dianne owes Warren her best summation of the issues and impacts of the two alignment options. Written DOT policies and project policies may differ on the finer points and there are several possible reasons to replace the water main now; for instance, possible disruption of the water main during construction, protection of the investment in new pavement by replacing utilities now, ease of permitting, etc. There is the potential for joint benefit with cost sharing to be negotiated.

9. I feel that Option 6 is the best course of action. If I had to be one of the people involved in this problem, I would want to be able to reach an agreement in such a way as to benefit everyone. For the same reason, if this were in the newspapers it would make everyone look like they did their part and the project was a joint effort.

10. Ethics should not be pushed aside just to save money, even if it is to help out a town or other organization. There is always another solution to help people out.

11. Dianne should talk with Warren and tell him her concerns. If the project cannot be done in accordance with DOT policy then Dianne should request to be removed from the project.

Comments from Board of Review Members

1. OPTIONS FOR DIANNE:

Option A:

Dianne asks Warren for an opportunity to discuss comments of the preceding day. Warren says, “Sure. What’s on your mind?”

Dianne says, “Could we step into your office?”

Warren answers, “Sure.”

Dianne clears her throat and says, with some hesitancy, “What is your expectation of what a “really thorough job” ought to yield for our project layout, especially regarding the routing of the closed-pipe drainage system? You’ve had a lot more experience than I in how to deal with the type of issues we talked about yesterday. Could you provide me some written notes and mark up the drawings of the 30% design I submitted last week? That would be really helpful and could save me from overrunning my budgeted design hours.”

Warren says, “No problem. I’ll get them to you this afternoon.”

Later that day, after Dianne had finished her brown-bag lunch and returned to her windowless office from the “staff” picnic table, just outside her office there sat the rolled-up drawings and her “Preliminary Design” brief. There was a big yellow Post-It Note saying, “Looks great! Proceed with 50% Design for review on September 15th. W” The note bore no date.

On a standard Department Intra-Office Memorandum form, Dianne wrote out, “Warren, In accordance with the instructions you returned today with the “30% Design” package, I will proceed with the 50% Design drawings and specifications for your review on September 15th.” She dated and signed the memo. She stuck Warren’s yellow note on her memo and photocopied it. Then after removing the note, she put her memo in the mail cart. She then made photocopies of Warren’s note stuck on the cover sheets of her Preliminary Design report and drawings and filed the copies in the lockable center drawer of her desk.

About 3:30 PM, she peddled her 10-speed home, warned up some leftovers, wrote out a check for the third payment of her college tuition loan, and leafed through her Advanced Highway Design text while pondering the next forty years of her life.

Option B:

Dianne logs onto Monster.com

2. Dianne should study all the reasonable options of designing the project – including cost estimates and other related factors. Her decision should be based on the facts she develops. She should prepare a summary report of her investigations, conclusions and recommendations, and she should review the report with Warren. Assuming she concludes that the drainage system should be on the opposite side from the water mains, she should so advise Warren and explain how she reached her conclusion. Warren has several options:

   (1) agree with Dianne's conclusions and recommendations,

   (2) show deficiencies in her approach, or

   (3) instruct her to change her conclusions and recommendations.

   On the assumption that Warren will find deficiencies in her approach, if Dianne agrees with Warren's concerns about deficiencies and this changes the conclusion, she should make the change to her design. If she does not agree with Warren, then she should not make the change in her conclusions and recommendations. If Warren instructs her to change her recommendation, she should decline and offer the let Warren submit the project over his signature.

   Dianne should not change her conclusions and recommendations that are not supported by the facts she develops. She has the ethical responsibility to base her recommendations on her engineering studies even if it may jeopardize her receiving a favorable recommendation for the P.E. license.

   Warren, hopefully, will do the ethical thing and provide an appropriate reference for the license.

Survey Results For Selected Applied Ethics Cases

Now You Have It, Now You Don't
(Percentage of votes agreeing)

Question 1 (President of Nearby Engineers)

1 .... 4%
2 .... 1%
3 .... 11%
4 .... 7%
5 .... 11%
6 .... 13%
7 .... 31%
8 .... 11%
9 .... 0%
10 ..... 6%
11 ..... 2%
12 ..... 3%

Question 2 (Professional Engineer on County Public Works Department staff)

13 ..... 3%
14 ..... 50%
15 ..... 12%
16 ..... 1%
17 ..... 3%
18 ..... 15%
19 ..... 12%
20 ..... 1%
21 ..... 3%

Question 3 (Director of Public Works for the County)

22 ..... 2%
23 ..... 5%
24 ..... 5%
25 ..... 3%
26 ..... 22%
27 ..... 14%
28 ..... 38%
29 ..... 8%
30 ..... 1%
31 ..... 2%
32 ..... 1%

Question 4 (President of Truehart Engineers)

33 ..... 10%
34 ..... 6%
35 ..... 3%
36 ..... 18%
37 ..... 7%
38 ..... 8%
39 ..... 6%
40 ..... 5%
41 ..... 35%
42 ..... 3%

Between a Buck and a Hired Place
(Percentage of votes agreeing)

1 .... 3%
2. .... 3%
3 .... 25%
4 .... 3%
5 .... 21%
6 .... 11%
7 .... 20%
8 .... 9%
9 .... 4%
10 .... 1%

Was It "Piracy" Or "Privacy"?
(Percentage of votes agreeing)

1 .... 10%
2 .... 6%
3 .... 2%
4 .... 13%
5 .... 7%
6 .... 28%
7 .... 13%
8 .... 15%
9 .... 3%
10 .... 2%

Don't Ask, Don't Tell…
(Percentage of votes agreeing)

1 .... 0%
2 .... 7%
3 .... 18%
4 .... 4%
5 .... 15%
6 .... 1%
7 .... 10%
8 .... 15%
9 .... 29%
10 .... 1%

Ye Olde Water Main
(Percentage of votes agreeing)

1 .... 2%
2 .... 6%
3 .... 5%
4 .... 21%
5 .... 14%
6 .... 26%
7 .... 7%
8 .... 13%
9 .... 2%
10 ...... 2%

Public Health And Safety--Code Enforcement

NSPE Opinions of the Board of Ethical Review - Case No. 98-5

Facts:

Engineer A serves as a director of a building department in a major city. Engineer A has been concerned that as a result of a series of budget cutbacks and more rigid code enforcement requirements, the city has been unable to provide a sufficient number of qualified individuals to perform adequate and timely building inspections. Each code official member of Engineer A's staff is often required to make as many as 60 code inspections per day.

Engineer A believes that there is no way even the most conscientious code official can make 60 adequate, much less thorough, inspections in one day, particularly under the newer, more rigid code requirements for the city. These new code requirements greatly enhance and protect the public's health and safety. The code officials are caught between the responsibility to be thorough in their inspections and the city's desire to hold down costs and generate revenue from inspection fees. Engineer A is required to sign off on all final inspection reports.

Engineer A meets with the chairman of the local city council to discuss his concerns. The chairman indicates that he is quite sympathetic to Engineer A's concerns and would be willing to issue an order to permit the hiring of additional code officials for the building department. At the same time, the chairman notes that the city is seeking to encourage more businesses to relocate into the city in order to provide more jobs and a strengthened tax base.

In this connection, the chairman seeks Engineer A's concurrence on a city ordinance that would permit certain specified buildings under construction to be "grandfathered" under the older existing enforcement requirements and not the newer, more rigid requirements now in effect. Engineer A agrees to concur with the chairman's proposal, and the chairman issues the order to permit the hiring of additional code officials for the building department, which Engineer A believes the city desperately needs.

Question:

Was it ethical for Engineer A to agree to concur with the chairman's proposal under the facts?

References:

Code of Ethics Section I.1. - Engineers, in the fulfillment of their professional duties, shall hold paramount the safety, health and welfare of the public.

Code of Ethics Section II.1.b. - Engineers shall approve only those engineering documents which are in conformity with applicable standards.

Code of Ethics Section II.3.b. - Engineers may express publicly technical opinions that are founded upon knowledge of the facts and competence in the subject matter.

Code of Ethics Section III.1.b. - Engineers shall advise their clients or employers when they believe a project will not be successful.

Discussion:

The duty to hold paramount the public health, safety, and welfare is among the most basic and fundamental obligations to which an engineer is required to adhere. While in many instances, the obligation is often clear and obvious, in other instances, there could be an obligation on the part of the engineer to balance competing or concurrent obligations or responsibilities to protect the public health and safety. The facts of this case are in many ways a classic ethical dilemma faced by many engineers in their professional lives. Engineers have a fundamental obligation to hold paramount the safety, health, and welfare of the public in the performance of their professional duties (See Code Section I.1.). Moreover, the Code provides guidance to engineers who are confronted with circumstances where their professional reputations are at stake. Sometimes engineers are asked by employers or clients to sign off on documents about which they may have reservations or concerns (See Code Section II.1.b.).

The Board has addressed public health and safety issues in the code and approval process on numerous occasions. In BER Case 92-4, Engineer A, an environmental engineer employed by the state environmental protection division, was ordered to draw up a construction permit for construction of a power plant at a manufacturing facility. He was told by a superior to move expeditiously on the permit and "avoid any hang-ups" with respect to technical issues. Engineer A believed the plans as drafted were inadequate to meet the regulation requirements and that outside scrubbers to reduce sulfur dioxide emissions were necessary and without them the issuance of the permit would violate certain air pollution standards as mandated under the 1990 Clean Air Act. His superior believed that the plans, which involved limestone mixed with coal in a fluidized boiler process that would remove 90% of the sulfur dioxide, will meet the regulatory requirements.

Engineer A contacted the state engineering licensure board and was informed, based upon the limited information provided to the board, that suspension or revocation of his engineering license was a possibility if he prepared a permit that violated environmental regulations. Engineer A refused to issue the permit and submitted his findings to his superior. The department authorized the issuance of the permit. The Board concluded that (a) it would not have been ethical for Engineer A to withdraw from further work in this case, (b) it would not have been ethical for Engineer A to issue the permit and (c) it would be ethical for Engineer A to refuse to issue the permit.

Specifically, the Board determined that it would not have been ethical for Engineer A to withdraw from further work on the project, because Engineer A had an obligation to stand by his position consistent with his obligation to protect the public, health, safety, and welfare and refuse to issue the permit. Said the Board, "Engineers have an essential role as technically-qualified professionals to 'stick to their guns' and represent the public interest under the circumstances where they believe the public health and safety is at stake."

As early as BER Case 65-12, the Board dealt with a situation in which a group of engineers believed that a product was unsafe. The Board then determined that as long as the engineers held to that view, they were ethically justified in refusing to participate in the processing or production of the product in question. The Board recognized that such action by the engineers would likely lead to loss of employment.

In BER Case 82-5, where an engineer employed by a large defense industry firm documented and reported to his employer excessive costs and time delays by subcontractors, the Board ruled that the engineer did not have an ethical obligation to continue his efforts to secure a change in the policy after his employer rejected his reports, or to report his concerns to proper authority, but has an ethical right to do so as a matter of personal conscience.

The Board noted that the case did not involve a danger to the public health or safety, but related to a claim of unsatisfactory plans and the unjustified expenditure of public funds. The Board indicated that it could dismiss the case on the narrow ground that the Code does not apply to a claim not involving public health and safety, but that was too narrow a reading of the ethical duties of engineers engaged in such activities.

The Board also stated that if an engineer feels strongly that an employer's course of conduct is improper when related to public concerns, and if the engineer feels compelled to blow the whistle to expose facts as he sees them, he may well have to pay the price of loss of employment. In this type of situation, the Board felt that the ethical duty or right of the engineer becomes a matter of personal conscience, but the Board was unwilling to make a blanket statement that there is an ethical duty in these kinds of situations for the engineer to continue the campaign within the company and make the issue one for public discussion.

More recently, in BER Case 88-6, an engineer was employed as the city engineer/director of public works with responsibility for disposal plants and beds and reported to a city administrator. After (1) noticing problems with overflow capacity, which are required to be reported to the state water pollution control authorities, (2) discussing the problem privately with members of the city council, (3) being warned by the city administrator to report the problem only to him, (4) discussing the problem again informally with the city council, and (5) being relieved by the city administrator of responsibility for the disposal plants and beds, the engineer continued to work in the capacity as city engineer/director of public works. In ruling that the engineer failed to fulfill her ethical obligations by informing the city administrator and certain members of the city council of her concern, the Board found that the engineer was aware of a pattern of ongoing disregard for the law by her immediate supervisor, as well as by members of the city council. After several attempts to modify the views of her superiors, the engineer knew, or should have known, that "proper authorities" were not the city officials, but more probably, state officials. The Board could not find it credible that a city engineer/director of public works for a medium-sized town would not be aware of this basic obligation. The Board said that the engineer's inaction permitted a serious violation of the law to continue and made the engineer an "accessory" to the actions of the city administrator and others.

Turning to the facts of the present case, Engineer A is faced with a predicament with a variety of options and alternatives. First, Engineer A could interpret the situation presented as one involving "trade-offs," in which Engineer A must weigh one "public good" (a better building inspection process) against a competing or concurrent "public good" (a consistent code enforcement process).

In such a situation, Engineer A could arguably rationalize a decision to permit the inconsistent application of a building code in order to accomplish the larger objective of obtaining the necessary resources to hire a sufficient number of code enforcement officials to provide proper protection to the public health and safety.

On the other hand, Engineer A's decision to permit developers to avoid compliance with the newer, updated building code enforcement requirements might potentially cause a real danger to the public health and safety if the new facility causes harm to the public because of its failure to comply with the more updated code requirements. In addition, agreeing to the chairman's arrangement has the appearance of compromising the public health and safety for political gain.

While this case presents a difficult dilemma for Engineer A, on balance, the Board believes that previous BER cases provide sufficient guidance for Engineer A. Each of the earlier cases discussed present a constant theme that the engineer must hold the public health and safety paramount and that the engineer has an responsibility to insist, however strongly and vociferously, that public officials and decision-makers take steps and corrective steps if necessary to see that this obligation is fulfilled.

The Code of Ethics makes it clear that engineers have an obligation to advise their clients or employers when they believe a project will not be successful.

In this case, Engineer A should make it plain and clear to the chairman that "righting a wrong with another wrong," does grave damage to the public health and safety (See Code Section III.1.b.).

Engineer A should insist that the public will be seriously damaged in either case and that if the integrity of the building code enforcement process is undermined for short-term gain, the city, its citizens, and its businesses will be harmed in the long term.

Conclusion:

It was not ethical for Engineer A to agree to concur with the chairman's proposal under the facts. Additionally, it was not ethical for Engineer A to sign inadequate inspection reports. (See Code Section II.1.b.).

Board of Ethical Review

Lorry T. Bannes, P.E.
James G. Fuller, P.E.
Donald L. Hiatte, P.E.
Joe Paul Jones, P.E.
Paul E. Pritzker, P.E.
Richard Simberg, P.E.
C. Allen Wortley, P.E., Chairman

NOTE: The NSPE Board of Ethical Review (BER) considers ethical cases involving either real or hypothetical matters submitted to it from NSPE members, other engineers, public officials and members of the public. The BER reviews each case in the context of the NSPE Code of Ethics and earlier BER opinions. The facts contained in each case do not necessarily represent all of the pertinent facts submitted to or reviewed by the BER.

Each opinion is intended as guidance to individual practicing engineers, students and the public. In regard to the question of application of the NSPE Code of Ethics to engineering organizations (e.g., corporations, partnerships, sole-proprietorships, government agencies, university engineering departments, etc.), the specific business form or type should not negate nor detract from the conformance of individuals to the NSPE Code. The NSPE Code deals with professional services — which services must be performed by real persons. Real persons in turn establish and implement policies within business structures.

These opinions are for educational purposes only. They may be reprinted without further permission, provided that this statement is included before or after the text of the case and that appropriate attribution is provided to the National Society of Professional Engineers' Board of Ethical Review.

Visit www.niee.org to learn how to obtain complete volumes that include all NSPE Opinions (or call 1-806-742-6433).

Gifts To Engineers

NSPE Opinions of the Board of Ethical Review - Case No. 81-4

Facts:

Engineers A, B, and C are principals or employees of a consulting engineering firm which does an extensive amount of design work for private developers. The engineers are involved in recommending to the developers a list of contractors and suppliers to be considered for selection on a bidding list for construction of the projects. Usually, the contractors and suppliers recommended by the engineers for the selected bidding list obtain most of the contracts from the developers. Over a period of years the officers of the contractors or suppliers developed a close business and personal relationship with the engineers of the firm.

From time to time, at holidays or on birthdays of the engineers with whom they dealt, the contractors and suppliers would give Engineers A, B, and C personal gifts of substantial value, such as home furnishings, recreational equipment, gardening equipment, etc.

Question:

Was it ethical for Engineers A, B, and C to accept gifts from the contractors and suppliers?

References:

Code of Ethics-Section II.4.c. - Engineers shall not solicit or accept financial or other valuable consideration, directly or indirectly, from contractors, their agents, or other parties in connection with work for employers or clients for which they are responsible.

Code of Ethics Section II.5.b. - Engineers shall not offer, give, solicit, or receive, either directly or indirectly, any political contribution in an amount intended to influence the award of a contract by public authority, or which may be reasonably construed by the public of having the effect or intent to influence the award of a contract. They shall not pay a commission, percentage, or brokerage fee in order to secure work except to a bona fide employee or bona fide established commercial or marketing agencies retained by them.

Code of Ethics Section III.5.b. - Engineers shall not accept commissions or allowances, directly or indirectly, from contractors or other parties dealing with clients or employers of the Engineer in connection with work for which the Engineer is responsible.

Discussion:

The previous cases dealing with gifts have been under what may be called "reverse facts" in that the gifts were given by the engineers to those in the position of being able to influence the award of contracts for engineering services.

In Case 79-8, an engineer in private practice gave a gift to an engineer in a public agency, and we held that it was unethical for the one engineer to make the gift, and unethical for the other engineer to accept it. But the primary thrust of the discussion revolved around the Code section barring the use of gifts by engineers to secure work. In that emphasis we referred to the criteria established in Case 60-9 on the issue of whether the gift was of a nature which might influence the award of work to engineers.

In Case 76-6 we dealt again with a gift by an engineer to an official of a foreign country, and also ruled that, notwithstanding the practice in the foreign country for officials to receive gifts from those who do business with the agency of the foreign country, it was unethical for the engineer to offer the gift.

In the first case involving gifts (Case 60-9), we looked at three circumstances, one of which involved engineering employees of an industrial company, who were in a position to recommend for or against the purchase of products, accepting nominal cash gifts from salesmen of particular products being offered their employer. In that case we also held that acceptance of even nominal gifts raised a question of integrity and hence was unethical.

The emphasis in the case before us raises more pertinently the idea of engineers accepting, rather than giving, gifts. Applying the principles of the earlier cases, and the language of the Code, as cited above in several sections, it seems clear that there would be, at least, a reasonable suspicion to others, and particularly to other contractors and suppliers, that acceptance of the gifts by Engineers A, B, and C would imply favoritism.

The language of Section II.4.c. covers acceptance of gifts, as well as their solicitation by engineers, and extends to the impact of such action on clients. Thus, the clients (developers) of the engineering firm may be led to question whether the recommendation of particular contractors or suppliers is totally unbiased and represents the independent judgment of the consulting firm.

The first part of Section II.5.b. deals with political contributions, but applies equally to offering any gift in order to secure work. While under that language, Engineers A, B, and C did not in this case "offer" a gift, the section represents the same philosophy as Section II.5.c. And we have also cited Section III.5.b., even though its emphasis is on acceptance of commissions or allowances from contractors dealing with clients of the engineer, because it is a reflection of the same ethical concept noted above.

When read in the context of the thrust of the entire Code on the matter of gifts, and in line with the ethical precept we have stated in the earlier cited cases, we affirm the overriding principle that engineers should "lean over backward" to avoid acceptance of gifts from those with whom they, or their firm, do business. We leave aside for this case the related issue of when a gift is permissible in terms of an expression of friendship or social custom, such as a calendar, a cigar, or picking up the check at a luncheon meeting. For general guidance on that point we refer the reader to the discussion of Case 60-9, namely that the guideline is that a gift of that nature be limited to those which will not raise any inference of compromising the independent professional judgment of the engineer, or that the giving or acceptance of such a gift be governed by the normal standards of good taste and acceptable custom.

Conclusion:
It was not ethical for Engineers A, B, and C to accept gifts from the contractors and suppliers.

Board of Ethical Review

Louis A Bacon, P.E.
Robert R. Evans, P.E.
James G. Johnstone, P.E.
Lawrence E. Jones, P.E.
Robert H. Perrine, P.E.
Alfred H. Samborn, P.E.
F. Wendell Beard, P.E., chairman

Note: This opinion is based on data submitted to the Board of Ethical Review and does not necessarily represent all of the pertinent facts when applied to a specific case. This opinion is for educational purposes only and should not be construed as expressing any opinion on the ethics of specific individuals.

Gifts

NSPE Opinions of the Board of Ethical Review - Case No. 60-9

Facts:

The following situations are consolidated into one case because they involve the same ethical principles:

Situation A: A consulting engineer who has done considerable work for a public body makes it a practice to take certain staff engineering employees of the agency to lunch or dinner three or four times a year, at an average cost of $5 per person. He also makes it a practice to give certain members of the engineering staff Christmas presents at an average cost of $10 each.

Situation B: Certain engineering employees of an industrial firm, who are in a position to recommend for or against the purchase of products used by the company, regularly receive cash gifts ranging from $25 to $100 from salesmen for particular products.

Situation C: Upon completion of a major engineering contract held by a consulting engineer, the chief engineer of the client who worked directly and intimately with the consultant receives a new automobile of the value of approximately $4,000 from the consultant with a letter stating that the gift is in appreciation of his close and friendly cooperation and assistance in the successful performance of the work.

Questions:

1. Was it ethical for the consulting engineer in Situation A to offer any of the gifts to the employees?
2. Was it ethical for the staff engineers in Situation A to accept any of the gifts tendered them?
3. Was it ethical for the engineer employees in Situation B to accept any of the gifts tendered them?
4. Was it ethical for the consultant in Situation C to offer any of the gifts to the employees?
5. Was it ethical for the Chief Engineer in Situation C to accept any of the gifts tendered him?

References:

Code of Ethics Code C8: "The engineer will act in professional matters for each client or employer as a faithful agent or trustee."

Code of Ethics Code C9:"He will act with fairness and justice between his client or employer and the contractor when dealing with contracts."

Code of Ethics Code C16: "He will not accept commissions or allowances, directly or indirectly, from contractors or other parties dealing with his clients or employer in connection with work for which he is responsible."

Code of Ethics Code R1:1: "He will be guided in all his relations by the highest standards."

Code of Ethics Code R1:4: "He will not offer to pay, either directly or indirectly, any commission, political contribution, or a gift, or other consideration in order to secure work, exclusive of securing salaried positions through employment agencies."

Code of Ethics Code R17:29: "He will not accept personal consideration in any form. This assures that his recommendations for the award of a contract cannot be influenced."

Discussion:

The question of when a gift is intended to or becomes an inducement to influence one's impartial decision, as distinguished from an expression of friendship or a social custom, has remained a perplexing one over the Years. No blanket rule covering all situations has been discovered. The size of the gift is usually a material factor, but must be related to the circumstances of the gift. It would hardly be felt a token gift, such as a cigar, a desk calendar, etc., would be prohibited. It has been customary in the business world for friends and business associates to tender such tokens of recognition or appreciation, and "picking up the tab" at a business luncheon or dinner is commonplace and well accepted in the mores of our society.

Recognizing the difficulties inherent in passing judgment on each instance, we believe the Canons and Rules state, in substance, that an engineer may neither offer nor receive a gift which is intended to or will influence his independent professional judgment. The full application of this principle requires the impossible-that we read the state of mind of the donor or donee. Therefore, we must apply a criterion which reasonable men might reasonably infer from the circumstances; that the giving or acceptance of the benefit be a matter of "good taste," and such that among reasonable men it might not be of a nature which raises suspicions of favoritism.

Applying these general principles to the situations at hand, we think that an occasional free luncheon or dinner, or a Christmas or birthday present when there is a personal relationship is acceptable practice. On the other hand, cash payments to those in a position to influence decisions favorable or unfavorable to the sever are not in good taste and do immediately raise a suspicion that there is an ulterior motive. Likewise, a very expensive gift has a connotation of placing the recipient in a position of obligation.

Conclusions:

1. Situation A: Occasional free luncheons or dinners and Christmas presents of relatively small value do not exceed the bounds of ethical behavior by the giver.
2. Situation A: Occasional free luncheons or dinners and Christmas presents of relatively small value do not exceed the bounds of ethical behavior by the receiver.
3. Situation B: Inasmuch as it does not appear that the employees performed any services for the cash gifts, and in view of their position of responsibility and trust to the employer, the receipt of such cash gifts immediately raises a doubt as to the integrity of the employees and leaves an impression that the intent is to influence their discretion in favor of the giver. Acceptance of cash gifts in such circumstances is an unethical act.
4. Situation C: It is unethical to offer such a gift.
5. Situation C: It is unethical to accept such a gift. It is conceivable that the expensive gift was tendered in good faith, honestly expressing appreciation for the cooperation and assistance of the chief engineer. Nor is there any direct evidence or implication that the consulting engineer would receive preferential treatment in any future work for the company. On the other hand, the chief engineer presumably had a duty to cooperate with the consulting engineer, in the interest of an efficient job by the consultant. And it might be inferred by reasonable men that one who had received an expensive gift would be under at least moral pressure in any future consideration of awarding additional work to the consultant. Here, we apply the concept of "good taste" and conclude that the chief engineer should decline the gift, thereby protecting his good name and reputation. and that of his profession, from any shadow of suspicion. In accepting such a gift he might not sully his character (what he is) but he would run a definite risk of besmirching his reputation (what others think he is). For the same reasons the consultant should not tender gifts of such magnitude, even though he sincerely acts with the best motives.

Board of Ethical Review

L. R. Durkee, P. E.
Phil T. Elliott, P. E.
Wylie W. Gillespie, P. E.
A. C. Kirkwood, P. E.
Marvin C Nichols, P. E.
Ezra. K. Nicholson, P E.
Pierce G. Ellis, P. E., Chairman

Affirming the Right Choice - Qualitatively Qualified

Public Administration at the University of Arkansas - Revised 7/19/96

Used by permission. Contact Dr. William Miller, Associate Professor of Political Science, University of Arkansas — http://plsc.uark.edu/book/books/ethics/index.htm

Context

Last year a policy analyst for the state Bureau of Economic Development resigned to take a position in the private sector. Bob Everly, the Bureau director, quickly initiated a hiring process according to the state hiring procedures. Now three candidates have completed the interview process and Bob sits down with the personnel director to make a decision.

All three candidates look good. Each has had experience in industrial development. Each met the basic requirements on the written civil-service test. One candidate, however, is requiring a salary level that is too high for the position classification. He will have to be excluded from the final list. The other two candidates are very comparable.

The first candidate, Jane Smith, has had two years of experience in a similar position in a neighboring state. She scored well on the entry exams. In the interviews for the position, she seemed professional and competent. Joe Miller, the second candidate, has had three years of experience in a similar position in a neighboring state. He also scored well on the entry exams. He interviewed well and seemed competent.

The fact that he has had more experience might give him the edge in the selection process. There is, however, another factor to be considered. All the employees in the Bureau are white and male. This has been duly noted in complaints to the state affirmative action board.

The Dilemma

Director Everly discusses this with his personnel manager. Joe is more experienced. Jane, who is both African-American and female, fits their need to increase diversity on the staff. Director Everly feels that the agency could benefit from more diverse viewpoints among its staff. Both applicants are competent and qualified.

The Decision

Director Everly, in spite of a disagreement with the personnel director, decides to hire Jane and meet their need for both competence and diversity.

Do you agree?

Public Welfare
Duty Of Government Engineer

NSPE Opinions of the Board of Ethical Review — Case 92-4

Facts:

Engineer A, an environmental engineer employed by the state environmental protection division, is ordered to draw up a construction permit for construction of a power plant at a manufacturing facility. He is told by a superior to move expeditiously on the permit and "avoid any hang-ups" with respect to technical issues. Engineer A believes the plans as drafted are inadequate to meet the regulation requirements and that outside scrubbers to reduce sulfur dioxide emissions are necessary and without them the issuance of the permit would violate certain air pollution standards as mandated under the 1990 Clear Air Act. His superior believes that plans, which involve limestone mixed with coal in a fluidized boiler process would remove 90% of the dioxide, will meet the regulatory requirements.

Engineer A contacts the state engineering registration board and is informed, based upon the limited information provided to the board, that suspension or revocation of his engineering license was a possibility if he prepared a permit that violated environmental regulations. Engineer A refused to issue the permit and submitted his findings to his superior. The department authorized the issuance of the permit. The case had received widespread publicity in the news media and is currently being investigated by state authorities.

Questions:

1. Would it have been ethical for Engineer A to withdraw from further work in this case?
2. Would it have been ethical for Engineer A to issue the permit?
3. Was it ethical for Engineer A to refuse to issue the permit?

References:

Code of Ethics Preamble - Engineering is an important and learned profession. The members of the profession recognize that their work has a direct and vital impact on the quality of life for all people. Accordingly, the services provided by engineers require honesty, impartiality, fairness and equity, and must be dedicated to the protection of the public health, safety and welfare. In the practice of their profession, engineers must perform under a standard of professional behavior which requires adherence to the highest principles of ethical conduct on behalf of the public, clients, employers and the profession.

Code of Ethics Section I.1. - Hold paramount the safety, health and welfare of the public in the performance of their professional duties.

Code of Ethics Section II.1.a. - Engineers shall at all times recognize that their primary obligation is to protect the safety, health, property and welfare of the public. If their professional judgment is overruled under circumstances where the safety, health, property or welfare of the public are endangered, they shall notify their employer or client and such other authority as may be appropriate.

Code of Ethics Section II.1.b. - Engineers shall approve only those engineering documents which are safe for public health, property and welfare in conformity with accepted standards.

Code of Ethics Section II.3.a. - Engineers shall be objective and truthful in professional reports, statements or testimony. They shall include all relevant and pertinent information in such reports, statements or testimony.

Discussion:

The facts of this case are in many ways a classic ethical dilemma faced by many engineers in their professional lives. Engineers have a fundamental obligation to hold paramount the safety, health and welfare of the public in the performance of their professional duties (Code Section I.1.). Moreover, the Code provides guidance to engineers who are confronted with circumstances where their professional reputation is at stake. Sometimes engineers are asked by employers or clients to sign off on documents in which they may have reservations or concerns.

The Board of Ethical Review has examined this issue over the years in differing contexts. As early as case BER 65-12, the Board dealt with a situation in which a group of engineers believed that a product was unsafe. The Board then determined that as long as the engineers held to that view, they were ethically justified in refusing to participate in the processing or production of the product in question. The Board recognized that such action by the engineers would likely lead to loss of employment.

In BER Case 82-5, where an engineer employed by a large defense industry firm documented and reported to his employer excessive costs and time delays by sub-contractors, the Board ruled that the engineer did not have an ethical obligation to continue his efforts to secure a change in the policy after his employer rejected his reports, or to report his concerns to proper authority, but has an ethical right to do so as a matter of personal conscience. The Board noted that the case did not involve a danger to the public health or safety, but related to a claim of unsatisfactory plans and the unjustified expenditure of public funds. The Board indicated that it could dismiss the case on the narrow ground that the Code does not apply to a claim not involving public health and safety, but that was too narrow a reading of the ethical duties of engineers engaged in such activities. The Board also stated that if an engineer feels strongly that an employer's course of conduct is improper when related to public concerns, and if the engineer feels compelled to blow the whistle to expose facts as he sees them, he may well have to pay the price of loss of employment. In this type of situation, the Board felt that the ethical duty or right of the engineer becomes a matter of personal conscience, but the Board was unwilling to make a blanket statement that there is an ethical duty in these kinds of situations for the engineer to continue the campaign within the company, and make the issue one for public discussion.

More recently, in BER Case 88-6, an engineer was employed as the city engineer/director of public works with responsibility for disposal plants and beds and reported to a city administrator. After (1) noticing problems with overflow capacity which are required to be reported to the state water pollution control authorities, (2) discussing the problem privately with members of the city council, (3) being warned by the city administrator to only report the problem to him, (4) discussing the problem again informally with the city council and (5) being relieved by the city administrator of responsibility for the disposal plants and beds by a technician, the engineer continued to work in the capacity as city engineer/director of public works. In ruling that the engineer failed to fulfill her ethical obligations by informing the city administrator and certain members of the city council of her concern, the Board found that the engineer was aware of a pattern of ongoing disregard for the law by her immediate supervisor as well as by members of the city council. After several attempts to modify the views of her superiors, the engineer knew or should have known that "proper authorities" were not the city officials, but more probably state officials. The Board could not find it credible that a city engineer/director of public works for a medium-sized town would not be aware of this basic obligation. Said the Board, the engineer's inaction permitted a serious violation of the law to continue and made the engineer an "accessory" to the actions of the city administrator and others.

Turning to the facts of this case, we believe the situation involved in this case is in many ways similar to the situation involved in BER Case 88-6. This, unlike BER Case 82-5 did not involve a matter of personal conscience, but rather a matter which had a direct impact upon the public health and safety. Yet unlike the circumstances involved in BER Case 88-6 where the issues were hidden from public note, here, the case involves facts which have received coverage in the media.

In view of this fact, we do not believe it is incumbent upon Engineer A to bring this issue to the attention of the "proper authorities". As we see it, such officials are already aware of the situation and have begun an investigation.

The reason for our position in BER Case 88-6 was that the engineer's failure to bring the problems to the attention of the "proper authorities" made it more probable that danger would ultimately result to the public health, safety and welfare. Here, the circumstances are presumably already known to appropriate public officials. To bring the matter to their attention is a useless act.

However, we believe it would not have been ethical for Engineer A to withdraw from further work on the project because Engineer A had an obligation to stand by his position consistent with his obligation to protect the public, health, safety and welfare and refuse to issue the permit. Engineers have an essential role as technically-qualified professionals to "stick to their guns" and represent the public interest under the circumstances where they believe the public health and safety is at stake.

We would also note that this case also raises another dimension which involves the role of the state licensing board in determining the ethical conduct of licensees.

Under the facts, Engineer A affirmatively sought the opinion of the state as to whether his approval of the permit could violate the state engineering registration law. We believe Engineer A's actions in this regard constitute appropriate conduct and actions are consistent with Section II.1.a. of the Code.

This case involves a question of public health and welfare and Engineer A's decision to disassociate himself from further work on this project avoids having Engineer A being placed in a professionally compromising situation.

Conclusions:

1. It would not have been ethical for Engineer A to withdraw from further work on the project.

2. It would not have been ethical for Engineer A to issue the permit.

3. It was ethical for Engineer A to refuse to issue the permit.

Board of Ethical Review

William A. Cox, Jr., P.E.
William W. Middleton, P.E.
William E. Norris, P.E.
William F. Rauch, Jr., P.E.
Jimmy H. Smith, Ph.D., P.E.
Otto A. Tennant, P.E.
Robert L. Nichols, P.E., Chairman

Note: In regard to the question of application of the Code to corporations vis-a-vis real persons, business form or type should not negate nor influence conformance of individuals to the Code. The Code deals with professional services, which services must be performed by real persons. Real persons in turn establish and implement policies within business structures. The Code is clearly written to apply to the Engineer and it is incumbent on a member of NSPE to endeavor to live up to its provisions. This applies to all pertinent sections of the Code.

A Well-Deserved Vacation

Public Administration at the University of Arkansas - Revised 7/19/96
Used by permission. Contact Dr. William Miller, Associate Professor of Political Science, University of Arkansas -- http://plsc.uark.edu/book/books/ethics/index.htm

Context

You are an administrator in the state economic development council. In your job, you often travel for the state. It feels like you have been sent everywhere from Japan to the smallest towns in your state. Next week you are scheduled to go to Hawaii for a conference of trade representatives. You are looking forward to this trip because the conference is lightly scheduled and recreation will be a big fringe for you. You have worked hard and deserve it.

The Dilemma

As you begin sifting through the day's mail, your secretary interrupts you with a phone call from the organizer of the conference. He informs you that the conference is officially canceled. The looming trade war has taken all the big hitters away. There will be no conference presentations, no important meetings and no substantial reason to go.

However….the conference coordinator says that he and a group of others are still planning on going. The motel has a block of rooms already paid for. The air fares have been paid and are not refundable.

Why not use these resources and have a good time? The conference coordinator will be glad to vouch that this is a business event if that is necessary.

You struggle to consider these things. You have no vacation time left, but this feels too good to pass up. No business. No pressures. Fun in the sun. The job takes more of your time than it should. You deserve a break and would hate to see all that money wasted.

The Question

What should you do?

Supplanting Another Engineer—Employee Participation—Registration— Adverse Comments on Applicant

NSPE Opinions of the Board of Ethical Review - Case No. 72-4

Facts:

John Adams, a registered engineer, had a formal agreement to perform preliminary design of a project for the ABC Company. Richard Edwards, a non registered engineer employee of Adams, performed a substantial part of the preliminary design. The working drawing phase of the project was to be performed only when and if authorized by the ABC Company. Prior to any decision being made on the working drawing phase, Edwards voluntarily terminated his employment with Adams and was immediately employed by George Barton, a registered engineer in private practice.

Thereafter the ABC Company notified Adams that it did not desire him to proceed with the working drawing phase and paid him the fees due for the preliminary design work. Barton was retained within a few days thereafter to furnish the working drawing engineering services to the ABC Company and assigned to continue work on the project. Neither the ABC Company, Barton, nor Edwards contacted Adams regarding these arrangements and events subsequent to the cessation of the arrangement between Adams and the ABC Company. Adams alleged that Barton and Edwards had conspired to supplant him in this engagement with the ABC Company.

Questions:

1. Was Edwards unethical in transferring his services from Adams to Barton under the circumstances stated?
2. Was Barton unethical for participating in the arrangement to transfer the work to his firm?
3. If Edwards had acted unethically prior to obtaining his registration as a professional engineer, would it be ethical for a registered engineer to withhold his recommendation that Edwards be registered, and to submit unfavorable comments on Edwards' character to the state registration board?

References:

Code of Ethics Section 7a - While in the employ of others, he will not enter promotional efforts or negotiations for work or make arrangements for other employment as a principal or to practice in connection with a specific project for which he has gained particular and specialized knowledge without the consent of all interested parties.

Code of Ethics Section 8 - The engineer will endeavor to avoid a conflict of interest with his employer or client, but when avoidable, the engineer shall fully disclose the circumstances to his employer or client.

Code of Ethics Section 11 - The engineer will not compete unfairly with another engineer by attempting to obtain employment or advancement or professional engagements by competitive bidding, by taking advantage of a salaried position, by criticizing other engineers, or by other improper or questionable methods.

Code of Ethics Section 11a - The engineer will not attempt to supplant another engineer in a particular employment after becoming aware that definite steps have been taken toward the other's employment.

Code of Ethics Section 12 - The engineer will not attempt to injure, maliciously or falsely, directly or indirectly, the professional reputation, prospects, practice, or employment of another engineer, nor will he indiscriminately criticize another engineer's work in public. If he believes that another engineer is guilty of unethical or illegal practice, he shall present such information to the proper authority for action.

Discussion:
The facts presented to us suggest, but do not necessarily prove, that Barton and Edwards had entered into some form of prearranged relationship to shift the continued work on the project from Adams to Barton after the preliminary phase had been completed by Adams.

We have consistently held that an engineer does not have an exclusive right to perform engineering services for a particular client and that a client has a right to change from one consulting engineer to another. In Case 62-18 we dealt in some detail with the supplanting question under circumstances in which several years had elapsed between the time of preliminary studies for a project by one engineer and the retention of another engineer for a new report and design of a project, concluding that there had not been any unethical act of supplanting. (See also Case 64-9 in which we distinguished a related situation and held that the second engineer had unethically supplanted the first.)

Assuming, as we do for the purpose of this case, that Barton and Edwards had conspired to have the work transferred from Adams to Barton by utilizing Edwards' intimate knowledge of the project based on his connection with and work during the preliminary design, we believe that both are in violation of the mandate of Section 11 and Section 11a in that such action was an attempt to advance their respective interests by taking advantage of a salaried position (as to Edwards) and by supplanting Adams for the balance of the project (as to Barton). Edwards was also, by this arrangement with Barton, in violation of Section 7a. Although that portion of the code refers to promotional efforts or negotiations for work " ... as a principal ...," it also forbids " ... practice in connection with a specific project for which he has gained particular and specialized knowledge without the consent of all interested parties." Adams was clearly an interested party. Section 8 is also cited because Edwards' conduct generated a conflict of interest with his employer, Adams. When the code language says that a conflict of interest may be tolerated when unavoidable, provided "... the engineer shall fully disclose the circumstances to his client or employer," it means that the disclosure must be made before the fact and not after the damage has been done, as in this case.

As to the ethical duty of an engineer called upon or volunteering to comment on Edwards' qualifications for registration as a professional engineer, Section 12 imposes a duty on all engineers to advise the "proper authority" if he believes that another engineer has engaged in unethical acts. We can imagine no more clear-cut application of this duty than in these circumstances and conclude that an engineer having knowledge of the facts must disclose them to the state registration board. It is not necessary for an engineer to know as a certainty that an applicant for registration had acted unethically (that is a matter of opinion), but if, as in this case, there is reasonable belief that an applicant for registration may have acted unethically, it is required that the basis for the belief be submitted to the registration board. Whether such action would constitute a sound basis for the registration board to refuse to register an applicant is within the discretion of the registration board in accordance with the qualification standards in the registration law and the interpretation of the facts under those standards.

Conclusions:

1. Edwards was unethical in transferring his services from Adams to Barton under the circumstances stated, insofar as the transfer of his services was to improperly shift the work on the project from Adams to Barton.
2. Barton was unethical in conspiring with Edwards to supplant Adams.
3. It would be ethical for an engineer to withhold his recommendation that Edwards be registered and to submit unfavorable comments on Edwards' character to the state registration board.

Board of Ethical Review

W.R. Gibbs, P.E.
Sherman Smith, P.E.
Joseph H. Littlefield, P.E
Albert L. Wolfe, P.E
James D. Maloney, P.E.
Frank H. Bridgers, P.E., chairman

Note—This opinion is based on data submitted to the Board of Ethical Review and does not necessarily represent all of the pertinent facts when applied to a specific case. This opinion is for educational purposes only and should not be construed as expressing any opinion on the ethics of specific individuals. This opinion may be reprinted without further permission, provided that this statement is included before or after the text of the case.

Board of Ethical Review Case Reports

The Board of Ethical Review was established to provide service to the membership of the NSPE by rendering impartial opinions pertaining to the interpretation of the NSPE Code of Ethics.

At a Distance

Public Administration at the University of Arkansas - Revised 7/22/98
Used by permission. Contact Dr. William Miller, Associate Professor of Political Science, University of Arkansas -- http://plsc.uark.edu/book/books/ethics/index.htm

Context

Betty Reid is the city manager in a rather small rural town. Like many small towns in her region, growth from larger cities is spreading out into her area. This creates new needs and expectations for her budget. For the five years she has worked in this city, John Maxwell has been her budget officer. John is a decent person, but has one limitation which is beginning to cause a bit of a problem. John hates computers. He does all of the city budget work on accounting paper. Five years from retirement, he doesn't see why he should retrain. The paper system, he says, works quite well.

The Dilemma

Betty is not so sure. She feels that some form of computerization is needed. The city council had made an occasional disparaging remark about the slowness of reporting. Betty has to spend more time entering budget data into reports than she likes. Though they are a small town, they have growing needs.

Betty has recently learned of a service that can do, on line over the Internet, what John does. This service has a variety of reporting formats and will give her economic forecasting and budget analysis. The reports include comparisons with other cities like hers. She could "wow" the city council with these reports and would feel more competent in addressing the growth needs of the city. All this for less than she is paying John to do the books.

The Decision

Betty appreciates what John has done for the city in the past. Though Betty likes John, she sees no options. Though she could wait a few years until he retires, Betty feels that the city would be the loser.

What should Betty do?

Conflict Of Interest — Community Service

NSPE Opinions of the Board of Ethical Review — Case 92-5

Facts:

In the early 1980s a community service corporation was established for the purpose of revitalizing a city's downtown area. One of the actions was the hiring of a consultant. After making a survey of the retail area influenced by retailers and doing an economic feasibility study, the consultant concluded that a significant catalyst needed to be established in the downtown area. The consultant suggested an "off-price mall". The group, using the statistical data developed, put together a brochure to entice developers to undertake the project. Approximately 120 brochures were sent to developers but none responded.

In 1986, Engineer A, a principal in a small structural, environmental and civil engineering firm, was elected president of the community service corporation. He receives no compensation for any of his services. The corporation concentrates on persuading the state government to build a state office building in the downtown area and to concentrate a number of its offices in the building. The effort is successful and the state office building is built. A second effort, to encourage the federal government to fund the construction of a federal courthouse and office building is also successful. A third effort, the construction of a high rise county office building with an underground connection to the federal and a city building along with a central fire station, police station and city hall is also begun. Engineer A has had a high profile and has generally been acknowledged as a leader in these efforts.

Selections are in the process of being made for the design work of the federal courthouse and office building. Larger design firms are beginning to contact smaller local consulting firms, including the firm headed by Engineer A's son, Engineer B, who is the president and chief executive officer. Engineer A is the chairman, principal stockholder of the firm. A major design firm submitting a proposal to the federal government to lead the design effort asks Engineer B to perform civil, structural and environmental engineering services in connection with the project. Engineer B agrees to perform the services.

Question:

Would it be ethical for the firm of Engineer A and Engineer B to agree to perform services in connection with the federal project?

References:

Code of Ethics Preamble - Engineering is an important and learned profession. The members of the profession recognize that their work has a direct and vital impact on the quality of life for all people. Accordingly, the services provided by engineers require honesty, impartiality, fairness and equity, and must be dedicated to the protection of the public health, safety and welfare. In the practice of their profession, engineers must perform under a standard of professional behavior which requires adherence to the highest principles of ethical conduct on behalf of the public, clients, employers and the profession.

Code of Ethics Section II.4.d. - Engineers in public service as members, advisors or employees of a governmental or quasi-governmental body or department shall not participate in decisions with respect to professional services solicited or provided by them or their organizations in private or public engineering practice.

Code of Ethics Section II.4.e. - Engineers shall not solicit or accept a professional contract from a governmental body on which a principal or officer of their organization serves as a member.

Code of Ethics Section III.5. - Engineers shall not be influenced in their professional duties by conflicting interests.

Discussion:

Fundamental to engineering ethics is the principle that engineers should not use public positions that they hold in a manner that will benefit themselves in their private dealings (Code Section II.4.d.). In particular, a basic ethical principle for engineers to be mindful of is that engineers should not seek or accept work from public agencies where a principal or an officer serves as a member (II.4.e.).

The Board of Ethical Review has on numerous occasions had an opportunity to address these two key ethical principles and the policy considerations relating to these principles. For example, in BER Case 62-7, the Board reviewed a case concerning an engineering consultant who had been retained by a county commission to perform all necessary engineering and advisory services. The commission did not have an engineering staff so the engineer acted as the staff for the commission in the preparation of sewage and water studies, the financing of sanitary districts, and the approval of plans submitted by others. The engineer was also retained by a private company to perform engineering design for the development of several thousand housing units which involved extensive contract negotiations between the commission and the developer. The Board found that the engineer was in a position of passing engineering judgment on behalf of the commission on work or contract arrangements which the engineer performed or in which he participated. This obviously involved the self-interest of the engineer and divided his loyalties. Even if the engineer acted with the best of intentions, he was put into a position of assessing his recommendations to two clients with possibly opposing interests. Given these realities, the Board concluded that a conflict of interest existed.

In BER Case 74-2, a case in which a state law required every municipality to retain a municipal engineer with that engineer's firm usually retained for engineering services for capital improvements needed by the municipality, the Board found that the engineer was not a bona fide "employee" of the municipality but a consultant, thus it was not unethical for him to serve as "municipal engineer" and participate in a consulting firm providing engineering services to the municipality. The Board reasoned that the public interest was best served by providing to small municipalities the most competent engineering services that they could acquire. It was assumed that the state law was intended to achieve that end.

In BER Case 75-7, the Board examined the question of whether an engineer who serves as a member of local governmental boards and commissions that involve some aspects of engineering may provide engineering services through his firm to the board and commissions. There the Board concluded that an engineer serving on a commission could ethically provide services to the private client because the engineer had abstained from the discussion and vote on certain permit applications. The Board cautioned, however, that care must be taken that the engineer in such a situation not have taken any action to influence a favorable decision on the permit.

In BER Case 82-4, the Board, in reviewing the aforementioned decisions, ruled that an engineer who served as a city engineer and a county engineer for a retainer fee may not ethically provide or render judgment on behalf of the city and county relative to projects on which the engineer had furnished services through a private client. "To do so", noted the Board, "is a useless act because it is basic to the Code of Ethics that an engineer will not submit plans or other work which he does not believe represents the best interests of the client". The Board could not see how an engineer could wear two hats and still represent the best interest of his client; to do so would constitute a conflict of interest. "If the county or city wishes to obtain a recommendation on the merits of the work", the Board stated, "it should retain another engineer for review."

In BER Case 85-2, a county hospital board owned a hospital facility and contracted with a private health care provider to manage, administer, and generally operate a hospital facility. An engineer, principal in a local engineering firm, served on the board of directors of the private health care provider. Certain engineering and surveying work needed to be performed at the facility. The engineer sought and received a contract from the provider to perform the engineering and surveying work at the hospital. The decision to select the engineer was made by the private health care provider's board of directors and the engineer participated in the decision. In deciding that it was unethical for the engineer to seek a contract with the private health care provider to provide engineering and surveying services, the Board, relying on BER Cases 75-7 and 82-4, noted that the engineer's services on the board was of a quasi-governmental nature and that the engineer had an obligation under Code Section II.4.d. and II.4.e. not to participate in the decision to award the engineer the contract. Following BER Case 85-2, the NSPE Code of Ethics Section II.4.d. was modified to include activities of a "quasi-governmental" nature.

Most recently, in BER Case 89-6, which involved an engineer who was chairman of a condominium association, the essential principles contained in BER Case 85-2 relating to Code Section II.4.d. were reaffirmed.

Turning to the case at hand, we believe the facts involved can be readily distinguished from the line of cases discussed above. In fact we are of the view that based upon the plain meaning of the language, neither Code Section II.4.d. nor II.4.e. would be particularly pertinent to this case.

As applied to the fact in this case, these two Code provisions are essentially intended (1) to prevent an engineer who serves as a member, advisor or employee of a governmental or similar entity from serving, for example, on the design selection, oversight, review committee of that entity (e.g., federal government) where the engineer's firm is competing for the work and (2) to prevent an engineering firm whose officer or principal serves as a member of a governmental body from soliciting or accepting a professional contract with that governmental body.

Based on the facts, there is no indication that either Engineer A or Engineer B are serving as members, advisors, or employees of the governmental body procuring the design services. Under the facts, the design services are being procured, overseen and reviewed by an agency of the federal government and not by the community services corporation on which Engineer A served.

Also, importantly Engineer A and B had no direct relationship with the federal agency procuring the A/E services.

Conclusion:
It would be ethical for the firm of Engineer A and Engineer B to agree to perform services in connection with the federal project.

Board of Ethical Review

William A. Cox, Jr., P.E.
William W. Middleton, P.E.
William E. Norris, P.E.
William F. Rauch, Jr., P.E.
Jimmy H. Smith, P.E.
Otto A. Tennant, P.E.
Robert L. Nichols, P.E., Chairman

Note: In regard to the question of application of the Code to corporations vis-a-vis real persons, business form or type should not negate nor influence conformance of individuals to the Code. The Code deals with professional services, which services must be performed by real persons. Real persons in turn establish and implement policies within business structures. The Code is clearly written to apply to the Engineer and it is incumbent on a member of NSPE to endeavor to live up to its provisions. This applies to all pertinent sections of the Code.

The City Owes Me: Oppression and the IGA

Public Administration at the University of Arkansas - Revised 4/27/98
Used by permission. Contact Dr. William Miller, Associate Professor of Political Science, University of Arkansas — http://plsc.uark.edu/book/books/ethics/index.htm

Context

Jan Sims is a housing inspector with the city housing code enforcement department. Her job is very demanding. She reports to work at 8:00 in the morning and seldom leaves before 6:00 or 6:30 in the evening. She is also a single mother whose sister watches her two children while Jan works. She likes her job, but rightfully feels that the city is exploiting her. They need to hire another inspector. Unfortunately, the city can't afford to hire another inspector until at least the next fiscal year.

The Dilemma

About 6:00 Jan finishes her last inspection and heads home. On the way, she pulls her city owned car into the grocery store and loads up with what she will need for supper. Another day, thank God, is over.

The next day Jan finds a memo on her desk from the housing director's office. A citizen complained about the use of a city car for private purposes. The memo reminds her of the city policy forbidding the use of city property for non-work related activities. The memo clearly indicates that if Jan is caught doing this again, she may be terminated.

The Decision

Moments after she reads the memo, the director visits Jan in her office. "I had to write the memo", he says. "I understand that we work you overtime and that you are not compensated for that time. Let's privately consider the city car a bit of a fringe benefit. Use the city car when you need to. Just keep a low profile. We don't want any irate citizens causing trouble."

What do you think of this decision?

Academic Qualifications

NSPE Opinions of the Board of Ethical Review - Case 79-5

Facts:

Engineer A received a Bachelor of Science degree in 1940 from a recognized engineering curriculum, and subsequently was registered as a professional engineer in two states.

Later, he was awarded an earned "Professional Degree" from the same institution. In 1960 he received a Ph.D. degree from an organization which awards degrees on the basis of correspondence without requiring any form of personal attendance or study at the institution, and is regarded by state authorities as a "diploma mill."

Engineer A has since listed his Ph.D. degree among his academic qualifications in brochures, correspondence, and otherwise, without indicating its nature.

Question:

Was Engineer A ethical in citing his Ph.D. degree as an academic qualification under these circumstances?

References:

Code of Ethics Section 3(a) - The Engineer shall not make exaggerated, misleading, deceptive, or false statements or claims about his professional qualifications, experience, or performance in his brochures, correspondence, listings, advertisements, or other public communications.

Code of Ethics Section 3(b) - The above prohibitions include, but are not limited to, the use of statements containing a material misrepresentation of fact or omitting a material fact necessary to keep the statement from being misleading; statements intended or likely to create an unjustified expectation; statements containing prediction of future success; statements containing an opinion as to the quality of the Engineer's services; or statements intended or likely to attract clients by the use of showmanship, puffery, or self-laudation, including the use of slogans, jingles, or sensational language or format.

Discussion:

In Case 72-11 we dealt with a related question under a slightly different wording in the Code which declared it unethical for an engineer to allow himself to be listed for employment using exaggerated statements of his qualifications. In that case, however, the alleged offense related to statements for employment in which the engineer played down his major technical design experience to emphasize his lesser managerial and administrative experience in order to meet employment opportunities then available. We concluded that such action was not unethical under the "exaggerated" standard because the engineer had in fact some degree of competence in the managerial and administrative areas, and because his action was not intended to deceive a prospective employer by an untruthful statement.

Now the language of the Code has been revised and extended beyond the "exaggerated" standard to embrace misleading, deceptive, or false statements regarding professional qualifications. The mandate has been buttressed by the requirement that statements prohibited should not omit a material fact necessary to keep the statement from being misleading.

We believe that this case should be resolved on the basic question of whether a claim to a Ph.D. degree under these circumstances is "misleading." Ordinarily, employers, clients, and the engineering profession generally understand that a statement of academic achievement in the form of a degree means an earned degree from an accredited educational institution. Thus, it is customary for those who may list an honorary degree, for instance, to indicate that it is not an earned degree by use of "Hon." after the citation of the degree.

The state engineering registration laws usually refer to academic qualifications as a basis for examination or otherwise to qualify for registration by reference to a degree approved by the state board. Most state boards apply that type of language to refer to a degree awarded through completion of an engineering program approved by the Engineer's Council for Professional Development, which is the nationally recognized accrediting agency for engineering education.

It is beyond our purview to comment broadly on the pros and cons of correspondence courses, or the awarding of certain degrees by mail. Whatever other merits this type of education may have for an individual, it is enough for our purpose to say that mail order degrees are not the accepted norm in the engineering profession and uniformly will not be recognized by state registration boards or the profession itself as meeting required educational standards.

Under these conditions we must conclude that Engineer A is charged with knowledge of the accepted standards of the profession. In stating that he had a Ph.D. degree he should have been aware that those who received his communications would be deceived.

There is some flexibility allowed for state registration boards to decide which educational attainments meet the standards for registration purposes, and there is some flexibility allowed to members of the profession in listing academic degrees from institutions or curricula not recognized by state boards. But the bounds of such flexibility are exceeded when the basis for the claimed educational achievement is a mail order procedure not involving recognition by any recognized accrediting body.

Finally, we add that it would not be sufficient for Engineer A to overcome the objective by merely listing the name of the organization which awarded the mail order degree.

Such organizations may have impressive names which would lead the person receiving the communication to assume that it is a recognized educational institution, and that the claimed degree is one to be taken at face value.

Conclusion:
Engineer A was unethical in citing his Ph.D. degree as an academic qualification under these circumstances.

Board of Ethical Review

Louis A. Bacon, P.E.
Robert R. Evans, P.E.
James G. Johnstone, P.E.
Robert H. Perrine, P.E.
James F. Shivler, Jr., P.E.
L.W. Sprandel, P.E.
Donald C. Peters, P.E., chairman

*Note—This opinion is based on data submitted to the Board of Ethical Review and does not necessarily represent all of the pertinent facts when applied to a specific case. This opinion is for educational purposes only and should not be construed as expressing any opinion on the ethics of specific individuals.

This opinion may be reprinted without further permission, provided that this statement is included before or after the text of the case.

Appearances

Public Administration at the University of Arkansas - Revised 4/27/98
Used by permission. Contact Dr. William Miller, Associate Professor of Political Science, University of Arkansas — http://plsc.uark.edu/book/books/ethics/index.htm

Context

John Swane works in the contracts department at NASA. For twenty years he has demonstrated competence and integrity in his work. Now he is considering a move. What with his salary being rather low compared to the private sector and the cutbacks, he is thinking this might be a good time to move.

At a recent conference on the future of aerospace technology, he had an interesting conversation with the Big Boom Rocket company CIO. They are looking for a person with his qualifications. The CIO pretty much offered him a job at higher pay and great benefits. A move is involved, so John is talking with his family. They seem supportive and it looks like he will accept.

The Dilemma

Meanwhile, Big Boom has made a bid for a new booster rocket for the space station program. John is in charge of reviewing the bids. In his professional opinion, the bid by Big Boom is the best of the bids. In normal circumstances he would simply sign his name recommending acceptance of Big Boom's bid.

The problem is that some might see such an action, followed closely by his leaving to work for Big Boom, as a case of conflict of interest. John knows that this is not true. The bid speaks for itself and Big Boom should get the contract. He could remove himself from the situation claiming the possible appearance of conflict of interest. This would slow down the bidding process and result in a loss of money for Big Boom. Moreover, Big Boom would be hesitant to hire him after discussions of conflict of interest.

On the other hand, if he signs the bid and later takes the job with Big Boom, though there may be talk of conflict of interest, he will be secure in his new job. In either case, he feels he is doing nothing illegal.

The Decision

John contemplates the issues. He decides to sign the contract and then leave for Big Boom.

Do you agree with this decision? Why? or Why Not?

Ethical Association with Other Engineers

NSPE Opinions of the Board of Ethical Review - Case No. 75-3

Facts:

Engineer A, president of an engineering consulting firm, was charged with and found to have violated a section of the state society code of ethics and was publicly reprimanded for the offense. Subsequently, an engineering consulting firm headed by Engineer B proposes to engage in a joint venture with the firm headed by Engineer A, but raises the question whether such an association would conflict with the Code of Ethics.

Question:

May the firm headed by Engineer B ethically engage in a joint venture with the firm headed by Engineer A?

Reference:

Code of Ethics Section 13 - The Engineer will not associate with or allow the use of his name by an enterprise of questionable character, nor will he become professionally associated with engineers who do not conform to ethical practices, or with persons not legally qualified to render the professional services for which the association is intended.

Discussion:

Our previous discussions of Section 13 of the code have all related to various factual situations generally involving the offer or furnishing of engineering services by persons or companies not authorized to engage in engineering practice (Cases 61-4, 63-10,69-6 and 69-9). We have not heretofore been required to pass on the meaning of Section 13 when an engineer has been disciplined for unethical conduct.

Section 13 actually embraces three different aspects in its entirety. The first deals with an enterprise of a questionable character (see Case 69-6 as an example). The third part involves relationships with persons not legally qualified to offer or provide professional services (see Case 61-4 as an example).

We are now confronted with the second portion of Section 13, which on the face of the language would appear to absolutely rule out an association with any engineer who has violated the Code of Ethics.

However, we do not believe that such a harsh and unyielding interpretation of the language is required and justified in all circumstances. One semantic problem to be first resolved is whether the words "who do not conform to ethical practices" were intended to mean that an engineer found guilty of one violation of the code, no matter of what degree of severity, should be "read out" of the profession or considered an unethical engineer for all time to the extent that ethical engineers must shun him forever. Such a reading would be contrary to the spirit of our laws and traditions that redemption is a cherished virtue and that a person found to have violated the mores of society should "go forth and sin no more." Even the hardened criminal under our moral concepts may be accepted back into society as a useful citizen after he has paid the penalty for his transgressions. We believe that a proper reading of the language on this point should be construed to mean that an ethical engineer will not associate with an engineer who is known to habitually violate the code and who has shown no evidence of avoiding such unethical conduct as he may have engaged in previously after he has been duly "brought to book" for his past action.

Another troublesome aspect of the question before us involved the oft noted fact that the code is written in terms of personal conduct, whereas in the real world a large part of engineering practice is carried on by firms, whether partnerships or corporations, often comprising many hundreds of engineers as officers, partners, or employees. We cannot believe that the framers of the code could have meant to ban association with an entire firm because one of its officers or partners (to say nothing of engineering employees) had at one time violated some part of the code. The effect of such an interpretation would be to penalize many other engineers who were unaware of and not a party to any such violation. This would be a form of "guilt by association."

But if these readings of the language we must deal with are correct, what is the true meaning of the restriction? Does it mean that the firm of Engineer B may engage in the joint venture with the firm of Engineer A only if there is some assurance that Engineer A will not personally participate in the endeavor? Or does it mean that Engineer B must satisfy himself that Engineer A has undertaken to cease and desist from further unethical conduct in order to have him participate? Suppose that Engineer A was a partner, a member of the board of directors if the firm is incorporated, an individual practitioner, or an employee of a large firm in a management-level position, or an employee in a subordinate position? Does the relative rank of Engineer A in the firm make a difference?

We must confess that there are no easy or simple answers to the kinds of questions we have asked ourselves, and we are loath to conclude that the middle portion of Section 13 is without substantive meaning. On balance, we are constrained to read into the language the following principles:

1. A single transgression of the code by itself does not necessarily bar an engineer from future association with other engineers, depending upon the severity of the transgression.

2. A pattern of violations of the code by the engineer is grounds for other engineers to shun him in professional associations.

3. If it is determined that an engineer has committed a serious offense under principle No. 1, or has engaged in a pattern of violations under principle No. 2, he is deemed to be an unethical engineer, and his status in the firm is a material factor. If he is in a position to influence or control operational policies of the firm, ethical engineers and their firms should not associate

with the firm of the offending engineer. It would be misleading to distinguish between unethical individual engineers and their firms so as to rationalize that ethical engineers could associate with firms under the policy control or influence of unethical engineers.

Applying the above principles to the facts before us, and on the stated record that Engineer A was found to have engaged in an ethical violation on only one occasion for which he received a reprimand and in the absence of any evidence that he had engaged in a pattern of unethical conduct, we believe that the firm of Engineer B may ethically engage in a joint venture with the firm of Engineer A, but is required to maintain a careful scrutiny of the operation of the firm of Engineer A to assure itself to the extent possible that further unethical conduct will not develop during and with respect to the joint venture.

Conclusion:

Subject to the constraints noted above, the firm headed by Engineer B may ethically engage in a joint venture with the firm headed by Engineer A.

Board of Ethical Review

William J. Deevy, P.E.
William R. Gibbs, P.E.
Joseph N. Littlefield, P.E.
Donald C. Peters, P.E.
James F. Shivler, Jr. P.E.
Louis W. Sprandel, P.E.
Robert E. Stiemke, P.E., chairman

Note—This opinion is based on data submitted to the Board of Ethical Review and does not necessarily represent all of the pertinent facts when applied to a specific case. This opinion is for educational purposes only and should not be construed as expressing any opinion on the ethics of specific individuals.

Public Welfare — Hazardous Waste

NSPE Opinions of the Board of Ethical Review — Case 92-6

Facts:

Technician A is a field technician employed by a consulting environmental engineering firm. At the direction of his supervisor Engineer B, Technician A samples the contents of drums located on the property of a client. Based on Technician A's past experience, it is his opinion that analysis of the sample would most likely determine that the drum contents would be classified as hazardous waste. If the material is hazardous waste, Technician A knows that certain steps would legally have to be taken to transport and properly dispose of the drum including notifying the proper federal and state authorities.

Technician A asks his supervisor Engineer B what to do with the samples. Engineer B tells Technician A only to document the existence of the samples. Technician A is then told by Engineer B that since the client does other business with the firm, Engineer B will tell the client where the drums are located but do nothing else. Thereafter, Engineer B informs the client of the presence of drums containing "questionable material" and suggests that they be removed. The client contacts another firm and has the material removed.

Questions:

1. Was it ethical for Engineer B to merely inform the client of the presence of the drums and suggest that they be removed?

2. Did Engineer B have an ethical obligation to take further action?

References:

Code of Ethics Preamble - Engineering is an important and learned profession. The members of the profession recognize that their work has a direct and vital impact on the quality of life for all people. Accordingly, the services provided by engineers require honesty, impartiality, fairness and equity, and must be dedicated to the protection of the public health, safety and welfare. In the practice of their profession, engineers must perform under a standard of professional behavior which requires adherence to the highest principles of ethical conduct on behalf of the public, clients, employers and the profession.

Code of Ethics Section I.1. - Hold paramount the safety, health and welfare of the public in the performance of their professional duties.

Code of Ethics Section II.1. - Engineers shall hold paramount the safety, health and welfare of the public in the performance of their professional duties.

Code of Ethics Section II.1.a. - Engineers shall at all times recognize that their primary obligation is to protect the safety, health, property and welfare of the public. If their professional judgment is overruled under circumstances where the safety, health, property or welfare of the public are endangered, they shall notify their employer or client and such other authority as may be appropriate.

Code of Ethics Section II.1.c. - Engineers shall not reveal facts, data or information obtained in a professional capacity without the prior consent of the client or employer except as authorized or required by law or this Code.

Code of Ethics Section II.3.a. - Engineers shall be objective and truthful in professional reports, statements or testimony. They shall include all relevant and pertinent information in such reports, statements or testimony.

Code of Ethics Section III.1. - Engineers shall be guided in all their professional relations by the highest standards of integrity.

Code of Ethics Section III.3. - Engineers shall avoid all conduct or practice which is likely to discredit the profession or deceive the public.

Code of Ethics Section III.3.a. - Engineers shall avoid the use of statements containing a material misrepresentation of fact or omitting a material fact necessary to keep statements from being misleading or intended or likely to create an unjustified expectation; statements containing prediction of future success; statements containing an opinion as to the quality of the Engineers' services; or statements intended or likely to attract clients by the use of showmanship, puffery, or self-laudation, including the use of slogans, jingles, or sensational language or format.

Code of Ethics Section III.4. - Engineers shall not disclose confidential information concerning the business affairs or technical processes of any present or former client or employer without his consent.

Discussion:

The extent to which an engineer has an obligation to hold paramount the public health and welfare in the performance of professional duties (Section I.1.) has been widely discussed by the Board of Ethical Review over the years. In many of these cases this basic duty has frequently intersected with the duty of engineers not to disclose confidential information concerning the business affairs, etc., of clients (Section III.4.)

For example, in BER Case 89-7 an engineer was retained to investigate the structural integrity of a 60 year old occupied apartment building which his client was planning to sell. Under the terms of the agreement with the client, the structural report written by the engineer was to remain confidential. In addition, the client made it clear to the engineer that the building was being sold "as is" and the client was not planning to take any remedial action to repair or renovate any system within the building. The engineer performed several structural tests on the building and determined that the building was structurally sound. However, during the course of providing services, the client confided in the engineer that the building contained deficiencies in the electrical and mechanical systems which violated applicable codes and standards. While the engineer was not an electrical nor mechanical engineer, he did realize that those deficiencies could cause injury to the occupants of the building and so informed the client. In his report, the engineer made a brief mention of his conversation with the client concerning the deficiencies; however, in view of the terms of the agreement, the engineer did not report the safety violations to any third parties. In determining that it was unethical for the engineer not to report the safety violations to appropriate public authorities, the Board, citing cases decided earlier, noted that the engineer "did not force the issue but instead went along without dissent or comment. If the engineer's ethical concerns were real, the engineer should have insisted that the client take appropriate action or refuse to continue work on the project." The Board concluded that the engineer had an obligation to go further particularly because the Code uses the term "paramount" to describe the engineer's obligation to protect the public safety health and welfare.

More recently, in BER Case 90-5, the Board reaffirmed the basic principle articulated in BER Case 89-7. There, tenants of an apartment building sued its owner to force him to repair many of the building's defects. The owner's attorney hired an engineer to inspect the building and give expert testimony in support of the owner. The engineer discovered serious structural defects in the building which he believed constituted an immediate threat to the safety of the tenants. The tenants' suit had not mentioned these safety-related defects. Upon reporting the findings to the attorney, the engineer was told he must maintain this information as confidential as it is part of the lawsuit. The engineer complies with the request. In deciding it was unethical for the engineer to conceal his knowledge of the safety-related defect, the Board discounted the attorney's statement that the engineer was legally bound to maintain confidentiality, noting that any such duty was superseded by the immediate and imminent danger to the building's tenants. While the Board recognized that there may be circumstances where the natural tension between the engineer's public welfare responsibility and the duty of non-disclosure may be resolved in a different manner, the Board concluded that this clearly was not the case under the facts.

Turning to the facts in this case, we believe the basic principles enunciated in BER Cases 89-7 and 90-5 are applicable here as well except in a different context. Unlike the facts in the earlier cases, Engineer B made no oral or written promise to maintain the client's confidentiality. Instead, Engineer B consciously and affirmatively took actions that could cause serious environmental danger to workers and the public, and also a violation of various environmental laws and regulations.

Under the facts, it appears that Engineer B's primary concern was not so much maintaining the client's confidentiality as it was in maintaining good business relations with a client. In addition, it appears that as in all cases which involve potential violations of the law, Engineer B's actions may have had the effect of seriously damaging the long-term interests and reputation of the client. In this regard, we would also note that under the facts it appears that the manner in which Engineer B communicated the presence of the drums on the property must have suggested to the client that there was a high likelihood that the drums contained hazardous materials.

We believe that this subterfuge is wholly inconsistent with the spirit and intent of the Code of Ethics because it makes the engineer an accomplice to what may amount to an unlawful action.

Clearly, Engineer B's responsibility under the facts was to bring the matter of the drums possibly containing hazardous material to the attention of the client with a recommendation that the material be analyzed. To do less would be unethical. If analysis demonstrates that the material is indeed hazardous, the client would have the obligation of disposing of the material in accordance with applicable federal, state and local laws.

Conclusions:

1. It was unethical for Engineer B to merely inform the client of the presence of the drums.

2. It was unethical for Engineer B to fail to advise his client that he suspected hazardous material and provide a recommendation concerning removal and disposal in accordance with federal, state and local laws.

Board of Ethical Review

William A. Cox, Jr., P.E.
William W. Middleton, P.E.
William E. Norris, P.E.
William F. Rauch, Jr., P.E.
Jimmy H. Smith, P.E.
Otto A. Tennant, P.E., and
Robert L. Nichols, P.E., Chairman

Note: In regard to the question of application of the Code to corporations vis-a-vis real persons, business form or type should not negate nor influence conformance of individuals to the Code. The Code deals with professional services, which services must be performed by real persons. Real persons in turn establish and implement policies within business structures. The Code is clearly written to apply to the Engineer and it is incumbent on a member of NSPE to endeavor to live up to its provisions. This applies to all pertinent sections of the Code.

Who Pays for the Groceries

Public Administration at the University of Arkansas - Revised 7/3/96
Used by permission. Contact Dr. William Miller, Associate Professor of Political Science, University of Arkansas -- http://plsc.uark.edu/book/books/ethics/index.htm

Context

John Sampson is a policy analyst and planner for the state highway department. Much of his time is spent developing a master plan for the state's growing network of the highways. John is a mid-level bureaucrat who does not set policy, but his analysis certainly does have a role in how the policy agenda shapes up.

He has recently been working on route planning for a new regional feeder highway. This is a controversial development. The region is environmentally sensitive, yet has experienced rapid growth. There seem to be two viable alternatives in route planning.

- The least expensive in cost, the Speedy Route, is also the quickest. It will save about 8 minutes off the other route.
- The Green Route would be less environmentally destructive, but would cost the state about 10 percent more.

The Dilemma

The highway department has always been sympathetic to the needs of the transportation industry. Truckers and shippers are already advocating for the least expensive and quickest route over the mountain. The transport lobbyists have been in constant dialogue with the planners and know the issues well.

Environmentalists and preservationists are out of the loop. John knows that it is the Green Route that the environmentalists would favor. The environmentalists would certainly advocate for the less environmentally harmful route if they had the opportunity, but they may never really know the issues.

As John develops his report, he feels pressured to minimize the possibility of the more expensive and longer route. Upper management in the transportation department wants to keep the agenda free of "unnecessary" controversy. They have said that they believe the Green route is not as financially feasible. They encourage John to leave it out of his report. Why present an option that is not the best? John is not so sure.

The Question

What are John's options? What should John do?

Credit For Engineering Work — Research Data

NSPE Opinions of the Board of Ethical Review — Case 92-7

Facts:

The XYZ Company headed by Engineer A offered to provide funding to professors in the chemistry department of a major university for research on removing poisonous heavy metals (copper, lead, nickel, zinc, chromium) from waste streams. The university then agreed to contract with XYZ company to give the company exclusive use of the technology developed in the field of water treatment and waste water stream treatment. Under the agreement, XYZ Company will provide a royalty to the university from profits derived from the use of the technology. Also, a group of the university professors organized QRS, a separate company to exploit applications of the technology other than the treatment of water and waste water.

At the same time that the university research was being conducted, XYZ continued to conduct research in the same area. Performance figures and conclusions were developed. XYZ freely shared the figures and conclusions with QRS organized by the university professors.

At the university, Engineer B, a professor of civil engineering, wanted to conduct research and develop a paper relating to the use of the technology to treat sewage. Engineer B contacted the professors in the university's chemistry department. The chemistry professors provided XYZ's data to Engineer B for use in the research and paper. The professors did not reveal to Engineer B that the data was generated by Engineer A and XYZ company.

Engineer B's paper was published in a major journal. Engineer A's data was displayed prominently in the paper and the work of XYZ constituted a major portion of the journal. The paper credits two of the chemistry professors as major authors along with Engineer B. No credit was given to Engineer A or XYZ as the source of the data, the funds that supported the research. After publication Engineer B learns about the actual source of the data and its finding.

Question:

Does Engineer B have an obligation under the Code of Ethics to clarify the source of the data contained in the paper?

References:

Code of Ethics Section III.10. - Engineers shall give credit for engineering work to those to whom credit is due, and will recognize the proprietary interests of others.

Code of Ethics Section III.10.a. - Engineers shall, whenever possible, name the person or persons who may be individually responsible for designs, inventions, writings, or other accomplishments.

Discussion:

The issue of providing credit for research work performed by others is a vital matter in this day and age. Its importance is more than merely crediting contributions of individuals who have performed work in an area of engineering and scientific research. In actual fact, funding decisions for research and development of various technologies are vitally affected by the credit and acknowledgments.

Over the years, the Board has examined these issues in a variety of contexts. In BER Case 75-11, Engineer A performed certain research and then prepared a paper on an engineering subject based on that research which was duly published in an engineering magazine under his byline. Subsequently, an article on the same subject under the name of Engineer B appeared in another engineering magazine. A substantial portion of the text of Engineer B's article was identified word-for-word with the article authored by Engineer A. Engineer A contacted Engineer B and requested an explanation. Engineer B replied that he had submitted with his article a list of six references, one of which identified the article by Engineer A, but that the list of references had been inadvertently omitted by the editor. He offered his apology to Engineer A for the mishap because his reference credit was not published as intended. In ruling that Engineer B did not act ethically by his actions, we distinguished research from plagiarism. We offered that a "quotation from many sources is research" and "quotation from a single or limited number of sources is plagiarism." However, in either event, it is contemplated that the author will identify and give credit to his sources — single or many. In addition, we noted the important belief of Engineer B that he would have been without fault if the list of references had been published at the end of the article. This belief represented a lack of understanding of the requirements of the Code. Merely listing the work of Engineer A in a list of references to various articles only tells the reader that Engineer B had consulted and read those cited articles of other authors. It no way tells the reader that a large portion of his text is copied from the work of another.

More recently, in BER Case 83-3, Engineer B submitted a proposal to a county council following an interview concerning a project. The proposal included technical information and data that the council requested as a basis for the selection. Smith, a staff member of the council, made Engineer B's proposal available to Engineer A. Engineer A used Engineer B's proposal without Engineer B's consent in developing another proposal, which was subsequently submitted to the council. The extent to which Engineer A used Engineer B's information and data is in dispute between the parties. In finding that it was unethical for Engineer A to use Engineer B's proposal without Engineer B's consent, we indicated that Engineer A had an obligation to refuse to accept the proposal from Smith and also noted that Engineer A's actions constituted unfair competition by improper and questionable methods in violation of Code Section III.7.

Taking BER Cases 75-11 and 83-3 together, we believe that the instant case can be distinguished from the two earlier cases. Unlike the facts in BER Cases 75-11 and 83-3, Engineer B did not knowingly fail to credit Engineer A or XYZ corporation for its contributions to the research which formed the basis of his paper. Instead, Engineer B assumed that the material he received from the other professors was developed solely by those professors.

We conclude that Engineer B did not knowingly and deliberately fail to credit Engineer A or XYZ for their contributions to the research. However, we believe that had Engineer B made more of an effort to substantiate the sources contained in his paper, he may have been able to identify those sources. We would also emphasize our deep concern over the conduct of the chemistry professors who for whatever reason(s) misled Engineer B by failing to reveal the sources of the data. While not technically covered by this Code, the conduct of the chemistry professors is clearly deplorable and is unacceptable under the philosophical standards embodied in the Code of Ethics.

Finally, we would suggest that Engineer B prepare and request that the journal publish a clarification of the matter explaining how the matter occurred along with an apology for any misunderstanding which may have arisen as a result of the publication of the paper.

Conclusion:

Engineer B has an obligation to request that the journal publish a clarification of the matter explaining how the matter occurred along with an apology for any misunderstanding which may have arisen as a result of the publication of the paper.

Board of Ethical Review:

William A. Cox, Jr., P.E.
Jimmy H. Smith, P.E.
William W. Middleton, P.E.
Otto A. Tennant, P.E.
William E. Norris, P.E.
William F. Rauch, Jr., P.E.
Robert L. Nichols, P.E., Chairman

Note: This opinion is for educational purposes only and should not be construed as expressing any opinion on the ethics of specific individuals.

Expert Testimony Report
And Redesign by Another Engineer

NSPE Opinions of the Board of Ethical Review - Case 71-4

Facts:

Engineer A designed a facility for a client who, after construction of the project, filed a lawsuit against him claiming 1) that the cost grossly exceeded the preliminary estimate and 2) that there were numerous design errors.

The client terminated the services of Engineer A upon the filing of the lawsuit and retained Engineer B to study the work performed by Engineer A and to testify on the basis of his study as an expert witness at the trial on the client's behalf. Engineer B prepared a report prior to the trial listing many alleged deficiencies in the work of Engineer A—some dealing with overall design philosophy which are matters of judgment and opinion and some alleged factual defects in the design of construction.

Pending trial of the lawsuit the client retained Engineer B to redesign the project in accordance with the findings of his report and to be in general charge of the reconstruction of the facility. The reconstruction will destroy most of the actual physical evidence of the alleged defects prior to the trial.

Questions:

1. Is it ethical for Engineer B to offer expert testimony at the trial on a mixture of opinion on design philosophy and alleged factual errors in the design?

2. Was it ethical for Engineer B to undertake a contract to redesign and be in charge of reconstruction of the project in view of his critical report and pending testimony at the trial?

References:

Code of Ethics Section 5 - The Engineer will express an opinion of an engineering subject only when founded on adequate knowledge and honest conviction.

Code of Ethics Section 11 - The Engineer will not compete unfairly with another engineer by attempting to obtain employment or advancement or professional engagements by competitive bidding, by taking advantage of a salaried position, by criticizing other engineers, or by other improper or questionable methods.

Code of Ethics Section 12 - The Engineer will not attempt to injure, maliciously or falsely, directly or indirectly, the professional reputation, prospects, practice, or employment of another engineer, nor will he indiscriminately criticize another engineer's work in public. If he believes that another engineer is guilty of unethical or illegal practice, he shall present such information to the proper authority for action.

Discussion:

Before dealing with the primary issues in this case, we dispose of the collateral question raised by the statement of facts that the reconstruction of the facility will destroy most of the actual physical evidence of the alleged defects prior to the trial. This is a matter of legal procedure, and the attorneys for Engineer A may take steps to prevent prejudice to their client through the use of affidavits of actual conditions before construction starts, depositions of prospective witnesses, stipulations by the parties, the use of certified photographs, plans and other documents, and, if necessary, the use of restraining orders to prevent the reconstruction before vital physical evidence is destroyed.

In Case 63-6, dealing with language in the Canons of Ethics similar to that in the Code of Ethics as cited above, we noted that "there may ... be honest differences of opinion among equally qualified engineers on the interpretation of the known physical facts." We cannot, therefore, fault Engineer B for preparing a report critical of the work of Engineer A when retained to do just that by the client. However, if Engineer B's criticism of the design philosophy of Engineer A is merely a difference of opinion as to application of another valid solution he is outside of his right of ethical criticism. That type of criticism must be considered "indiscriminate" and is not ethically permissible under Section 12. Engineer B should offer his testimony and opinion at the trial only on the alleged physical defects in the work of Engineer A.

The facts do not tell us the extent to which Engineer B may have actively solicited the contract with the client to redesign the project, or whether the client took the sole initiative in seeking out the services of Engineer B. In the absence of such facts and for the purpose of this case, we shall assume that Engineer B did not obtain his assignment by criticizing Engineer A before being retained or by improper or questionable methods. Nor do we find anything in the code which could reasonably be construed to say that Engineer B may not proceed with his redesign work while the question of Engineer A's alleged negligence remains unresolved.

If a client is dissatisfied with the design work of an engineer to the extent of filing a lawsuit, it would be unreasonable to expect him to suspend the project until the lawsuit is settled, which might take several years. To proceed with his project he requires the services of another engineer, and he is entitled to obtain such services.

We might ponder the wisdom of Engineer B agreeing to handle the redesign and reconstruction after he has developed a critical report of the work of the previous engineer and while the validity of his critical report is pending trial and judgment by a judge or jury. To do so Engineer B exposes himself to suspicion that his criticism was prompted by the expectation that he would achieve an economic benefit by being selected to perform the redesign. And in retaining Engineer B for the redesign the client may impair the acceptance of the forthcoming critical expert testimony of Engineer B.

Despite these pragmatic considerations of the choices made by Engineer B and the client, we cannot support any finding of unethical conduct on the part of Engineer B in agreeing to undertake the redesign.

Conclusions:

1. It would be ethical for Engineer B to offer expert testimony at the trial on alleged factual errors in the design by Engineer A, but it would be unethical for Engineer B to offer opinion on the design philosophy (merely a difference of opinion as to application of another valid solution) of Engineer A.

2. It was ethical for Engineer B to undertake a contract to redesign and be in charge of reconstruction of the project prior to his testimony at the trial.

Board of Ethical Review

Frank H. Bridgers, P.E.
W.R. Gibbs, P.E.
C.C. Hallvik, P.E.
James D. Maloney, P.E.
Robert E. Stiemke, P.E.
Albert L. Wolfe, P.E.
Sherman Smith, P.E., chairman

Note—This opinion is based on data submitted to the Board of Ethical Review and does not necessarily represent all of the pertinent facts when applied to a specific case. This opinion is for educational purposes only and should not be construed as expressing any opinion on the ethics of specific individuals.

Attempt To Influence Prospective City/Client During Relocation Negotiations

NSPE Opinions of the Board of Ethical Review — Case 92-8

Facts:

Engineer A is the principal of a large engineering firm that provides civil engineering services to state, county and local governments and agencies. The firm is planning to relocate one of its regional offices to a medium-sized city. Part of the relocation involves the construction of a large office building. The relocation will greatly benefit the city, by among other things, creating needed first class office space, enhancing the city's tax base and providing needed additional employment in construction as well as in other areas. Having the city employ the engineering services offered by the firm would be an added incentive to the firm's selection of the city. Engineer A verbally suggests to city officials during the relocation negotiations with the city that he "hoped the city would consider employing the services of his firm in the future for part of the engineering services requirements." City officials at the meeting do not respond specifically to Engineer A's verbal suggestion. Ultimately, Engineer A's firm agrees to relocate to the city.

Question:

Would it be unethical for Engineer A to verbally suggest to city officials during the relocation negotiations that they consider employing the services of his firm?

References:

Code of Ethics Section II.5.b. - Engineers shall not offer, give, solicit or receive, either directly or indirectly, any political contribution in an amount intended to influence the award of a contract by public authority, or which may be reasonably construed by the public of having the effect or intent to influence the award of a contract. They shall not offer any gift, or other valuable consideration in order to secure work. They shall not pay a commission, percentage or brokerage fee in order to secure work except to a bona fide employee or bona fide established commercial or marketing agencies retained by them.

Code of Ethics Section III.2.a. - Engineers shall seek opportunities to be of constructive service in civic affairs and work for the advancement of the safety, health and well-being of their community.

Discussion:

The practice of engineering has become increasingly competitive in recent years. Firms have been forced to become far more aggressive in marketing their services. Some firms have used innovative methods to accelerate their marketing techniques in order to secure work. While such methods are frequently appropriate and ethical, at times such practices may go beyond the bounds of the Code of Ethics.

The NSPE Code of Ethics makes clear that engineers shall not "offer any gift or valuable consideration in order to secure work." (Code Section II.5.b.) This section of the Code has been interpreted on numerous occasions by the Board of Ethical Review, but mostly in the context of political contributions. A review of some of the more recent political contributions and similar ethics cases may be helpful in evaluating the facts in the present case.

BER Case 76-6 involved gifts to foreign officials. There an engineer whose firm did overseas work in a foreign county was advised by a high ranking official of that country that it was established practice to make personal gifts to government officials awarding contracts. The engineer's failure to adhere to the practice would result in the engineer's firm receiving no future contracts. In finding it was unethical for the engineer to accept the contract and make the gifts as described, the Board noted that such practices are not dissimilar to the arguments advanced by those who had at that time been revealed as offering financial payments to public officials to influence the award of contracts for architectural-engineering services. The Board noted that engineers should decline being drawn into such seamy procedures for self-gain.

Later in BER Case 78-4, a local group of business and community leaders banded together and organized a fund raising committee to support a bond issue to finance public works. Many extensive engineering and architectural projects were to be financed by the bond issue. In ruling it would be ethical for engineering firms to contribute to the promotional fund in the expectation or possibility that those firms might later seek design commissions arising from the public works programs, the Board noted that Section III.2.a. is one of many provisions that is not susceptible to precise construction or enforcement. However, the Board concluded in the context of the case, that the proposed public works program for the community would be constructive and would advance the well-being of the citizenry. Applying its views to the facts of the case, the Board noted that as with many situations, the question of "motivation" may be a mixed one, but that even if there is some degree of self-interest motivation, the Board concluded that it was sufficiently remote and removed from undue influence to eliminate any substantial concern that the Code is offended.

Turning to the facts of the immediate case, it is our view that it can easily be distinguished from the earlier-cited cases. Unlike those earlier cases, this case neither involves the requesting, offering or acceptance of a gift or a political contribution. Instead, this case involves an engineering firm that is in the process of selecting a site for the relocation of one of its regional offices. There does not appear to be any "quid pro quo" involved under which an understanding or agreement to provide something of value in exchange for some other thing of value. Here, Engineer A is merely suggesting to city officials that at the appropriate point in time and under the proper circumstance, the city consider retaining Engineer A's firm to provide engineering services to the city. While we are not entirely comfortable with the context in which the suggestion is made, we do not believe that the suggestion during the relocation negotiations rises to a level of impropriety or Code violation.

We would note that the fact that the city officials present during the meeting did not respond specifically to Engineer A's verbal suggestion indicates to us that the city officials were in no way being coerced to make any promises of future work to Engineer A's firm.

In closing, we would merely note that in view of the fact that many if not most state, county and local governments employ qualifications-based selection procedures for the procurement of architectural and engineering services, Engineer A's comments should be viewed in that context. If one was to assume that the city here had a QBS procedure in place, Engineer A's comments could easily be understood as an expression of interest in submitting proposals to the city and a request that such proposals be given due consideration.

Conclusion:

It would not be unethical for Engineer A to verbally suggest to city officials during the relocation negotiations with the city that they consider employing the services of his firm.

Board of Ethical Review

William A. Cox, Jr., P.E.
William W. Middleton, P.E.
William E. Norris, P.E.
William F. Rauch, Jr., P.E.
Jimmy H. Smith, P.E.
Otto A. Tennant, P.E.
Robert L. Nichols, P.E., Chairman

Note: In regard to the question of application of the Code to corporations vis-a-vis real persons, business form or type should not negate nor influence conformance of individuals to the Code. The Code deals with professional services, which services must be performed by real persons. Real persons in turn establish and implement policies within business structures. The Code is clearly written to apply to the Engineer and it is incumbent on a member of NSPE to endeavor to live up to its provisions. This applies to all pertinent sections of the Code.

Use Of Disadvantaged Firm After Learning Of Impropriety

NSPE Opinions of the Board of Ethical Review, Case 92-9

Facts:

Engineer A is a principal in a large consulting engineering firm specializing in civil and structural engineering. Engineer A's firm does a large percentage of its engineering work for public agencies at the state, federal and local level. Engineer A is frequently encouraged by representatives of those agencies to consider retaining the services of small, minority, or women-owned design firms as sub-consultants to the firm, particularly on publicly funded projects.

For about a year, Engineer A's firm has retained the services of Engineer B's firm, a disadvantaged firm of a type described above, on several public and private projects. Engineer A's firm has gotten a good deal of public relations benefit as a result of its retention of Engineer B's firm, particularly among its public and private clients. The work of Engineer B's firm is adequate but not of high quality. In addition, Engineer B suddenly began charging Engineer A much higher charges and fees in recent months, particularly after an article appeared in a local publication that was very complementary of Engineer A's efforts to retain disadvantaged firms.

Question:

What would be the proper action for Engineer A to take under the circumstances?

References:

Code of Ethics Preamble - Engineering is an important and learned profession. The members of the profession recognize that their work has a direct and vital impact on the quality of life for all people. Accordingly, the services provided by engineers require honesty, impartiality, fairness and equity, and must be dedicated to the protection of the public health, safety and welfare. In the practice of their profession, engineers must perform under a standard of professional behavior which requires adherence to the highest principles of ethical conduct on behalf of the public, clients, employers and the profession.

Code of Ethics Section II.2.a. - Engineers shall undertake assignments only when qualified by education or experience in the specific technical fields involved.

Code of Ethics Section III.6. - Engineers shall uphold the principle of appropriate and adequate compensation for those engaged in engineering work.

Discussion:

Over the past several years a significant amount of socio-economic legislation and regulation has been enacted at the federal, state and local levels to promote the retention of businesses that had been heretofore under-represented in the procurement process. As a result, many engineering firms have been encouraged both by public and private clients to establish goals to retain qualified employees and consultants representative of such under-represented groups.

This Board has never had occasion to examine a case in the context of such a program. As a general proposition, we believe the Code of Ethics is generally supportive of the establishment of voluntary programs that provide engineers with the opportunity to be of constructive service in community affairs and to work for its advancement and well-being. We should also note that many governmental and private procurement procedures take into account such factors consistent with their procurement requirements and standards.

Having made these general observations, we turn to the case before us. It appears that while the philosophy of establishing voluntary targets or goals for the retention of disadvantaged firms is not inconsistent with the objective of the Code of Ethics, we believe that the continued retention of a firm that is abusing its relationship with its client may be at odds with the intent of the Code.

As noted in BER Case 75-3, which involved the question of whether it was ethical for an engineer to joint venture with another engineer that had earlier been publicly reprimanded for an ethics violation, the Board concluded that in order for the engineer to ethically engage in the joint venture, the engineer must maintain a careful scrutiny of the operation of the firm of the other engineer to assure itself to the extent possible that further unethical conduct will not develop during and with respect to the joint venture.

Obviously, the facts in BER Case 75-3 were somewhat different because there the ethical violation had occurred at an earlier time under different circumstances and the question for the Board related to a future association with the unethical firm. However, we believe that a logical extension of that case should be that if an engineer's scrutiny of the operation of the firm reveals improper action, the engineer has an ethical obligation to disassociate with that firm in a manner that would not be prejudicial to his client.

We would also note BER Case 78-2 to reinforce our earlier point regarding the Code's traditional concern and application to larger societal interests and affairs.

Specifically with respect to the particular issue in the instant case regarding Engineer B's unjustified escalation of his firm's fees and charges, we would note the discussion in BER Case 77-3. There the Board noted in the context of a fee dispute between two engineers that Section III.6. is merely intended as generally descriptive and cannot be specifically defined or stated in any all inclusive manner.

As the Board noted in BER Case 69-11, the key to avoiding misunderstanding in this area is through careful negotiation and discussion and through a "give-and take" procedure.

In the context of the present case, we believe this type of negotiations was lacking as it appears under the facts, Engineer B unilaterally imposed an escalation of his firm's fees and charges. Instead, Engineer B had an obligation to negotiate any future increases in his fees and charges with Engineer A's firm.

We would note that there may also be contractual issues involved in this case, but we do not pass judgment as to any legal questions that may have arisen as a result of Engineer B's conduct.

Conclusion:

Engineer A has an obligation to discuss and negotiate with Engineer B in an effort to improve the quality and relative value of Engineer B's services. If a mutual agreement cannot be reached concerning the terms and conditions of service, Engineer A should terminate his relationship with Engineer B and in the future continue to strive to retain qualified employees and consultants representative of such under-represented groups.

Board of Ethical Review

William A. Cox, Jr., P.E.
William W. Middleton, P.E.
William E. Norris, P.E.
William F. Rauch, Jr., P.E.
Jimmy H. Smith, P.E.
Otto A. Tennant, P.E.
Robert L. Nichols, P.E., Chairman

Note: In regard to the question of application of the Code to corporations vis-a-vis real persons, business form or type should not negate nor influence conformance of individuals to the Code. The Code deals with professional services, which services must be performed by real persons. Real persons in turn establish and implement policies within business structures. The Code is clearly written to apply to the Engineer and it is incumbent on a member of NSPE to endeavor to live up to its provisions. This applies to all pertinent sections of the Code. This opinion is for educational purposes only and should not be construed as expressing any opinion on the ethics of specific individuals.

Padding the Budget Case — Let Me Be Honest...So You Can Cut Me

Public Administration at the University of Arkansas - Revised 11/7/96
Used by permission. Contact Dr. William Miller, University of Arkansas

Context

John Savage is the director of the street department in a large metropolitan city. He is new at the job. A recent graduate of a well-known public administration masters program, he is enthusiastic and excited about making the streets in his city more user friendly.

One of his first tasks is to work on a budget request for the next fiscal year. This is not a task he looks forward to. He calls his administrative staff together to make some crucial decisions regarding the next year.

The Dilemma

After many hours of hard work, John and the staff arrive at what they feel is a good budget. There is no fat and yet there are increases for what John believes are vital programs which will improve urban transportation. The budget has also included estimated increases in materials and other unknowns. John sits back and smiles. He says, "So this is it! Two and one half million dollars."

After a moment of silence, Ed Seitz, a twenty year veteran in the department, speaks up. "John, you can't submit the budget we just worked out. You have to go through and raise portions of it enough to allow for the city budget director and the city council to cut some off. If you leave it as it is, we will not receive enough to do our job."

John is flabbergasted. His staff wants him to pad the budget. They want him to submit for more than the department needs in order to get what they need. What a crazy ritual! It is dishonest and he won't do it!

The Decision

Overnight John settles down. He considers the valuable projects that may get cut in the name of honesty. He can't let that happen. The mission of his agency is too important to let his personal needs and values interfere. He will not propose a budget of two and a half million. He will submit for two and three quarter million and get what he really needs.

What do you think of this decision?

Is this practice a common part of our "culture?" (See the end of "Perspectives on Customs in Different Cultures" on page 100)

Credit For Engineering Work — Design Competition

NSPE Opinions of the Board of Ethical Review — Case 92-1

Facts:

Engineer A is retained by a city to design a bridge as part of an elevated highway system. Engineer A then retains the services of Engineer B, a structural engineer with expertise in horizontal geometry, superstructure design and elevations to perform certain aspects of the design services. Engineer B designs the bridge's three curved welded plate girder spans which were critical elements of the bridge design.

Several months following completion of the bridge, Engineer A enters the bridge design into a national organization's bridge design competition. The bridge design wins a prize. However, the entry fails to credit Engineer B for his part of the design.

Question:

Was it ethical for Engineer A to fail to give credit to Engineer B for his part in the design?

References:

Code of Ethics Section I.3. - Issue public statements only in an objective and truthful manner.

Code of Ethics Section II.3.a. - Engineers shall be objective and truthful in professional reports, statements or testimony. They shall include all relevant and pertinent information in such reports, statements or testimony.

Code of Ethics Section III.3. - Engineers shall avoid all conduct or practice which is likely to discredit the profession or deceive the public.

Code of Ethics Section III.5.a. - Engineers shall not accept financial or other considerations, including free engineering designs, from material or equipment suppliers for specifying their product.

Code of Ethics Section III.10.a. - Engineers shall, whenever possible, name the person or persons who may be individually responsible for designs, inventions, writings, or other accomplishments.

Discussion:

Basic to engineering ethics is the responsibility to issue statements in an objective and truthful manner (Section I.3.) The concept of providing credit for engineering work to those to whom credit is due is fundamental to that responsibility. This is particularly the case where an engineer retains the services of other individuals because the engineer may not possess the education, experience and expertise to perform the required services for a client. The engineer has an obligation to the client to make this information known (Section II.3.a.) As noted in BER Case 71-1, the principle is not only fair and in the best interests of the profession, but it also recognizes that the professional engineer must assume personal responsibility for his decisions and actions.

In BER Case 71-1, a city department of public works retained Firm A to prepare plans and specifications for a water extension project. Engineer B, chief engineer of the department having authority in such matters, instructed Firm A to submit its plans and specifications without showing the name of the firm on the cover sheets but permitted the firm to show the name of the firm on the working drawings. It was also the policy of the department not to show the name of the design firm in the advertisements for construction bids, in fact, the advertisements stated "plans and specifications as prepared by the city department of public works." The Board noted that the policy of the department is, at best, rather unusual in normal engineering practices and relationships between retained design firms and clients. The Board surmised on the basis of the submitted facts that the department policy was intended to reflect the idea that the plans and specifications when put out to construction bid are those of the department. In concluding that Engineer B acted unethically in adopting and implementing a policy which prohibited the identification of the design firm on the cover sheets for plans and specification, the Board noted that Engineer B, in carrying out the department policy, denied credit to Firm A for its work. The Code of Ethics Section III.10.a. states that engineers shall, whenever possible, name the person or persons who may be individually responsible for designs, inventions, writings, or other accomplishments. The Board concluded that under the circumstances, it was possible for Engineer B to name the persons responsible for the design.

While each individual case must be understood based upon the particular facts involved, we believe that Engineer A had an ethical obligation to his client, to Engineer B as well as to the public to take reasonable steps to identify all parties responsible for the design of the bridge.

Conclusion:

It was unethical for Engineer A to fail to give credit to Engineer B for his part in the design.

Board of Ethical Review

William A. Cox, Jr., P.E.
William W. Middleton, P.E.
William E. Norris, P.E.
William F. Rauch, Jr., P.E.
Jimmy H. Smith, P.E.
Otto A. Tennant, P.E.
Robert L. Nichols, P.E., Chairman

Note: This opinion is for educational purposes only and should not be construed as expressing any opinion on the ethics of specific individuals.

Advertising — Misstating Credentials

NSPE Opinions of the Board of Ethical Review — Case 92-2

Facts:

Engineer A is an EIT who is employed by a medium-sized consulting engineering firm in a small city. Engineer A has a degree in mechanical engineering and has performed services almost exclusively in the field of mechanical engineering. Engineer A learns that the firm has begun a marketing campaign and in its literature lists Engineer A as an electrical engineer. There are other electrical engineers in the firm. Engineer A alerts the marketing director, also an engineer, to the error in the promotional literature, and the marketing director indicates that the error will be corrected. However, after a period of six months, the error is not corrected.

Question:

Under the circumstances, what actions, if any, should Engineer A take?

References:

Code of Ethics Section I.3. - Issue public statements only in an objective and truthful manner.

Code of Ethics Section I.5. - Avoid deceptive acts in the solicitation of professional employment.

Code of Ethics Section II.3. - Engineers shall issue public statements only in an objective and truthful manner.

Code of Ethics Section II.5.a - Engineers shall not falsify or permit misrepresentation of their, or their associates', academic or professional qualifications. They shall not misrepresent or exaggerate their degree of responsibility in or for the subject matter of prior assignments. Brochures or other presentations incident to the solicitation of employment shall not misrepresent pertinent facts concerning employers, employees, associates, joint venturers or past accomplishments with the intent and purpose of enhancing their qualifications and their work.

Code of Ethics Section III.3.a. - Engineers shall avoid the use of statements containing a material misrepresentation of fact or omitting a material fact necessary to keep statements from being misleading or intended or likely to create an unjustified expectation; statements containing prediction of future success; statements containing an opinion as to the quality of the Engineers' services; or statements intended or likely to attract clients by the use of showmanship, puffery, or self-laudation, including the use of slogans, jingles, or sensational language or format.

Discussion:

Engineers, as with all professionals, are admonished to perform professional services only in areas of their competence. As part of the engineer's relations with his client, employer and the general public the engineer has a fundamental obligation to issue public statements in an objective and truthful manner (Code Section I.3.) In addition, where the engineer is seeking professional engagements, the engineer must always take all reasonable steps to avoid misleading and deceptive acts in the solicitation of professional employment.

Taken together, these three basic principles suggest that in making offers of professional services to clients or potential clients, as well as in communications with employers, engineers have a basic ethical responsibility to take appropriate steps to ensure that such offers avoid misleading, deceptive and untruthful language. While we freely acknowledge that this is sometimes a difficult line to draw, we believe that the best method of determining whether the line is violated is by reviewing each individual matter on a case-by-case basis. Over the years, the Board of Ethical Review has had the opportunity to review at least two similar cases as the one presently before it. In BER Case 83-1, the Board considered the ethical conduct of an engineer who, as a principal in an engineering firm, terminated an engineer but continued to distribute a previously printed brochure listing the terminated engineer, who stayed on for a period of time as one of his "key employees", and continued to use a previously printed brochure with the terminated engineer's name in it well after the terminated engineer left the firm. The Board considered whether it was the "intent and purpose" of the engineer to "enhance the firm's qualifications and work" by including the terminated engineer's name in the promotional brochure after the terminated engineer left the firm. The Board found that the facts presented in the case demonstrated that the engineer acted with "intent and purpose" in distributing the misleading brochure. The engineer was aware of the impending termination of the terminated engineer. The engineer had distributed the brochure while the terminated engineer was still employed but had been given notice of the termination. The Board noted that this could easily mislead potential clients into believing that the terminated engineer, highlighted as a "key employee", would be available in the firm for consultation on future projects. Moreover, since the engineer distributed the brochure after the terminated engineer left the firm, the Board concluded that it would be a clear misrepresentation of a pertinent fact with the intent to enhance the firm's qualifications and as such constituted a violation of the Code.

More recently, in BER Case 90-4, a case involving similar issues, an engineer, one of a few engineers in a medium-sized firm with expertise in hydrology, gave two weeks notice of intent to move to another firm. Nevertheless, a principal in the firm continued to distribute a brochure identifying the engineer as an employee of the firm and listed the departing engineer on the firm resume. In finding it was not unethical for the principal to continue to represent the engineer as an employee of the firm under the circumstances described, we distinguished BER Case 90-4 from BER Case 83-1. The Board noted that in BER Case 83-1, the terminated engineer was highlighted in the firm's promotional brochure as a "key employee".

Under the totality of the facts and circumstances of the case, it was apparent that the engineer's continued inclusion of the terminated engineer's name in the brochure constituted an overt misrepresentation of an important fact concerning the overall make-up of the firm. However, in BER Case 90-4, there was no suggestion that any of the brochures or other promotional material describe the departing engineer as a "key employee" in the firm. Nor was there any effort or attempt on the part of the firm to highlight the activities or achievements of the departing engineer in the field of hydrology.

While the facts revealed that the departing engineer was one of the few engineers in the firm with expertise in the field of hydrology, the departing engineer was not the only engineer in the firm who possessed such expertise. In addition, it appeared that this area of practice did not constitute a significant portion of the services provided by the firm; therefore it seemed to the Board that the inclusion of the name of the departing engineer in the firm's brochure and resume did not constitute a misrepresentation of "pertinent facts".

In addition, in BER Case 90-4, the Board was reluctant to conclude that the actions of the firm and the engineer in including the name of the departing engineer in the firm's brochure and resume demonstrated an intent to "enhance the firm's qualifications and work". Under the facts, there did not appear to be the same motive on the part of the principal engineer or the firm to act in a manner which will materially benefit the firm as was so in BER Case 83-1. In BER Case 90-4, the action by the firm and engineer appear more in the manner of an oversight without malice or intent. While the Board has in the past found that unethical conduct occurred in the absence of intentional actions, the Board did not consider the facts of BER Case 90-4 to be of a nature to make such a finding.

Significantly, in BER Case 90-4, the Board noted that it was in no way condoning the failure of an engineering firm to correct material (brochures, resumes, etc.) which might have the unintentional effect of misleading clients, potential clients and others. While the Board recognized the realities of firm practice and the logistical problems involved in marketing and promotion, the Board noted it was important for firms to take actions to expeditiously correct any false impressions which might exist. In this regard, the Board indicated that engineering firms that use printed material as part of their marketing efforts should take reasonable steps to assure that such written material is as accurate and up-to-date as possible. In the case of marketing brochures and other similar materials, errata sheets, cover letters, strike-outs and, if necessary, reprints should be employed within a reasonable period of time in order to correct inaccuracies, particularly where a firm has reason to believe that a misunderstanding might occur. Firms that fail to take such measures run the risk of breaching ethical behavior.

We believe that the instant case presents a clear illustration of the last point raised earlier by the Board in BER Case 90-4. Under the facts, the firm's marketing director has been informed by the engineer in question that the firm's marketing brochure contains inaccurate information that could mislead and deceive a client or potential client. Under the reasoning in BER Case 90-4, the marketing director has an ethical obligation to take expeditious action to correct the error.

Again, while we recognize basic logistical problems involved in distributing, correcting and reprinting brochures and other promotional material, we believe the marketing director, a professional engineer, has an ethical obligation both to the clients and potential clients as well as to Engineer A to expeditiously correct the misimpression which may have been created. As we noted in BER Case 90-4, this could take the form of a simple and inexpensive errata sheet inserted into the brochure.

In closing, while we believe Engineer A has taken an appropriate step in alerting the marketing director to the error in the brochure, we believe that after a period of six months during which the error was not corrected, Engineer A should raise the issue with a principal in the firm. While there is no indication that what has occurred under these facts is anything other than a negligent oversight, continued inaction by the firm in light of actual knowledge of the error could easily raise questions of improper and unethical conduct.

Conclusion:

Engineer A should raise the issue of the error with a principal in the firm and note the appropriate requirements under the state board's rules of professional conduct in writing.

Board of Ethical Review

William A. Cox, Jr., P.E.
William W. Middleton, P.E.
William E. Norris, P.E.
William F. Rauch, Jr., P.E.
Jimmy H. Smith, P.E.
Otto A. Tennant, P.E.
Robert L. Nichols, P.E., Chairman

Note: In regard to the question of application of the Code to corporations vis-a-vis real persons, business form or type should not negate nor influence conformance of individuals to the Code. The Code deals with professional services, which services must be performed by real persons. Real persons in turn establish and implement policies within business structures. The Code is clearly written to apply to the Engineer and it is incumbent on a member of NSPE to endeavor to live up to its provisions. This applies to all pertinent sections of the Code.

Participation in Production of Unsafe Equipment

NSPE Opinions of the Board of Ethical Review - Case No. 65-12

Facts:

Engineers of Company A prepared plans and specifications for machinery to be used in a manufacturing process and Company A turned them over to Company B for production. The engineers of Company B in reviewing the plans and specifications came to the conclusion that they included certain miscalculations and technical deficiencies of a nature that the final product might be unsuitable for the purposes of the ultimate users, and that the equipment, if built according to the original plans and specifications, might endanger the lives of persons in proximity to it. The engineers of Company B called the matter to the attention of appropriate officials of their employer who, in turn, advised Company A of the concern expressed by the engineers of Company B. Company A replied that its engineers felt that the design and specifications for the equipment were adequate and safe and that Company B should proceed to build the equipment as designed and specified. The officials of Company B instructed its engineers to proceed with the work.

Question:

What are the ethical obligations of the engineers of Company B under the stated circumstances?

References:

Code of Ethics Section 1c - He will advise his client or employer when he believes a project will not be successful.

Code of Ethics Section 2 - The Engineer will have proper regard for the safety, health, and welfare of the public in the performance of his professional duties. If his engineering judgment is overruled by non-technical authority, he will clearly point out the consequences. He will notify the proper authority of any observed conditions which endanger public safety and health.

Code of Ethics Section 2c - He will not complete, sign, or seal plans and/or specifications that are not of a design safe to the public health and welfare and in conformity with accepted engineering standards. If the client or employer insists on such unprofessional conduct, he shall notify the proper authorities and withdraw from further service on the project.

Discussion:

The engineers of Company B fulfilled their obligation under Section 1c of the Code by notifying their employer that they did not believe the project would be successful as designed by the engineers of Company A. They also met the requirements of Section 2 in pointing out the consequences to be expected from proceeding under the original plans and specifications. By their actions the engineers of Company B regarded their "duty to the public welfare as paramount," as required by Section 2a.

The further and more difficult question, however, is whether the engineers of Company B are required or ethically permitted to refuse to proceed with the production on the basis of plans and specifications which they continue to regard as unsafe.

In BER Case 61-10, we held that engineers assigned to the redesign of a commercial product of lower quality should not question the company's business decision, but had an obligation to point out any safety hazards in the new design. In that case, however, the redesign of the product involved only a question of a lower quality product and did not raise the problem of the product endangering public health or safety.

Section 2c of the Code is specific in holding that engineers will not complete, sign, or seal plans and/or specifications that are not of a design safe to the public health and welfare. In this situation the engineers of Company B have not been requested, or required, to "sign, or seal plans and/or specifications" at all. This has been done by the engineers of Company A. A literal construction of the Code language may, therefore, indicate that the engineers of Company B may ethically proceed with their role in the production process. But we think that this is too narrow a reading of the Code and that the purpose and force of Section 2c is that the engineer will not participate in any way in engineering operations which endanger the public health and safety.

The last sentence of Section 2c is likewise clear in requiring that the engineers not only notify proper authority of the dangers which they believe to exist, but that they also "withdraw from further service on the project." This mandate applies to engineers serving clients or employers.

Where, as in this case, there is an apparent honest difference of opinion as to the safety features of the machinery between the engineers of Company A and the engineers of Company B it would be appropriate for the question to be referred to an impartial body of experts, such as a technical engineering society in the particular field of practice, for an independent determination.

So long as the engineers of Company B hold to their opinion that the machinery as originally designed and specified would be unsafe to the public they should refuse to participate in its processing or production under the mandate of Section 2c. While such refusal to comply with the instruction of their employer may cause a most difficult situation, or even lead to the loss of employment, we must conclude that these considerations are subordinate to the requirements of the Code.

<u>**Conclusion**</u>:

The ethical obligations of the engineers of Company B are to notify their employer of possible dangers to the public safety and seek to have the design and specifications altered to make the machinery safe in their opinion; if the opinions cannot be reconciled they should propose submission of the problem to an independent and impartial body of experts: unless and until the engineers of Company B are satisfied that the machinery would not jeopardize the public safety they should refuse to participate in any engineering activity connected with the project.

Board of Ethical Review

T.C. Cooke, P.E.
James Hallett, P.E.
W.S. Nelson, P.E.
N.O. Saulter, P.E.
K.F. Wendt, P.E.
A.C. Kirkwood, P.E., Chairman

*Note—This opinion is based on data submitted to the Board of Ethical Review and does not necessarily represent all of the pertinent facts when applied to a specific case. This opinion is for educational purposes only and should not be construed as expressing any opinion on the ethics of specific individuals.

This opinion may be reprinted without further permission, provided that this statement is included before or after the text of the case.

Participation in Professional and Technical Societies - Ethical Duty of Employer and Employee

NSPE Opinions of the Board of Ethical Review - Case No. 82-7

Facts:

Engineer A has been employed by an organization for more than 20 years. During his early years of employment he was encouraged by his superiors to join and participate in the activities of both a technical society and a professional society. Within those societies, Engineer A held several board and committee positions, of which, entry into key positions was approved by his superiors. He presently holds a committee position.

Engineer A's immediate supervisor, Engineer B, opposes Engineer A's participation in activities of his professional society on any other than annual leave basis, although existing organization rules encourage the use of excused leave for such purposes. It is Engineer B's view that such participation does not result in "benefits for the employer"; he feels that such participation does not constitute "employer training." Engineer B has refused to permit written communications from Engineer A asking for administrative leave to attend professional society meetings to go through Engineer B to higher leave personnel.

When summoned by the chief executive officer (CEO) on another matter, Engineer A took the opportunity to ask his opinion of attendance and participation in technical and professional society meetings by his engineers. The CEO reaffirmed the organization policy.

When Engineer A prepared a travel request to go through his superior, Engineer B, to the CEO, Engineer B refused to forward the travel request and told Engineer A that he did not appreciate Engineer's A going over his head to discuss attendance and participation in technical and professional societies with his superior.

Questions:

1. Was it ethical for Engineer A to discuss attendance and participation in technical and professional societies with the CEO without first notifying his superior?

2. Was it ethical for Engineer B to hinder Engineer A's efforts to obtain excused leave in order to attend technical and professional society meetings?

References:

Code of Ethics Section I.4. - Engineers, in the fulfillment of their professional duties, shall ... Act in professional matters for each employer or client as faithful agents or trustees.

Code of Ethics Section III.1.f. - Engineers shall avoid any act tending to promote their own interest at the expense of dignity and integrity of the profession.

Code of Ethics Section III.11.a. - Engineers shall encourage engineering employees' efforts to improve their education.

Code of Ethics Section III.11.b. - Engineers shall encourage engineering employees to attend and present papers at professional and technical society meetings.

Discussion:

The two questions posed are best addressed by reference to four Code sections. Section I.4. requires engineers to act professionally and faithfully in dealing with their employer. Section III.1.f. cautions engineers to avoid any act tending to promote their own interests at the expense of the profession. Sections III.11.a. and b. admonish engineers to encourage their engineer employees to improve their knowledge through education and in particular through attendance and participation in professional and technical society meetings. With that background we will proceed to evaluate the actions of Engineers A and B.

It is possible for this Board to review the actions of Engineer A and to conclude that as a factual matter he was disloyal and promoting his own self-interests by going beyond his immediate superior to obtain permission to attend and participate in professional and technical society activities. However, if we were to do so, we would be ignoring the basic underlying philosophy of engineering—professionalism. The essence of professionalism is the unique service a practitioner renders to a client by virtue of having developed special capabilities. In line with that view we believe an employer of engineers has an obligation to treat engineers as professional individuals. It is incumbent on the employer of any employed professional engineer to create an environment conducive to the continued development of professional capabilities. Of course it is the professional obligation of the practitioner to expend some time and effort to continuous expansion of his or her knowledge and capabilities. Such expansion of knowledge may be gained in a variety of ways. We believe one of those ways is by participating in the activities of a professional society. In particular, participation in the committee work of a professional society allows the practicing engineer the opportunity to gain a greater understanding of the new trends and advances in his profession, permits him to interact and exchange views and insights with other engineers, and provides the engineer with a better perspective as to the role of the engineer in society.

We are of the view that a fundamental issue was at stake when Engineer A discussed attendance and participation in technical and professional societies with the CEO. What was at stake was Engineers A's professional integrity and his obligation to expand his knowledge and capabilities.

In addition, we note that it was the general policy of the employer to encourage Engineer A's participation in the activities of technical and professional activities. It was only Engineer A's immediate supervisor, Engineer B, who hindered his efforts to participate. In view of those factors, we are of the view that Engineer A acted professionally and faithfully in his dealings with his employer.

Although it may have been more appropriate for Engineer A to first meet with his supervisor, Engineer B, to inform him of his intention to seek the CEO's permission to attend and participate in technical and professional organization's activities, we are not convinced that his failure to do so tended to promote his own self-interest at the expense of the dignity and integrity of the profession. Although his action might be characterized as a deception, given the intransigence of his supervisor, Engineer B, in not permitting him to communicate with his superior on the matter of participation in professional and technical society activities, one can better understand his decision to pursue this route. We find that Engineer A's failure to inform Engineer B of his intention to seek the CEO's permission to attend and participate in technical and professional society activities did not promote his own interest at the expense of the profession.

As for Engineer B, we are of the opinion that his opposition is neither in accord with the Code nor supported by experience. Sections III.11.a. and b. admonish engineers to encourage their employees to participate in a variety of activities in order to foster their professional growth and development. As Section III.11.b. plainly states, among these activities are professional and technical society meetings. Engineer B was of the view that Engineer A's participation in technical and professional societies did not constitute "employee training" and did not result in "benefits to the employer." Aside from the question of whether this was in fact an accurate assessment of Engineer A's society activities, there is the issue of whether standards such as "employee training" or "benefits to the employer" are the only yardsticks by which professional and technical society activities and continuing engineering education programs should be measured. We think not but leave the question for another day. It suffices to say that in the instant case, contrary to Engineer B's view, Engineer A's participation in professional and technical society meetings was of the type intended by Sections III.11.a. and b. of the Code.

We note, however, that our decision today must not be construed to mean that an engineer should as a matter of course be granted excused leave from his employment without due regard to the needs and requirements of his employer. We believe that Section I.4. mandates that an engineer must be sensitive to the needs and requirements of his employer. When an employer chooses to limit his employees' participation in technical and professional society activity because those employees' services are critical to the operation of his organization, Section I.4. requires the employee to accede to his employer's decision. Although an engineer has an obligation to further his professional growth and development, it should never be pursued in a manner that would be adverse to the interest of his employer.

Conclusions:

1. It was ethical for Engineer A to discuss attendance and participation in technical and professional societies with the CEO without first notifying his superior.

2. It was unethical for Engineer B to hinder Engineer A's efforts to obtain excused leave in order to attend technical and professional society meetings.

Board of Ethical Review

Ernest C. James, P.E.
Lawrence E. Jones, P.E.
Robert H. Perine, P.E.
James L. Polk, P.E.
J. Kent Roberts, P.E.
Alfred H. Samborn, P.E.
F. Wendell Beard, P.E., chairman

Note—This opinion is based on data submitted to the Board of Ethical Review and does not necessarily represent all of the pertinent facts when applied to a specific case. This opinion is for educational purposes only and should not be construed as expressing any opinion on the ethics of specific individuals.

This opinion may be reprinted without further permission, provided that this statement is included before or after the text of the case.

Whistleblowing

NSPE Opinions of the Board of Ethical Review - Case No. 82-5

Facts:

Engineer A is employed by a large industrial company which engages in substantial work on defense projects. Engineer A's assigned duties relate to the work of subcontractors, including review of the adequacy and acceptability of the plans for material provided by subcontractors. In the course of this work Engineer A advised his superiors by memoranda of problems he found with certain submissions of one of the subcontractors, and urged management to reject such work and require the subcontractor to correct the deficiencies he outlined. Management rejected the comments of Engineer A, particularly his proposal that the work of a particular subcontractor be redesigned because of Engineer A's claim that the subcontractor's submission represented excessive cost and time delays.

After the exchange of further memoranda between Engineer A and his management superiors, and continued disagreement between Engineer A and management on the issues he raised, management placed a critical memorandum in his personnel file, and subsequently placed him on three months' probation, with the further notation that if his job performance did not improve, he would be terminated.

Engineer A has continued to insist that his employer had an obligation to insure that subcontractors deliver equipment according to the specifications, as he interprets same, and thereby save substantial defense expenditures. He has requested ethical review and determination of the propriety of his course of action and the degree of ethical responsibility of engineers in such circumstances.

Question:

Does Engineer A have an ethical obligation, or an ethical right, to continue his efforts to secure change in the policy of his employer under these circumstances, or to report his concerns to proper authority?

References:

Code of Ethics Section II.1l.a. - Engineers shall at all times recognize that their primary obligation is to protect the safety, health, property, and welfare of the public. If their professional judgment is overruled under circumstances where the safety, health, property, or welfare of the public are endangered, they shall notify their employer or client and such other authority as may be appropriate.

Code of Ethics Section III.2.b. - Engineers shall not complete, sign, or seal plans and/or specifications that are not of a design safe to the public health and welfare and in conformity with accepted engineering standards. If the client or employer insists on such unprofessional conduct, they shall notify the proper authorities and withdraw from further service on the project.

Discussion:

In BER Case 65-12 we dealt with a situation in which a group of engineers believed that a product was unsafe, and we determined that so long as the engineers held to that view they were ethically justified in refusing to participate in the processing or production of the product in question. We recognized in that case that such action by the engineers would likely lead to loss of employment.

In BER Case 61-10 we distinguished a situation in which engineers had objected to the design of a commercial product, but which did not entail any question of public health or safety. On that basis we concluded that this was a business decision for management and did not entitle the engineers to question the decision on ethical grounds.

The Code section in point related to plans and specifications "that are not of a design safe to the public health and welfare," and ties that standard to the ethical duty of engineers to notify proper authority of the dangers and withdraw from further service on the project.

That is not quite the case before us; here the issue does not allege a danger to public health or safety, but is premised upon a claim of unsatisfactory plans and the unjustified expenditure of public funds. We could dismiss the case on the narrow ground that the Code does not apply to a claim not involving public health or safety, but we think that is too narrow a reading of the ethical duties of engineers engaged in activities having a substantial impact on defense expenditures or other substantial public expenditures that relate to "welfare" as set forth in Section III.2.b.

The situation presented here has become well known in recent years as "whistleblowing", and we note that there have been several cases evoking national interest in the defense field. As we recognized in earlier cases, if an engineer feels strongly that an employer's course of conduct is improper when related to public concerns, and if the engineer feels compelled to blow the whistle to expose the facts as he sees them, he may well have to pay the price of loss of employment. In some of the more notorious cases of recent years engineers have gone through such experiences and even if they have ultimately prevailed on legal or political grounds, the experience is not one to be undertaken lightly.

In this type of situation, we feel that the ethical duty or right of the engineer becomes a matter of personal conscience, but we are not willing to make a blanket statement that there is an ethical duty in these kinds of situations for the engineer to continue his campaign within the company, and make the issue one for public discussion. The Code only requires that the engineer withdraw from a project and report to proper authorities when the circumstances involve endangerment of the public health, safety, and welfare.

Conclusion:

Engineer A does not have an ethical obligation to continue his effort to secure a change in the policy of his employer under these circumstances, or to report his concerns to proper authority, but has an ethical right to do so as a matter of personal conscience.

Board of Ethical Review

Ernest C. James, P.E.
Lawrence E. Jones, P.E.
Robert H. Perrine, P.E.
James L. Polk, P.E.
J. Kent Roberts, P.E.
Alfred H. Samborn, P.E.
F. Wendell Beard, P.E., chairman

Note—This opinion is based on data submitted to the Board of Ethical Review and does not necessarily represent all of the pertinent facts when applied to a specific case. This opinion is for educational purposes only and should not be construed as expressing any opinion on the ethics of specific individuals.

Gifts to Foreign Officials

NSPE BER Case No. 76-6

Facts:

Richard Roe, P.E., is president and chief executive officer of an engineering firm which has done overseas assignments in various parts of the world. The firm is negotiating for a contract in a foreign country in which it has not worked previously. Roe is advised by a high-ranking government official of that country that it is established practice for those awarded contracts to make personal gifts to the governmental officials who are authorized to award the contracts, and that such practice is legal in that country. Roe is further advised that while the condition is not to be included in the contract, his failure to make the gifts will result in no further work being awarded to the firm and to expect poor cooperation in performing the first contract. He is further told that other firms have adhered to the local practice in regard to such gifts.

Question:

Would it be ethical for Roe to accept the contract and make the gifts as described?

References:

Code of Ethics Section 11b - He will not pay, or offer to pay, either directly or indirectly, any commission, political contribution, or a gift, or other consideration in order to secure work, exclusive of securing salaried positions through employment agencies.

Discussion:

On its face, the code is clear and direct to the point. There is no question under the factual situation that the gifts are a direct consideration for securing the work, and there is here no pretense about the intention of the foreign government officials.

In BER Case 60-9 we acted upon a domestic case under Rule 4 of the then-prevailing Rules of Professional Conduct, which was the same as the present. In that case we dealt with three levels of gifts, involving 1) taking employees of a public agency to luncheon or dinner a few times during the year at an average cost of $5 per person; 2) engineers of an industrial firm receiving cash gifts of from $25 to $100 from salesmen for certain products, which may be specified by the engineers; and 3) a consulting engineer giving the chief engineer of a client an automobile of the value of approximately $4000 at the completion of a project as an expression of appreciation for the cooperation of the recipient of the gift. (Realistically, these figures would now be much higher than those amounts on account of inflation since 1960.)

It is now worth repeating the language used at that time to indicate the basic principles which should govern the question:

"The question of when a gift is intended to or becomes an inducement to influence one's impartial decision, as distinguished from an expression of friendship or a social custom, has remained a perplexing one over the years. No blanket rule covering all situations has been discovered. The size of the gift is usually a material factor, but must be related to the circumstances of the gift. It would hardly be felt a token gift, such as a cigar, a desk calendar, etc., would be prohibited. It has been customary in the business world for friends and business associates to tender such tokens of recognition or appreciation, and 'picking up the tab' at a business luncheon or dinner is commonplace and well accepted in the mores of our society."

"Recognizing the difficulties inherent in passing judgment on each instance, we believe the canons and rules state, in substance, that an engineer may neither offer nor receive a gift which is intended to or will influence his independent professional judgment. The full application of this principle requires the impossible—that we read the state of mind of the donor or donee. Therefore, we must apply a criterion which reasonable men might reasonably infer from the circumstances; that the giving or acceptance of the benefit be a matter of 'good taste,' and such that among reasonable men it might not be of a nature which raises suspicions of favoritism."

"Applying these general principles to the situations at hand, we think that an occasional free luncheon or dinner, or a Christmas or birthday present when there is a personal relationship is acceptable practice. On the other hand, cash payments to those in a position to influence decisions favorable or unfavorable to the giver are not in good taste and do immediately raise a suspicion that there is an ulterior motive. Likewise, a very expensive gift has a connotation of placing the recipient in a position of obligation."

From those principles we then concluded that the practice in Situation 1 was ethically permissible, but those in situation 2 and 3 were unethical.

In the case before us we are not told the amount of proposed gifts, but we take it in context that they would be substantial. Accordingly, they would clearly be a violation of the code if offered in the United States.

The basic issue remaining is whether the flat prohibition in Section 11b applies to work in a foreign country where the laws and customs permit the gifts to government officials. It is worth noting in considering this point that in a different but related context, the NSPE Board of Directors in July 1966 adopted a so-called "When in Rome" clause to permit the submission of tenders for work in foreign countries when such is required by laws, regulations, or practices of the foreign country.

However, after further discussion and debate, the Board of Directors in January 1968 rescinded the "When in Rome" clause from the code as recommended by the Professional Engineers in Private Practice Section, which recognized a division of opinion on the policy but concluded, "... the profession should maintain a 'pure' position on competitive bidding; otherwise our opposition to competitive bidding will be chipped away, piece by piece."

The issue before us is a very current one. In recent months the press has been filled with reports of investigations of charges that certain industrial concerns have made improper gifts of large sums to foreign officials to secure contracts for their products. The defense to this activity has generally been that the companies making such gifts had "no choice," meaning that without such action they would not have been able to secure the contracts because competitors in other countries would have complied with the practice. There may be some appeal to this line of argument from a purely pragmatic standpoint, but it must of necessity fail in the final analysis.

Even though the practice may be legal and accepted in the foreign country, and even though some might argue on pragmatic grounds that United States commercial companies should "go along" to protect the jobs of employees in this country, we cannot accept it for professional services. No amount of rationalization or explanation will change the public reaction that the profession's claim of placing service before profit has been compromised by a practice which is repugnant to the basic principles of ethical behavior under the laws and customs of this country. Even if the "go along" philosophy is accepted as an exception only for foreign work, the result must be a "chipping away" of ethical standards, leading to contention that such conduct should also be accepted in the United States when and if it is argued that such is the local or area practice.

This approach is not dissimilar to the arguments advanced by those who have so recently been revealed as offering financial payments to public officials to influence the award of contracts for architect-engineer services. The rationale was "We had no choice. Others were doing it, and if we did not we would not be considered." The short answer is that there is a choice—the choice of declining to be drawn into a seamy procedure for self-gain.

We believe that the code must be read on this most basic point of honor and integrity, not only literally, but in the spirit of its purpose—to uphold the highest standards of the profession. Anything less is a rationalization which cannot stand the test of placing service to the public ahead of all other considerations.

Conclusion:
It would unethical for Roe to accept the contract and make the gifts as described.

Board of Ethical Review

William J. Deevy, P.E.
Joseph N. Littlefield, P.E.
James F. Shivler, Jr., P.E.
Robert E. Stiemke, P.E., chairman
William R. Gibbs, P.E.
Donald C. Peters, P.E.
L.W. Sprandel, P.E.

Note--This opinion is based on data submitted to the Board of Ethical Review and does not necessarily represent all the pertinent facts when applied to a specific case. This opinion is for educational purposes only and should not be construed as expressing any opinion on the ethics of specific individuals.

Public Criticism Of Bridge Safety

NSPE Opinions of the Board of Ethical Review Case No. 88-7

Facts:

Engineer A, a renowned structural engineer, is hired for a nominal sum by a large city newspaper to visit the site of a state bridge construction project, which has had a troubled history of construction delays, cost increases, and litigation primarily as a result of several well-publicized, on-site accidents. Recently the state highway department has announced the date for the opening of the bridge. State engineers have been proceeding with repairs based upon a specific schedule. Engineer A visits the bridge and performs a one-day visual observation. Her report identifies, in very general terms, potential problems and proposes additional testing and other possible engineering solutions.

Thereafter, in a series of feature articles based upon information gleaned from Engineer A's report, the newspaper alleges that the bridge has major safety problems that jeopardize its successful completion date. Allegations of misconduct and incompetence are made against the project engineers and the contractors as well as the state highway department. During an investigation by the state, Engineer A states that her report was intended merely to identify what she viewed were potential problems with the safety of the bridge and was not intended to be conclusive as to the safety of the bridge.

Question:

Was it ethical for Engineer A to agree to perform an investigation for the newspaper in the manner stated?

Reference:

Code of Ethics Section II.3.a. - Engineers shall be objective and truthful in professional reports, statements or testimony. They shall include all relevant and pertinent information in such reports, statements or testimony.

Code of Ethics Section II.3.b. - Engineers may express publicly a professional opinion on technical subjects only when that opinion is founded upon adequate knowledge of the facts and competence in the subject matter.

Code of Ethics Section II.3.c. - Engineers shall issue no statements, criticisms or arguments on technical matters which are inspired or paid for by interested parties, unless they have prefaced their comments by explicitly identifying the interested parties on whose behalf they are speaking, and by revealing the existence of any interest the engineers may have in the matters.

Code of Ethics Section III.2.a. - Engineers shall seek opportunities to be of constructive service in civic affairs and work for the advancement of the safety, health and well-being of their community.

Code of Ethics Section III.3.a. - Engineers shall avoid the use of statements containing a material misrepresentation of fact or omitting a material fact necessary to keep statements from being misleading or intended or likely to create an unjustified expectation; statements containing prediction of future success; statements containing an opinion as to the quality of the Engineers' services; or statements intended or likely to attract client by the use of showmanship, puffery, or self-laudation, including the use of slogans, jingles, or sensational language or format.

Discussion:

The technical expertise that engineers can offer in the discussion of public issues is vital to the interests of the public. We have long encouraged engineers to become active and involved in matters concerning the well-being of the public. Moreover, the NSPE Code of Ethics makes clear that engineers should "seek opportunities to be of constructive service in civic affairs and work for the advancement of the safety, health and well-being of their community." (Section III.2.a.)

Obviously, this important involvement must be appropriate to the circumstance of the situation. In situations where an engineer is being asked to provide technical expertise to the public discussion, the engineer should offer objective, truthful, and dispassionate professional advice that is pertinent and relevant to the points at issue. The engineer should only render a professional opinion publicly, when that opinion is (1) based upon adequate knowledge of the facts and circumstance involved, and (2) the engineer clearly possesses the expertise to render such an opinion. The Board has earlier visited situations in which engineers have publicly rendered professional opinions.

In BER Case 65-9, a consulting engineer who had performed the engineering work on a portion of an interstate highway to which a proposed controversial highway by-pass would connect, issued a public letter which was published in the local press, criticizing the cost estimates of the engineers of the state highway department, stating alleged disadvantages of the proposed route, and pointing out an alternative route. The newspaper story contained the full text of the letter from the consulting engineer. In deciding that it was ethical for the engineer to publicly express criticism of the proposed highway routes prepared by engineers of the state highway department, the Board stated: " . . . the whole purpose of engineering is to serve the public interest. When an engineering project has such a direct and substantial impact on the daily life of the citizenry as the location of a highway it is desirable that there be public discussion.

The Code does not preclude engineers, as citizens, from participating in such public discussion. Those engineers who have a particular qualification in the field of engineering involved may be said to even have a responsibility to present public comment and suggestions in line with the philosophy expressed in the Code."

Thereafter, in BER Case 79-2, the Board ruled that where an engineer had significant environmental concerns, it was not unethical for the engineer to criticize a town engineer and a consulting engineer with respect to findings contained in a report on a sanitary landfill for the town. Said the Board: "It is axiomatic that an engineer's primary ethical responsibility is to follow the mandate of the Code to place the public welfare over all other considerations." We noted that these issues in the public arena are subject to open public debate and resolution by appropriate public authority. Here the engineer was acting within the intent of the Code in raising his concern. We concluded by citing earlier decision BER Case 63-6 in which we noted: "There may also be honest differences of opinion among equally qualified engineers on the interpretation of the known physical facts. The Code does not prohibit public criticism; it only requires that the engineer apply due restraint in offering public criticism of the work of another engineer; BER Case 88-7 the engineering witness will avoid personalities and abuse, and will base his criticism on the engineering conclusions or application of engineering data by offering alternative conclusions or analyses."

It is clear, based upon the Code of Ethics and several interpretations of the Code by this Board that the engineer may and, indeed in some cases, must ethically provide technical judgment on a matter of public importance with the aforementioned considerations concerning expertise, adequacy of knowledge, and the avoidance of personality conflicts in mind. However, we must note that under the facts of this case, we are not merely dealing with a disinterested engineer who on her own has decided to come forward and offer her professional views.

Rather, we are dealing with an engineer who was retained by a newspaper to provide her professional opinion with the understanding that the opinion could serve as the basis for news articles concerning the safety of the bridge. This fact gives an added ethical dimension to the case and requires our additional analysis. In this regard, it is our view that as a condition of her retention by the newspaper involved, Engineer A has an ethical obligation to require that the newspaper clearly state in the articles that Engineer A had been retained for a fee by the newspaper in question to perform the one day observation of the bridge site.

We should also add that in circumstances such as here where an engineer is being retained by a newspaper to offer a professional opinion concerning a matter of public concern, the engineer must act with particular care, should exercise the utmost integrity and dignity, and should take whatever reasonable steps are necessary to enhance the probability that the engineer's professional opinions are reported completely, accurately, and not out of context.

While we recognize that there are limits to what an engineer can do in these areas, we believe that the engineer has an obligation to the public as well as to the profession to protect the integrity of her professional opinions and the manner in which those opinions are disseminated to the public.

Conclusion:

It was not unethical for Engineer A to agree to perform an investigation for the newspaper in the manner stated but Engineer A has an obligation to require the newspaper to state in the article that Engineer A had been retained for a fee by the newspaper to provide her professional opinion concerning the safety of the bridge.

Board of Ethical Review

Eugene N. Bechamps, P.E.
Robert J. Haefeli, P.E.
Robert W. Jarvis, P.E.
Lindley Manning, P.E.
Paul E. Pritzker, P.E.
Harrison Streeter, P.E.
Herbert G. Koogle, P.E., L.S., chairman

Note: This opinion is based on data submitted to the Board of Ethical Review and does not necessarily represent all of the pertinent facts when applied to a specific case. This opinion is for educational purposes only and should not be construed as expressing any opinion on the ethics of specific individuals.

Reconciling Design Drawings And Record Drawings

NSPE Opinions of the Board of Ethical Review - Case No. 00-02

Facts:

Engineer Bob prepares a set of drawings for a client for the design and construction of a building. Owner contracts with Contractor X, not an engineer, for construction, but does not retain Engineer Bob for construction phase services. Engineer Bob is paid in full for his work.

Engineer Bob's drawings are filed with town code officials and a building permit is issued. Contractor X builds the building, but does not follow Engineer Bob's design, relying upon Contractor X's own experience in construction. Following construction, Contractor X, with the assistance of Engineer Carl, prepares a set of record "as built" drawings based upon the actual construction of the building as reported by Contractor X.

Because the design and the construction drawings are not reconciled, the building official refuses to issue an occupancy permit to the Owner. Owner asks Engineer Bob to "reconcile" the original design and the record drawings. Engineer Bob, not wanting to perform additional studies, agrees to perform the "reconciliation."

Questions:

1. Was it ethical for Engineer Bob to perform the design reconciliation?
2. Was it ethical for Engineer Carl to prepare a set of record drawings based on the construction without notifying Engineer Bob?

References:

Code of Ethics Section II.1.b. - Engineers shall approve only those engineering documents which are in conformity with applicable standards.

Code of Ethics Section II.1.d. - Engineers shall not permit the use of their name or associate in business ventures with any person or firm which they believe is engaged in fraudulent or dishonest enterprise.

Code of Ethics Section II.2.b. - Engineers shall not affix their signatures to any plans or documents dealing with subject matter in which they lack competence, nor to any plan or document not prepared under their direction and control.

Code of Ethics Section II.3.b. - Engineers may express publicly technical opinions that are founded upon knowledge of the facts and competence in the subject matter.

Code of Ethics Section III.2.b. - Engineers shall not complete, sign or seal plans and/or specifications that are not in conformity with applicable engineering standards. If the client or employer insists on such unprofessional conduct, they shall notify the proper authorities and withdraw from further service on the project.

Code of Ethics Section III.7.a. - Engineers in private practice shall not review the work of another engineer for the same client, except with the knowledge of such engineer, or unless the connection of such engineer with the work has been terminated.

Code of Ethics Section III.8.a. - Engineers shall conform with state registration laws in the practice of engineering.

Discussion:

The facts in this case involve one of the most fundamental issues involved in the practice of engineering – the professional engineer's responsibility for the work which the engineer has signed and sealed.

Over the years, the NSPE Board of Ethical Review (BER) has had occasion to explore this issue in some detail. In BER Case No. 91-8, an engineer's firm was retained by a major fuel company to perform site investigations in connection with certain requirements under state and federal environmental regulations.

Under the procedures established by the engineer's firm, the site visits would be conducted by engineering technicians under direct supervision of Engineer B who would perform all observations, sampling, and preliminary report preparation. Engineering technicians would also take photographs of the sites. No professional engineers were present during the site visits. Following site visits, all pertinent information and material was presented to Engineer B, who was competent in this field. Following a careful review, Engineer B would certify that the evaluations were conducted in accordance with engineering principles.

In considering whether it was ethical for Engineer B to certify that the evaluations were conducted in accordance with engineering principles, the Board noted that the NSPE Code of Ethics is very clear concerning the requirements of engineers not to affix their signatures to any plans or documents dealing with subject matter in which the engineers lack competence, nor to any plan or document not prepared under their direction and control.

The Board concluded that it was ethical for the engineer to certify that the evaluations were conducted in accordance with engineering principles, so long as the engineer exercising direction and control performs a careful and detailed review of the material submitted by the engineer's staff and there has been full compliance with NSPE Code Section II.2.c.

In BER Case No. 86-2, an engineer was the chief engineer within a large engineering firm, and affixed his seal to some of the plans prepared by licensed engineers working under his general direction who did not affix their seals to the plans. At times, the Engineer also sealed plans prepared by unlicensed graduate engineers working under his general supervision.

Because of the size of the organization and the large number of projects being designed at any one time, the engineer found it impossible to give a detailed review or check of the design. He believed he was ethically and legally correct in not doing so because of his confidence in the ability of those he had hired and who were working under his general direction and supervision. By general direction and supervision, the engineer meant that he was involved in helping to establish the concept, the design requirements, and review elements of the design or project status as the design progressed.

The engineer was consulted about technical questions and he provided answers and direction in these matters. In evaluation of the facts and circumstances in this case, the Board focused on the language in NSPE Code Section II.2.b. relating to the obligation of engineers not to affix their signature to documents or plans...not prepared under their "direction and control." Following a careful review of the plain meaning of the terms "direction" and "control," the Board concluded that the terms have meaning which, when combined, would suggest that an engineer would be required to perform all tasks related to the preparation of the drawings, plans, and specifications in order for the engineer ethically to affix his seal.

The Board also noted at the time that the NCEES Model Law would require that an engineer must be in "responsible charge" - meaning "direct control and personal supervision of engineering work" - in order to affix his seal. After careful evaluation, the Board concluded that it would not be ethical for the engineer to seal plans that have not been prepared by him or which he has not checked and reviewed in detail.

In BER Case No. 90-6, the Board considered two separate fact situations involving the signing and sealing by an engineer of documents prepared using a CADD system. In considering the facts, the Board noted that the rendering of the Board's decision in BER Case No. 86-2 raised a considerable degree of discussion within the engineering community because to many, it appeared to be inconsistent with customary and general prevailing practices within the engineering profession and would therefore place a significant number of practitioners in conflict with the provisions of the NSPE Code.

The Board noted at the time that the NSPE Code is not a static document and must reflect and be in consonance with general prevailing practices within the engineering profession. Said the Board, "the Code must not impose an impossible or idealistic standard upon engineers, but rather must establish a benchmark of reasonable and rational methods of practice for it to maintain its credibility and adherence."

The Board determined that the conclusion in BER Case No. 86-2 should be modified to reflect actual practices which exist within engineering and not impose an impossible standard upon practice. Said the Board, "Were the Board to decide BER Case No. 86-2 today, the Board would conclude that it was not unethical for the engineer in that instance to seal plans that were not personally prepared by him as long as those plans were checked and reviewed by the engineer in some detail. The Board does not believe this represents a reversal of the Board's decision in BER Case No. 86-2, but rather a clarification, particularly for those who were troubled by the Board's discussion and conclusion in that case."

Turning to the facts in the present case, the Board is of the view that the facts and circumstances go beyond anything that would be permitted under the letter or the spirit of the NSPE Code of Ethics. The Board interprets the facts to suggest that Engineer B is being asked to adopt a design that was neither prepared by Engineer B, not under Engineer B's direct control or supervision, and does not reflect the professional judgment and intent of Engineer B. Instead, it appears that the Owner is seeking to have Engineer B seal the drawings in question to satisfy the requirements of the building official.

Unlike BER Case No. 91-8, the work in question was not performed under the responsible charge (direct control or personal supervision) of Engineer Bob. In fact, the work that was prepared by Engineer Bob was essentially ignored or rejected by Contractor X in favor of another solution chosen by the Contractor X. Since Engineer Bob was not retained for construction-phase services, Engineer Bob never had the opportunity to observe the work and provide guidance to Owner or Contractor X as to the relationship of the work to the design and construction documents prepared by Engineer Bob.

Moreover, the Board believes its earlier decisions in BER Case Nos. 86-2 and 90-6 are instructive under these facts because in both instances, the Board stressed the critical importance of the engineer being squarely involved either in the preparation of the work or being in responsible charge of the work which the engineer ultimately seals. In contrast, the facts in this case illustrate an example where an engineer is being asked to sign and seal work for which the engineer was neither in responsible charge nor which the engineer was involved in preparing.

In essence, it can be argued that the facts present the appearance that Engineer Bob's services were used by the Owner merely to gain approval for the project with no intent on the part of the Client or Contractor X to follow Engineer Bob's design intent.

Clearly, the facts and circumstances in the present case suggest a difficult situation for Engineer Bob's client. It is not entirely clear under the facts whether Owner knew or chose to accept Contractor X's decision to ignore Engineer Bob's designs. In unilaterally altering Engineer Bob's design, Contractor X may have engaged in the unlicensed practice of engineering.

However, since Engineer Bob's design was approved by the building official, and Contractor X's approach was at variance with the approved design, Owner may now find that Contractor X's approach will result in additional design and/or construction costs to obtain building official acceptance of the building.

Clearly there is a lesson here for clients that fail to appreciate the importance of publicly approved design drawings. There appears to be substantial reason that the structure might not be approvable as built since Contractor X is not an engineer.

Conclusions:

1 It was not ethical for Engineer Bob to reconcile his original design documents without extensive investigation to assure that all original design intent was followed.

2. The Owner is the ultimate client, therefore, it was not ethical for Engineer Carl to prepare a set of record drawings based on the construction without notifying Engineer Bob. Moreover, there is a possibility that Engineer Carl was aiding and abetting the unlicensed practice of engineering.

Board of Ethical Review

Lorry T. Bannes, P.E.
John W. Gregorits, P.E.
Louis L. Guy, Jr., P.E.
William J. Lhota, P.E.
Paul E. Pritzker, P.E.
Harold E. Williamson, P.E.
E. D. "Dave" Dorchester, P.E., Chairman

Note: This opinion is based on data submitted to the Board of Ethical Review and does not necessarily represent all of the pertinent facts when applied to a specific case. This opinion is for educational purposes only and should not be construed as expressing any opinion on the ethics of specific individuals.

Review By Engineer Of Work Of Design Engineer For Client

NSPE Opinions of the Board of Ethical Review - Case No. 00-12

Facts:

Customer X (a cellular phone company) asks Corporation Y (a tower manufacturer) to design and manufacture a 300' antenna tower. Engineer Adam, as an employee of Corporation Y, performs a structural design of the tower and provides signed/sealed drawings of the design.

Customer X wants to make sure it receives a good product, so it engages Corporation Z, a tower manufacturer that competes with Corporation Y, to analyze Corporation Y's tower design and manufacture. Engineer Bob, as an employee of Corporation Z, performs a structural analysis of Corporation Y's design and puts together a report of his findings. Engineer Bob then signs and seals his own report and submits it to Customer X.

Question:

Was it ethical for Engineer Bob to perform a structural analysis of Corporation Y's design and put together a report of his findings?

References:

Code of Ethics Section III.1.f. - Engineers shall not promote their own interest at the expense of the dignity and integrity of the profession.

Code of Ethics Section III.7 - Engineers shall not attempt to injure, maliciously or falsely, directly or indirectly, the professional reputation, prospects, practice or employment of other engineers. Engineers who believe others are guilty of unethical or illegal practice shall present such information to the proper authority for action.

Code of Ethics Section III.7.b. - Engineers in governmental, industrial or educational employ are entitled to review and evaluate the work of other engineers when so required by their employment duties.

Code of Ethics Section III.7.c. - Engineers in sales or industrial employ are entitled to make engineering comparisons of represented products with products of other suppliers.

Discussion:

Review of one engineer's work by another engineer has been the subject of previous Board of Ethical Review Opinions.

In BER Case No. 79-7, Engineer A was retained by the prime professional engineer to provide mechanical and electrical engineering services for a large housing project. The project was completed and occupied four years later, and Engineer A was fully paid for his services.

Approximately seven years after the original occupancy, ownership of the facility changed. The new owner informed Engineer A he had retained Engineer B to make an engineering inspection of the facility, and there were problems associated with the wiring.

At the owner's request, a joint inspection of the wiring was made by the two engineers and the city wiring inspector. The inspection did not reveal any defects in the wiring. The owner advised Engineer B of his complaint concerning the plumbing and heating systems. Engineer B thereafter conducted a further study and filed a report with the owner.

The report noted there was no problem with the design of the plumbing system, but concluded there were design inadequacies in the original sizing of the equipment for hot water and heating.

Engineer B recommended the installation of equipment of higher capacity. Engineer A thereafter filed a complaint with the state engineering registration board alleging that Engineer B had acted improperly in that the report was not objective and did not include all pertinent information, and further alleged that the actions of Engineer B were self-serving at the expense of the dignity and reputation of Engineer A.

Engineer A requested the registration board to find Engineer B guilty of "misconduct" in that Engineer B had obtained employment by a questionable method of criticizing Engineer A without his knowledge. In ruling that it was not unethical for Engineer B to take the assignment and render the report to the owner, the Board noted that it was apparent that Engineer A knew that Engineer B had been retained to make an engineer's inspection of the facility and that the resulting evaluation would necessarily entail a review of the original designs. Also, it is equally clear that the connection of Engineer A with the project had been terminated some years earlier.

The Board also noted that the purpose of the language in the code relating to reviewing the work of another engineer is intended to provide the engineer whose work is being reviewed by another engineer the opportunity to submit his comments or explanation for his technical decisions, thereby enabling the reviewing engineer to have the benefit of a fuller understanding of the technical considerations in the original design in framing his comments or suggestions for the ultimate benefit of the client. (See BER Case Nos. 68-6 and 68-11).

On the basis of the facts in the present case, there does not appear to be any indication that Engineer Bob had undertaken his review and subsequent report with the intent to injure the professional reputation or practice of Engineer Adam.

It is therefore the Board's view that Engineer Bob did not have any obligation to first discuss this matter with Engineer Adam to obtain some understanding of the reasons and justifications for Engineer Adam's initial design. Under the facts, Engineer Adam and his company, Corporation Y, had a professional relationship with Customer X. However, neither Engineer Adam nor Corporation Y had any ethical right to know what Customer X was planning or did consult with Engineer Bob. If indeed, Engineer Bob concluded that some changes were needed in the equipment originally specified, Engineer Bob was free to include that information in the report.

As a matter of ethics, Engineer Bob did not have a professional obligation to consult with Engineer Adam concerning the technical issues being addressed by Engineer Bob. NSPE Code of Ethics Sections III.7.b. and III.7.c. make it clear that industrial employees are free to evaluate the designs of other engineers when so required by their employment duties. Any other determination under the NSPE Code would, in fact, be contrary to the interests of the client and the language of the NSPE Code of Ethics.

Conclusion:
It was ethical for Engineer Bob (an employee of a tower design and manufacturing company) to perform a structural analysis of Engineer Adam's (an employee of another tower design and manufacturing company) design and put together a report of his findings without first consulting with Engineer Adam.

Board Of Ethical Review

Lorry T. Bannes, P.E.
John W. Gregorits, P.E.
Louis L. Guy, Jr., P.E.
William J. Lhota, P.E.
Paul E. Pritzker, P.E.
Harold E. Williamson, P.E.
E. D. "Dave" Dorchester, P.E., Chairman

Note: This opinion is based on data submitted to the Board of Ethical Review and does not necessarily represent all of the pertinent facts when applied to a specific case. This opinion is for educational purposes only and should not be construed as expressing any opinion on the ethics of specific individuals.

Brokerage Of Engineering Services - Building Inspection Services

NSPE Opinions of the Board of Ethical Review — BER Case 92-3

Facts:

Engineer Adam performs building inspection services. Engineer Adam is contacted by IJK, Inc., a firm that refers companies to professional engineers that perform building inspection services. IJK, Inc. and similar companies are involved in assisting relocating employees in the sale and purchase of residences. Typically IJK, Inc. makes contact with the client, takes an order for a job, and passes the order on to the professional engineer available in the geographic area.

Engineer Adam performs the services, prepares a report and submits the report to IJK, Inc. Engineer Adam has learned that IJK, Inc. has occasionally made modifications to the report without consulting with the engineer. Engineer Adam invoices IJK, Inc. for his services at half what he would normally charge to another client for the same services. IJK, Inc. invoices the client for its services, twice the amount that is charged by Engineer Adam, a fact later learned by Engineer Adam. IJK, Inc. has no exclusive contractual or business relationship with Engineer Adam and IJK, Inc. possesses no engineering expertise.

Questions:

1. Is it ethical for Engineer Adam to continue association with the referral firm after he learns that IJK, Inc. has a history of changing reports?
2. Is it ethical for Engineer Adam to continue association with the referral firm after learning that IJK, Inc. is indicating a fee for Engineer Adam's services to the IJK, Inc. client which is different from the fee charged by Engineer Adam?

References:

Code of Ethics Section II.1.d. - Engineers shall not permit the use of their name or firm name nor associate in business ventures with any person or firm which they have reason to believe is engaging in fraudulent or dishonest business or professional practices.

Code of Ethics Section II.5.b. - Engineers shall not offer, give, solicit or receive, either directly or indirectly, any political contribution in an amount intended to influence the award of a contract by public authority, or which may be reasonably construed by the public of having the effect or intent to influence the award of a contract. They shall not offer any gift, or other valuable consideration in order to secure work. They shall not pay a commission, percentage or brokerage fee in order to secure work except to a bona fide employee or bona fide established commercial or marketing agencies retained by them.

Code of Ethics Section III.3.a - Engineers shall avoid the use of statements containing a material misrepresentation of fact or omitting a material fact necessary to keep statements from being misleading or intended or likely to create an unjustified expectation; statements containing prediction of future success; statements containing an opinion as to the quality of the Engineers' services; or statements intended or likely to attract clients by the use of showmanship, puffery, or self-laudation, including the use of slogans, jingles, or sensational language or format.

Discussion:

The basic question posed by this case is whether it would be appropriate for an engineer to associate with a brokerage firm, particularly where there is evidence that the broker had made modifications to the report provided by the engineer.

Over the years, the Board of Ethical Review has had opportunities to discuss and offer opinion on the subject of brokering of engineering services. In BER Case 83-5, a local landscape architect, through a network of contacts, was able to locate engineering projects. The landscape architect contacted an engineer and proposed to refer these clients to the engineer in return for a fee over and above the value of the landscaping work which the landscape architect would presumably perform on the jobs.

Generally, little landscaping work was required on the projects. In ruling the arrangement was unethical, the Board, referring to Code Section II.5.b., noted that there was nothing to indicate that the landscape architect was a "bona fide marketing agency". Instead, it appeared that the landscape architect was wearing two hats and wearing those hats simultaneously.

The landscape architect proposed to act both as a marketing representative for the engineer and, at the same time, expected to perform services at an inflated rate in connection with the work that the landscape architect secured for the engineer. The Board ruled that such conduct did not demonstrate the requisite good faith, integrity of dealing and honesty implicit in the definition of a "bona fide marketing agency" as required by the Code.

Later, in BER Case 86-1, the Board considered two separate factual situations involving the solicitation of work by a business consortium consisting of an engineering firm, architectural firm, construction firm and financial firm.

In one case, to defray consortium expenses for promotion, publicity, overhead, etc., each firm was required to pay to the consortium an entrance fee plus a percentage of income derived from business successfully generated from referrals by other consortium members. In the other case, each firm was required to pay the entrance fee plus a referral fee directly to the consortium firm member which "found" the new business client. In finding the first arrangement to be proper but the second improper, the Board noted that both consortiums were being formed primarily for marketing purposes and represent, in effect, a "pooling" of individual firm marketing capabilities and efforts through an umbrella approach.

In this sense, the consortium is quite similar to the joint ventures where one firm learns of a potential project and forms liaisons with other firms having expertise complementary to the others. Marketing efforts are combined to secure the business and fee arrangements agreed to by all joint venture participants. The first consortium represented a relatively unique approach to marketing.

The second consortium involved a referral fee, a portion of which was exchanged between consortium firm members, constituted a payment for valuable consideration in order to secure work, prohibited by Code Section II.5.b.

In the instant case, Engineer Adam's performance of building inspection services under the circumstances appears to be inconsistent with the provisions of the Code of Ethics. Engineer Adam's involvement with IJK, Inc. appears in line with the facts in the second set of circumstances described in BER Case 86-1.

In view of the fact that IJK, Inc. typically makes contact with the employer, takes an order for a job, and passes the order on to the professional engineer available in the geographic area suggests that IJK, Inc. is acting purely as a broker under the facts and that Engineer Adam's forbearance of his full fee constitutes payment for valuable consideration in order to secure work, prohibited by Code Section II.5.b.

In addition, under the facts, it is clear that Engineer Adam performs all necessary services and prepares the report for the actual client. IJK, Inc. provides no benefit to the client other than simply passing the professional report prepared by Engineer Adam to the client. IJK, Inc. appears to be acting purely as a "go-between" and does not appear to be adding any value to the services purchased by clients even though such clients are paying a significant fee for the involvement of IJK, Inc.

We are disturbed by the fact that IJK, Inc. a company without any particular competence or expertise in the performance of building design services may occasionally make modifications to reports without consulting with the engineer. If such changes fall into the latter category, it would appear that IJK, Inc. may be entering the realm of performing engineering services in violation of the law.

In this regard, we would especially note that the Code of Ethics (Section II.1.d.) clearly admonishes engineers to avoid allowing their name or the name of their firm to be used in a joint venture where the co-venturer may be engaged in improper business activities and practices.

While it is not entirely clear from the facts whether IJK, Inc. is holding itself out to clients as capable of performing engineering services, we would simply give a note of caution that whatever line might exist in connection with the illegal practice of engineering, IJK, Inc. may be coming close to crossing it. If that were the case, Section II.1.d. would certainly provide a clear basis for Engineer Adam to avoid association with IJK, Inc.

Finally, we are also disturbed if Engineer Adam continues association with IJK, Inc. after learning that IJK, Inc. is indicating a fee for Engineer Adam's services on the IJK, Inc. invoice to its client which is different from the fee charged by Engineer Adam.

On the face of it, IJK's practice misrepresents Engineer Adam's actual fee and is a deceptive practice.

Conclusions:

1. It was unethical for Engineer Adam to continue association with the referral firm after he learns that IJK, Inc. has a history of changing reports.

2. It was unethical for Engineer Adam to continue association with the referral firm after learning that IJK, Inc. is indicating a fee for Engineer Adam's services to IJK's client which is different from the fee charged by Engineer Adam.

Board of Ethical Review

William A. Cox, Jr., P.E.
William W. Middleton, P.E.
William E. Norris, P.E.
William F. Rauch, Jr., P.E.
Jimmy H. Smith, P.E.
Otto A. Tennant, P.E.
Robert L. Nichols, P.E., Chairman

Note: In regard to the question of application of the Code to corporations vis-a-vis real persons, business form or type should not negate nor influence conformance of individuals to the Code. The Code deals with professional services, which services must be performed by real persons. Real persons in turn establish and implement policies within business structures. The Code is clearly written to apply to the Engineer and it is incumbent on a member of NSPE to endeavor to live up to its provisions. This applies to all pertinent sections of the Code.

Signing And Sealing Of Work – Making Changes

NSPE Opinions of the Board of Ethical Review - Case No. 02-2

Facts:

Engineer A is a professional engineer with expertise in electronics engineering and radio communications. Engineer A designs specialized antenna systems for broadcast stations in City X. A particular antenna system design was necessitated by the location of a large municipal highway department maintenance facility on the radio station's property.

Engineer A designs a specialized antenna system and signed and sealed those preliminary drawings. The construction of a new building for the highway department was about to begin. A meeting was called so that all parties involved could wrap up last minute details so the project could proceed expeditiously. During the meeting, Engineer A was asked to clarify some details about the antenna system's relationship to the foundation of the building and to address some other technical questions. Engineer A was unable to answer the questions about the building foundation because Engineer A was never provided with the final plans for the building— Engineer A was only provided preliminary drawings for the building and a site plan. Following the meeting, the project manager sent Engineer A a full set of drawings.

Engineer A's preliminary drawings were included with the final plans, but an unknown person had crossed out Engineer A's notes on each page of Engineer A's signed and sealed drawings without Engineer A's knowledge or permission. The project manager, not a licensed engineer, subsequently revealed to Engineer A that Engineer B, one of the prime consultants, made the changes to Engineer A's plans, and signed and sealed the drawings and that the changes should not have been made without Engineer A's approval. The project manager said that the changes were made to avoid a delay in distributing the bid documents.

Question:

Was it ethical for Engineer B, a prime consultant, to make changes to Engineer A's work?

References:

Code of Ethics Section II.2.b. - Engineers shall not affix their signatures to any plans or documents dealing with subject matter in which they lack competence, nor to any plan or document not prepared under their direction and control.

Code of Ethics Section II.2.c. - Engineers may accept assignments and assume responsibility for coordination of an entire project and sign and seal the engineering documents for the entire project, provided that each technical segment is signed and sealed only by the qualified engineers who prepared the segment.

Code of Ethics Section III.7.a. - Engineers in private practice shall not review the work of another engineer for the same client, except with the knowledge of such engineer, or unless the connection of such engineer with the work has been terminated.

Discussion:

The signing and sealing of engineering documents involves fundamental issues relating to the practice of engineering. The signature and seal on a set of engineering drawings is an indication that the signing and sealing engineer is taking personal and professional responsibility for the contents of the work.

The NSPE Code of Ethics addresses this issue in considerable detail and the NSPE Board of Ethical Review (Board) considered this issue on numerous occasions. In addition, state engineering licensure boards maintain strict rules and policies on the signing and sealing of engineering documents for the public health, safety, and welfare.

The Board has reviewed and considered issues relating to the signing and sealing of engineering work on a variety of occasions. For example, in BER Case 86-2, Engineer A was the chief engineer within a large engineering firm and affixed his seal to some of the plans prepared by licensed engineers working under Engineer A's general direction who did not affix their seals to the plans. At times, Engineer A also sealed plans prepared by non-registered, graduate engineers working under his general supervision. Because of the size of the organization and the large number of projects being designed at any one time, Engineer A found it impossible to give a detailed review or check of the design. He believed he was ethically and legally correct in not doing so because of his confidence in the ability of those he hired and who were working under his general direction and supervision. By general direction and supervision, Engineer A meant that he was involved in helping to establish the concept, the design requirements, and review elements of the design or project status as the design progressed. Engineer A was consulted about technical questions and he provided answers and direction in these matters. In finding that it was unethical for Engineer A to seal plans that were not prepared by him, or which he has not checked and reviewed in detail, the Board established criteria for "direction and control" in the NSPE Code, Section II.2.b. In BER Case 86-2, the Board said "It is clear that 'direction and control' have a meaning which when combined would suggest that an engineer would be required to perform all tasks related to the preparation of the drawings, plans, and specifications in order for the engineer to ethically affix his seal."

Even though the facts and circumstances in the two cases are quite different, the Board believes BER Case 86-2 is instructive in the present case because it identifies the requirements for the appropriate signing and sealing of work by an engineer and also describes situations where it is unethical for a professional engineer to sign and seal work prepared by others. In the preparation of engineering documents, there is a clear need for close collaboration between all parties involved in the design elements of a project in order for there to be good coordination. Every member of the design team brings different levels of design and management expertise to the process and having a clear and straightforward procedure for review and approval of the work is a responsibility of each party involved. While BER Case 86-2 presents a situation where an engineer's oversight did not reach the necessary threshold in order to ethically sign and seal the work as required by the NSPE Code of Ethics, the present case involves a basic disregard for the work product of another licensed professional engineer. Both situations are unacceptable under the language of the NSPE Code.

Specifically, under the language of the NSPE Code, while Engineer B may have had a general right to sign and seal a set of final drawings as the prime design engineer, it is also clear that Engineer B could not ethically sign and seal drawings that were prepared by another engineer, were preliminary in nature and then represent those drawings as final, regardless of the time constraints involved. While it is frequently a basic reality in today's engineering practice that time is of the essence, time considerations should never supersede to the need for competent engineering practice and the need for those with the appropriate level of knowledge and expertise to provide the necessary technical information as required in order to make the project successful for the benefit of the client and for the protection of the public. In the present case, the facts suggest that Engineer B may not have had the necessary level of technical competence and did not exercise appropriate direction and control over the work in order to assume responsibility for the work.

Conclusion:

It was not ethical for Engineer B, a prime consultant, to make changes to Engineer A's work.

Board of Ethical Review

E. Dave Dorchester, P.E.	Louis L. Guy, Jr., P.E.
William D. Lawson, P.E.	Robert L. Nichols, P.E.
Harold E. Williamson, P.E.	*William J. Lhota, P.E., Chair*

NOTE: The NSPE Board of Ethical Review (BER) considers ethical cases involving either real or hypothetical matters submitted to it from NSPE members, other engineers, public officials and members of the public. The BER reviews each case in the context of the NSPE Code and earlier BER opinions. The facts contained in each case do not necessarily represent all of the pertinent facts submitted to or reviewed by the BER.

The Following Engineering Ethics Cases are Excerpts Taken from the 2001 Opinions of the National Society of Professional Engineers Board of Ethical Review

Conclusions To These Cases Follow

Discussions and Conclusions of These And Other BER Cases Can Be Viewed At
http://www.niee.org/cases/index.htm

2001-2002 Members Of The Board Of Ethical Review

E. Dave Dorchester, P.E., NSPE
John W. Gregorits, P.E., F.NSPE
Louis L. Guy, Jr., P.E., F.NSPE
William D. Lawson, P.E., NSPE
Roddy J. Rogers, P.E., F.NSPE
Harold E. Williamson, P.E., NSPE
William J. Lhota, P.E., NSPE, Chair

The NSPE Board of Ethical Review (BER) considers ethical cases involving either real or hypothetical matters submitted to it from NSPE members, other engineers, public officials and members of the public. The BER reviews each case in the context of the NSPE Code and earlier BER opinions. The facts contained in each case do not necessarily represent all of the pertinent facts submitted to or reviewed by the BER.

Each opinion is intended as guidance to individual practicing engineers, students and the public. In regard to the question of application of the NSPE Code to engineering organizations (e.g., corporations, partnerships, sole-proprietorships, government agencies, university engineering departments, etc.), the specific business form or type should not negate nor detract from the conformance of individuals to the NSPE Code. The NSPE Code deals with professional services—which must be performed by real persons. Real persons in turn establish and implement policies within business structures.

Visit www.niee.org to view most of the BER cases free on the Internet and/or to learn how to obtain printed volumes that include NSPE Opinions.

(or call 806-742-NIEE)

Employment – Questioning Ability of Former Employer to Meet Client's Expectations

NSPE-BER Case No. 01-1

Facts:

Engineer Adam, a professional engineer working for a small private practice firm, leaves the employment of AB Inc. Engineer Adam had represented that he was going to start his own one-person consulting firm, Z-CORP, and that he would not be in the position of competing with AB Inc.

One month after Engineer Adam departs from AB Inc., Engineer Bob, a principal in AB Inc. learns that Engineer Adam has contacted one of AB Inc.'s employees, Engineer Carl, and offered him a position with Z-CORP.

Soon thereafter, Engineer Bob learns that Engineer Adam has contacted AB Inc.'s clients and is making representations that because Engineer Carl is going to be leaving AB Inc. to work for Z-CORP, AB Inc. will be "hard pressed" to perform successfully on its projects and that AB Inc.'s clients should hire Z-CORP to perform engineering services.

Questions:

1: Was it ethical for Engineer Adam to offer a position to Engineer Carl?

2: Was it ethical for Engineer Adam to make representations to AB Inc.'s clients that because Engineer Carl is going to be leaving AB Inc. to work for Z-CORP, AB Inc. will be "hard pressed" to perform successfully on its projects and that the clients should hire Z-CORP to perform engineering services?

Conflict Of Interest: Third Party Developer

NSPE-BER Case No. 01-2

Facts:

A developer, Mall Dev, has approached a town requesting approval to construct a development on a vacant site in Niceville. Based on the size of the development, Niceville is requesting that an environmental impact statement be prepared that will address traffic operations, as well as other issues.

Niceville requests an outside consultant, Engineer Adam, to assist the town in scoping out the necessary traffic analyses and to review and advise Niceville on possible traffic impacts of the proposed development. The development will be both retail and offices and will contain a supermarket.

The consultant, Engineer Adam, is also assisting other jurisdictions in review of proposals by Mall Dev. Engineer Adam has disclosed to the town all relationships, if any, with the proposed developer, Mall Dev with announced tenants, and with other customers that develop sites for retail development. Niceville is satisfied that there is no conflict of interest.

More specifically, Engineer Adam is not currently representing any other developers in the town, but in the past has prepared traffic impact studies for other developers on projects concerning other developments constructed in Niceville.

Engineer Adam is currently providing traffic impact studies to other developers in other jurisdictions, as well as services to Mall Dev. These have all been disclosed to Niceville.

Mall Dev, however, has informed Niceville that it believes the use of the consultant Engineer Adam is a conflict of interest and breaches the code of professional ethics.

Mall Dev bases its belief on the fact that Engineer Adam has worked in the past, and is currently working for, other developers who compete for the same tenants Mall Dev tries to attract to its developments.

Questions:

1. Would Engineer Adam's work for the Niceville constitute a conflict of interest?
2. Was it appropriate for Mall Dev to raise an ethical issue relating to Engineer Adam's actions?

Use of P.E. Designation – Not Licensed In State In Which Complaint Is Filed

NSPE -BER Case No. 01-3

Facts:

Engineer Adam is a safety engineer for a federal agency. He is responsible for independently overseeing the proper implementation of worker and nuclear safety programs in the agency's facilities, which are located in many different states, including the state in which Engineer Adam is licensed, State Y.

Engineer Adam is not required to be licensed by the federal agency, but has become licensed because of his personal commitment to the engineering profession.

Engineer Adam has never used his seal in the course of his employment. When Engineer Adam moves to State Z, he does not obtain an engineering license in State Z.

Engineer Adam reads a newspaper account about LMN Engineering, a subcontractor to the federal agency in which he works, having a conflict of interest with the agency.

Engineer Adam, acting on his ethical obligation to report violations of the NSPE Code of Ethics to a public authority, files a complaint against LMN Engineering. In the text of the complaint, Engineer Adam indicates that he is licensed in State Y but not licensed in State Z and signs the letter "Engineer Adam, P.E."

Engineer Adam is thereafter notified by the State Z engineering licensure board that his use of the title "P.E." in the letter is inappropriate because he is not licensed in State Z.

Questions:

1. Was it ethical for Engineer Adam to indicate in a State Z complaint letter, in which he had already indicated that he was not licensed in State Z, that he was a professional engineer?
2. Did Engineer Adam have an ethical obligation under the NSPE Code of Ethics to file a complaint in a state in which he was not licensed?

Patents – Dispute Over Right To Specify

NSPE-BER Case No. 01-4

Facts:

Engineer Adam, a structural engineer, designs structural systems for large developers on hotel projects.

Developer Jim would like to use a unique flooring system, but the system is patented by Inventor Charles, who is a professional engineer.

Developer Jim contacts Attorney Donald, who tells Developer Jim that Inventor Charles has a legitimate patent and recommends that Developer Jim negotiate with Inventor Charles to obtain a license for Inventor Charles's patent.

Developer Jim enters into negotiations with Inventor Charles, but the negotiations fail.

Thereafter, Developer Jim hires Attorney Edward, who reviews the patent and indicates that he disagrees with Attorney Donald, and also indicates that, in his professional view, there is a genuine dispute as to the legitimacy of Inventor Charles's patent.

Developer Jim tells Engineer Adam that he wants Engineer Adam to proceed with the project and have Engineer Adam specify the flooring system into the structural design of the project.

Question:

Would it be ethical for Engineer Adam to proceed with the project and reference the flooring system of the project's structural design?

Conflict Of Interest — Utility Audits For City

NSPE-BER Case No. 01-5

Facts:

Engineer Adam receives a "Request for Qualifications (RFQ)" from the City for the review of unbilled and mis-billed water and wastewater service records.

One paragraph of the RFQ reads as follows: "The consultant shall be entitled to receive X% of increased revenues generated.

If the consultant fails to identify and document unbilled or mis-billed water and wastewater sewer service records, the City shall be under no obligation to compensate the consultant."

Question:

Would it be ethical for Engineer Adam to enter into a contract under the circumstances described?

Confidentiality – Records Relating To Services To Former Client

NSPE-BER Case No. 01-6

Facts:

Several years ago Engineer Adam, a mechanical engineer, consulted for X-CORP, a pressure vessel manufacturer, on a specific pressure vessel problem relating to the design of a boiler system.

Engineer Adam's work focused on specific design and manufacturing defects that caused deterioration of the boiler system.

Engineer Adam completed his work and was paid for his services.

Ten years later, Engineer Adam was retained by Attorney Joe, plaintiff in a case involving the fatal explosion of a recently designed and manufactured pressure vessel at a facility previously owned by Engineer Adam's former client, X-CORP.

The facility was sold to Z-CORP seven years before the explosion. The litigation does not involve any of the issues related to the services Engineer Adam provided to X-CORP ten years earlier.

The defendant's attorney discovered through Engineer Adam's deposition and statements relating to his professional experience that Engineer Adam had worked for X-CORP on a pressure vessel problem.

Engineer Adam explains to the defendant's attorney that he is not relying upon any of his prior work for X-CORP in this case.

Nevertheless, the defendant's attorney requests that Engineer Adam provide his files from the previous work performed for X-CORP.

Question:

Would it be ethical for Engineer Adam to voluntarily release the files to defense counsel?

Conflict Of Interest – Privatization Of Plan Reviews

NSPE-BER Case No. 01-7

Facts:

A controversial new ordinance is developed by a local county Board of Supervisors to give property owners the option of hiring private engineers and architects to perform plan reviews and inspections normally performed by a building department.

The ordinance has stirred debate about who in the design and construction process is responsible for the code compliance of buildings.

According to county officials, the "affidavit ordinance" is intended to help the county encourage new development by streamlining the permitting process without compromising public safety.

The ordinance states that the private plan reviewer or inspector must be a licensed engineer or architect other than the design professional of record, and is required to carry liability insurance, without a deductible, of at least $1 million for residential projects and $2 million for commercial projects.

The ordinance also calls for an audit of 20% of all privately certified plans and 50% of all private inspections.

Contractors have criticized the ordinance, claiming that it will encourage private plan reviewers to take less personal responsibility.

Others have criticized the law because it will create a conflict of interest for the plan reviewers who are selected and paid by the property owners. It is recognized that most of the county plan reviews will still be performed by county plan reviewers and that the program is intended as an experiment.

Question:

Would it be ethical for engineers to participate in a private plan review under the circumstances described?

Associating With A Firm Not Authorized To Practice

NSPE-BER Case No. 01-8

Facts:

Engineer Adam is employed by Z-CORP in State X. Engineer Bob, the President of Z-CORP, passes away in July 2000. His widow, Widow Carol, a non-engineer, was the only other named corporate officer in Z-CORP. At the time of Engineer Bob's death, Engineer Adam is the only other professional engineer in Z-CORP.

Following discussions with Widow Carol, Engineer Adam tries to purchase Z-CORP from Widow Carol. However, negotiations break down and Engineer Adam decides to start his own firm in October 2000. Following a period of time, Widow Carol decides to run the engineering firm, Z-CORP.

As an "interim measure," Engineer Don, a personal friend of the deceased Engineer Bob, with a separate full-time practice, agrees to advise and help Widow Carol through a "transition period."

A year passes and there is still no professional engineer within the structure of Z-CORP who is in responsible charge of engineering work. Z-CORP continues to perform engineering services and take on new clients and new work through the limited advice of Engineer Don and two graduate engineer employees.

Under the laws of State X, a professional engineer employee must be in responsible charge of engineering work performed by an engineering company.

Questions:

1. What are Engineer Adam and Engineer Don's ethical obligations under the described facts?
2. Was it ethical for Engineer Don to assist Widow Carol through an undefined "transition period" and thereafter provide advice to Widow Carol to permit Z-CORP to continue to perform engineering services?

Reference – Quid Pro Quo

NSPE-BER Case No. 01-9

Facts:

Engineer Adam is licensed in State A, and State B and would like to become licensed in State C.

State C requires a recommendation of three licensed professional engineers licensed in any state.

Because Engineer Adam has worked for many years in a company with no other licensed engineers, Engineer Adam has not had very much exposure to licensed professional engineers.

Engineer Adam is able to obtain the recommendations of two licensed engineers.

His colleague, Engineer Bob, was licensed but has allowed his license to lapse.

Engineer Bob knows Engineer Adam well and respects Engineer Adam's professional judgment.

Engineer Adam offers to pay Engineer Bob to have Engineer Bob's license reinstated with the understanding that he will prepare a recommendation for Engineer Adam.

Question:

Was it ethical for Engineer Adam to offer to pay Engineer Bob to have Engineer Bob's license reinstated with the understanding that Engineer Bob will prepare a recommendation for Engineer Adam?

QUESTIONS AND CONCLUSIONS TO NSPE BER CASES FOR 2001

NSPE-BER Case No. 01-1 – Questions and Conclusions

Questions:

1: Was it ethical for Engineer Adam to offer a position to Engineer Carl?

2: Was it ethical for Engineer Adam to make representations to AB Inc.'s clients that because Engineer Carl is going to be leaving AB Inc. to work for Z-CORP, AB Inc. will be "hard pressed" to perform successfully on its projects and that the clients should hire Z-CORP to perform engineering services?

Conclusions:

Question 1: It was ethical for Engineer Adam to offer a position to Engineer Carl.

Question 2: It was not ethical for Engineer Adam to make representations that because Engineer Carl is going to be leaving AB Inc. to work for Z-CORP, that AB Inc. will be "hard pressed" to perform successfully on its projects and that AB Inc.'s clients should hire Z-CORP to perform engineering services. As an observation, the Board believes it was unethical for Engineer Adam to make misleading statements about Engineer Carl's future plans.

NSPE-BER Case No. 01-2 – Questions and Conclusions

Questions:

1: Would Engineer Adam's work for the Niceville constitute a conflict of interest?

2: Was it appropriate for Mall Dev to raise an ethical issue relating to Engineer Adam's actions?

Conclusions:

Question 1: Engineer Adam's work for the town would not constitute a conflict of interest since there was full disclosure. Based upon the language in the NSPE Code, it is clear that no conflict of interest exists and the engineer has fulfilled his ethical obligation under the NSPE Code.

Question 2: It was not appropriate for Mall Dev to raise an ethical issue relating to Engineer Adam's actions.

NSPE-BER Case No. 01-3 – Questions and Conclusions

Questions:

1. Was it ethical for Engineer Adam to indicate in a State Z complaint letter, in which he had already indicated that he was not licensed in State Z, that he was a professional engineer?
2. Did Engineer Adam have an ethical obligation under the NSPE Code of Ethics to file a complaint in a state in which he was not licensed?

Conclusions:

Question 1: It was ethical for Engineer Adam to indicate in a State Z complaint letter, in which he had already indicated that he was not licensed in the State Z, that he was a professional engineer. However, Engineer Adam has an obligation to become familiar with the legal requirements contained in applicable state engineering licensure statutes and regulations to avoid unintended violations of the law in the future.

Question 2: Engineer Adam did have an ethical responsibility under the NSPE Code of Ethics to file a complaint in a state in which he was not licensed.

NSPE-BER Case No. 01-4 – Question and Conclusion

Question:

Would it be ethical for Engineer Adam to proceed with the project and reference the flooring system of the project's structural design?

Conclusion:

It would be unethical for Engineer Adam to specify the flooring system into the project's structural design until the patent and proprietary rights of Inventor Charles are resolved.

NSPE-BER Case No. 01-5 – Question and Conclusion

Question:

Would it be ethical for Engineer Adam to enter into a contract under the circumstances described?

Conclusion:

It would be ethical for Engineer Adam to enter into a contract under the circumstances described.

NSPE-BER Case No. 01-6 – Question and Conclusion

Question:

Would it be ethical for Engineer Adam to voluntarily release the files to defense counsel?

Conclusion:

It would not be ethical for Engineer Adam to voluntarily release the files to the defense counsel.

NSPE-BER Case No. 01-7 – Question and Conclusion

Question:

Would it be ethical for engineers to participate in a private plan review under the circumstances described?

Conclusion:

It would be ethical for engineers to participate in a plan review with full disclosure under the circumstances described. However, the plan review process creates conflicts and questions that could be minimized if the local county, and not the developer, is responsible for retaining and paying the engineer for review and inspection.

NSPE-BER Case No. 01-8 – Questions and Conclusions

Questions:

1. What are Engineer Adam and Engineer Don's ethical obligations under the described facts?
2. Was it ethical for Engineer Don to assist Widow Carol through an undefined "transition period" and thereafter provide advice to Widow Carol to permit Z-CORP to continue to perform engineering services?

Conclusions:

Question 1: Engineer Adam and Engineer Don are ethically obligated to advise Widow Carol that Z-CORP may not be in full compliance with the laws of State X and suggest that Z-CORP self-report this fact to the state engineering licensure board. If Widow Carol refuses to take this action, it would appear that Engineer Adam and Engineer Don would have an obligation to personally bring this matter to the appropriate authorities (e.g., state engineering licensure board).

Question 2: It was ethical for Engineer Don to assist Widow Carol through a limited transition period and provide advice to Widow Carol, provided it is permissible under state laws and regulations.

NSPE-BER Case No. 01-9 – Question and Conclusion

Question:

Was it ethical for Engineer Adam to offer to pay Engineer Bob to have Engineer Bob's license reinstated with the understanding that Engineer Bob will prepare a recommendation for Engineer Adam?

Conclusion:

It was not ethical for Engineer Adam to offer to pay for Engineer Bob's licensure reinstatement with the understanding that Engineer Bob would prepare a recommendation for Engineer Adam.

NOTE:

A full discussion of all the 2001 NSPE-BER cases can be viewed and/or downloaded free at www.niee.org

APPENDICES

Appendix A

NSPE-BER CASES AVAILABLE ON THE INTERNET

For an Index linking to NSPE-BER cases, see www.niee.org. Click on "Ethics Cases," then click on "Consolidated Reference Table". Cases are listed by topic.

CASE # SUBJECT

79-5 .... Academic Qualifications
81-5 .... Advertising
73-2 .... Advertisement, Classified - Contract Work
65-13 .... Advertisement, Use of Engineer's Name in, to Validate Findings
65-7 .... Advertising - Use of Engineers' Creed in Political Advertisement
75-2 .... Advertising - Announcement Cards
62-8 .... Advertising - Billboard
72-1 .... Advertising - Bold Face in Telephone Directory
61-7 .... Advertising - Brochure at Convention
78-8 .... Advertising - Calendars - Pencils
72-3 .... Advertising - Direct Mail Solicitation - Supplanting Another Engineer
73-5 .... Advertising - Directory
72-8 .... Advertising - Distribution of Laudatory Article
66-9 .... Advertising - Engineers Week Section of Newspaper
63-7 .... Advertising - Full Page in Newspaper
66-4 .... Advertising - Good Will
75-16 .... Advertising - Group Advertisement by Engineering Firms
62-15 .... Advertising - Listing of Name
92-2 .... Advertising - Misstating Credentials
75-9 .... Advertising - Newsletter
64-8 .... Advertising of Engineering Services
84-2 .... Advertising Services of Engineering Staff
62-2 .... Advertising of Engineering Services - Brochure
61-3 .... Advertising of Engineering Services - Display
59-1 .... Advertising of Engineering Services - Text
60-1 .... Advertising - Press Release
68-9, 73-1, 77-2 Advertising - Professional Cards
96-11 .... Advertising - Promotional Reference to Work and Clients of Previous Employers
71-8 .... Advertising - Recruiting
63-3 .... Advertising - Repeated Use of Card
79-6 .... Advertising - Statement of Project Success
75-4 .... Advertising - Testing and Engineering Laboratory
71-11 .... Advertising - Use of Brochure Tied to Professional Directory Card
93-7 .... Agreement Not to Disclose Data, Findings, Conclusions
77-3 .... Appropriate Compensation for Engineering Services
93-3 .... Appropriate Notification and Review of Another Engineer's Work
69-9 .... Approval of Engineering Plans Related to Architectural Plans Prepared by Nonregistered Persons
65-14 .... Assistance in Preparation of Applications under Advance Planning Act
01-8 .... Associating with a Firm Not Authorized to Practice
61-4 .... Association with Nonregistered Engineers
92-8 .... Attempt to Influence Prospective City/ Client
75-15 .... Attempt to Restrain Employment of Engineer-Employees
95-7 .... Authorship of Article - Misleading Reference
71-5 .... Boycott of Public Agency Engineering Employment
76-10 .... Brochure - Distribution through Reader Service Card
80-2 .... Brochure of Subsidiary Firm
75-6 .... Brochure - Text and Build-in Reply Card
63-8, 74-8, 77-6 Brochure - Format and Content
71-2 .... Brokerage of Engineering Services
92-3 .... Brokerage of Engineering Services - Building Inspection Services
67-7 .... Certification of Plans Prior to Payment of Engineer
91-8 .... Certification of Work Performed by Technician
86-5 .... City Engineer Seeking to Retain Employees of Engineering Firm Independent of Their Firm
73-2 .... Classified Advertisement - Contract Work
96-10 .... Comments by One Engineer Concerning Another
78-7 .... Commission Basis of Payment Under Marketing Agreement
98-8 .... Competence to Certify Arms Storage Rooms
94-8 .... Competence to Perform Foundation Design
61-5 .... Compensation for Engineering Employment
60-2 .... Competitive Bidding - Professional Services

Appendix B

References Related To Engineering Ethics

Recently Published Full Textbooks

Charles E. Harris, Michael S. Pritchard, and Michael J. Rabins, Engineering Ethics: Concepts and Cases 2nd ed. ,Wadsworth/Thomson Learning Belmont, Calif. 2000

Mike W. Martin and Roland Schinzinger, Ethics in Engineering 4th ed., McGraw Hill, Boston 2005

Stephen H. Unger, Controlling Technology: Ethics and the Responsible Engineer 2nd ed., John Wiley & Sons, New York 1994

Caroline Whitbeck, Ethics in Engineering Practice and Research, Cambridge University Press, New York 1998

Recently Produced Engineering Ethics Videos

"Incident at Morales, an Engineering Ethics Story," National Institute for Engineering Ethics, 2003 (36 min.)

Recently Published Short Textbooks / Supplementary Books

Charle B. Fleddermann, Engineering Ethics 2nd ed., Pearson Prentice Hall, Upper Saddle River, N.J. 2004

Michael E. Gorman, Matthew M. Mehalik, and Patricia H. Werhane, Ethical, Environmental Challenges to Engineering, Prentice Hall, Upper Saddle River, N.J. 2000

Alastair S. Gunn and P. Aarne Vesilind, Hold Paramount: The Engineer's Responsibility to Society, Brooks/Cole—Thomson Learning, Pacific Grove, Calif. 2003

Joseph R. Herkert, ed., Social, Ethical, and Policy Implications of Engineering: Selected Readings, IEEE Press, New York 2000

Carl Mitcham and R. Shannon Duvall, Engineering Ethics, Prentice Hall, Upper Saddle River, N.J. 2000

Roland Schinzinger and Mike W. Martin, Introduction to Engineering Ethics, McGraw Hill, Boston 2000

Web Sites that Contain Links to Numerous Ethics Resources

National Institute for Engineering Ethics: www.niee.org

Online Ethics Center for Engineering and Science: www.onlineethics.org

Videos

1. ABC Video, "Inside the Third Reich," (from the autobiography of Albert Speer, 'Hitler's number one man'), American Broadcasting Companies, Inc, 1982, Color, 240 min. drama, 1982 (Embassy Home Entertainment, 1986)

2. American Society of Civil Engineers, "Ethics on Trial: The Case of Marvin L. Camper," ASCE, 1977 (60 min.)

3. Boisjoly, R., Presentation on the Challenger Accident at Duke University, Department of Civil Engineering, date? (90 minutes including Q&A) [Note: a 2-hour video of Mr. Boisjoly's presentation at MIT is also available, entitled, "Company Loyalty and Whistle Blowing: Ethical Decisions and the Space Shuttle Disaster"]

4. Ethics Resource Center, "A Matter of Judgment," An Ethics at Work Program, Ethics Resource Center, Washington, D.C., 1986 (5 vignettes of 5 or 6 min. duration, some with engineering topics; no commentary)

5. Fanlight Productions, "The Truesteel Affair," 47 Halifax St. Boston, MA 02130. 1983

6. Haggerty, T., "Engineering Ethics Dramatizations," (my title), University of Louisville Civil Engineering and Philosophy Departments, 1988 (about 90 minutes including two cases and discussion).

7. National Institute for Engineering Ethics, National Society of Professional Engineers, "Gilbane Gold: A Case Study in Engineering Ethics," 1989 (23 min.)

8. Petroski, H., "To Engineer is Human," BBC Discovery Series, 1986 (60 min.)

Books and Monographs

1. Alderman, F.E. and Schulz, R.A., Ethical Problems in Consulting Engineering, Report available from Alderman, 721 Fair Oaks Ave., South Pasadena, CA 91030, 1980.
2. Alger, et al, Ethical Problems in Engineering, Wiley, New York, 1965.
3. Anderson, R.M., Perucci, R., Schendel, D.E., and Trachtman, L.E., Divided Loyalties - Whistleblowing at BART, Purdue Research Foundation, 1980.
4. Baron, Marcia, The Moral Status of Loyalty, Center for the Study of Ethics in the Professions, Kendall/ Hunt Publishing, Dubuque, IO, 1'984, 36 pp.
5. Baum, R.J., Ethics and Engineering Curricula, The Teaching of Ethics VII, The Hastings Center, Hastings on Hudson, N.Y., 1980, 79 pp.
6. Baum, R.J. and Flores, A., eds, Ethical Problems in Engineering, Center for the Study of the Human Dimensions of Science and Technology, Rensellaer Polytechnic Institute, Troy, New York, 1978, 335 pp.

7. Baum, R.J., ed, Ethical Problems in Engineering, Second Edition, Volume Two: Cases, Center for the Study of the Human Dimensions of Science and Technology, Rensellaer Polytechnic Institute, Troy, New York, 1980, 259 pp.
8. Beabout, G.R., Wennemann, D.J., Applied Professional Ethics: A Developmental Approach for Use with Case Studies, University Press of America, Lanham, MD, 1994, 175 pp.
9. Beauchamp, T.L., Case Studies in Business, Society, and Ethics, 2nd Ed., Prentice-Hall, Englewood Cliffs, NJ 1989, 275 pp.
10. Bennett, F. L., The Management of Engineering: Human, Quality, Organizational, Legal, and Ethical Aspects of Professional Practice, John Wiley & Sons, Inc., 1996, 478 pp.
11. Blinn, K.W., Legal and Ethical Concepts in Engineering, Prentice-hall, 1989, 334 pp. (see pp. 11-18)
12. Board of Ethical Review, NSPE, Opinions of the Board of Ethical Review, Vols. I - VII, NSPE Publications, National Society of Professional Engineers, Alexandria, VA.
13. Bok, Sissela, Lying: Moral Choice in Public and Private Life, Vintage Books, 1979.
14. Buchanan, R.A., The Engineers: A History of the Engineering Profession in Britain, 1750-1914, Jessica Kingsley Publishers, London, 1989, 230 pp.
15. Burgunder, L.B., Legal Aspects of Managing Technology, South-Western Publishing, Cincinnati, OH, 1995, 482 pp.
16. Callahan, D. and Bok, S., Ethics Teaching in Higher Education, Plenum Press, New York, 1980, 315 pp.
17. Callahan, J.C., ed, Ethical Issues in Professional Life, Oxford University Press, New York, 1988, 470 pp.
18. Cameron R. and Millard, A.J., Technology Assessment: A Historical Approach, Center for the Study of Ethics in the Professions, Kendall/Hunt Publishing, Dubuque, IO 1985, 37 pp.
19. Center for the Study of Ethical Development, DIT Manual, University of Minnesota, Minneapolis, MN, 1986.
20. Center for the Study of Ethics in the Professions, Moral Issue sin Engineering: Selected Readings, V. Weil, Editor, Illinois Institute of Technology, 1988, 300 pp.
21. Chalk, R., Franke, M. and Chafer, S.B., AAAS Professional Ethics Project: Professional Ethics Activities of the Scientific and Engineering Societies, American Association for the Advancement of Science, Washington, D.C., December, 1980, 224 pp.
22. Chevron Corporation, Our Business Conduct: Principles and Practices, 1986, 31 pp.
23. Cohen, R.M. and Witcover, J., A Heartbeat Away: The Investigation and Resignation of Vice President Spiro T. Agnew, Viking Press, New York, 1974.
24. Curd, M and May, L., Professional Responsibility for Harmful Actions, Center for the Study of Ethics in the Professions, Kendall/Hunt Publishing, Dubuque, IO, 1984, 30 pp.
25. Dalcourt, G.J., The Methods of Ethics, University Press of America, Lanham, MD, 1983, 237 pp.
26. Dunham & Young, Contracts, Specifications, and Law for Engineers, 4th Ed, J. Bockrath, Editor, 1986.
27. Eddy, P., Potter, E., and Page, B., Destination Disaster: From the Tri-Motor to the DC-10, Quadrangle Press, New York, 1976.

28. Elbaz, S.W., Professional Ethics and Engineering: A Resource Guide, National Institute for Engineering Ethics, Arlington, VA, 1990, 50 pp.
29. Ethics Resource Center and Behavior Research Center, Ethics Policies and Programs in American Business, The Ethics Resource Center, Washington, D.C., 1990, 44 pp.
30. Firmage, D.A., Modern Engineering Practice: Ethical, Professional, and Legal Aspects, Garland STPM, New York, 1980.
31. Fishkin, J.S., The Limits of Obligation, Yale University Press, 1982, 184 pp.
32. Flores, A., ed, Ethical Problems in Engineering, Second Edition, Volume One: Readings, Center for the Study of the Human Dimensions of Science and Technology, Rensellaer Polytechnic Institute, Troy, New York, 1980.
33. Flores, A., Ethics and Risk Management in Engineering, Westview Press, Boulder, CO 1988.
34. Florman, S.C., The Existential Pleasures of Engineering, St. Martin's Press, New York, 1976, 160 pp.
35. Florman, S.C., Blaming Technology: The Irrational Search for Scapegoats, St. Martin's Press, New York, 1981, 207 pp.
36. Florman, S.C., The Civilized Engineer, St. Martin's Press, New York, 1987, 258 pp.9 [ch.1-9].
37. Ford, D.F., Three Mile Island: Thirty Minutes to Meltdown, Viking, New York, 1982.
38. Frankel, Mark, ed., Science, Engineering and Ethics: State of the Art and Future Directions, Report of a AAAS Workshop and Symposium, Feb., 1988, AAAS, 104 pp.
39. Fredrich, A.J., Sons of Martha: Civil Engineering Readings in Modern Literature, American Society of Civil Engineers, New York, 1989, 596 pp.
40. Garrett, T.M., et al, Cases in Business Ethics, Appleton Century Crofts, New York, 1968.
41. General Dynamics Corporation, The General Dynamics Ethics Program Update, St. Louis, MO, 1988, 32 pp.
42. Godson, J., The Rise and Fall of the DC-10, David McKay, New York, 1975.
43. Goldman, A.H., The Moral Foundations of Professional Ethics, Rowman and Littlefield, Totowa, NJ, 1979.
44. Gorlin, R.A., ed., Codes of Professional Responsibility, 2nd Ed., The Bureau of national Affairs, Washington, DC, 1990, 555 pp.
45. Gray, M. and Rosen, I., The Warning: Accident at Three Mile Island, W. W. Norton, New York, 1982.
46. Gunn, A.S., and Vesilind, P.A., Environmental Ethics for Engineers, Lewis Publishers, 1986, 153 pp.
47. Harris, C.A., Applying Moral Theories, Wadsworth Publishing, Belmont, CA 1986, 191 pp.
48. Harris, C.E., Prithcard, M.S., Rabins, M.J., Engineering Ethics: Concepts and Cases, Wadsworth Publishing, Belmont, CA 1995, 411 pp.
49. Hoffman, W.M., Lange, A.E., Fedo, D.A., Ethics and The Multinational Enterprise: Proceedings of the Sixth National Conference on Business Ethics, University Press of America, Lanham, MD, 1986, 530 pp.
50. Jackall, R., Moral Mazes: The World of Corporate Managers, Oxford University Press, New York, 1988, 249 pp.
51. Jaksa, J.A. and Pritchard, M.S., Communication Ethics: Methods of Analysis, Wadsworth Publishing, Belmont, CA 1988, 172 pp.

52. Johnson, D.G., Ethical Issues in Engineering, Prentice-Hall, Englewood Cliffs, N.J. 1991, 392 pp.
53. Johnson, D.G. and Snapper, J.W., eds, Ethical Issues in the Use of Computers, Wadsworth Publishing, Belmont, CA, 1985, 363 pp.
54. Josephson Institute of Ethics, The Ethics of American Youth: A Warning and a Call to Action, October, 1990.
55. Kamm, L.J., Successful Engineering: A Guide to Achieving Your Goals, 1989.
56. Kant, I., Grounding for the Metaphysics of Morals, Hackett Publishing Company, Indianapolis, IN, 1981 Ed. translated by J. W. Ellington, 72 pp.
57. Kemper, J.D., Engineers and Their Profession, 3rd Ed., Hlt, Reinhart & Winston, New York, 1982.
58. Ladenson, R.F., et al, A Selected Annotated Bibliography of Professional Ethics and Social Responsibility in Engineering, Center for the Study of Ethics in the Professions, Illinois Institute of Technology, Chicago, 1980.
59. Layton, E.T., Jr., The Revolt of the Engineers: Social Responsibility and the American Engineering Profession, Johns Hopkins University Press, Baltimore, 1971, 1986, 286 pp.
60. LeMaire, H.P., Personal Decisions, University Press of America, Inc., Lanham, MD, 1982, 200 pp.
61. Leopold, A., A Sand County Almanac and Sketches Here and There, Oxford University Press, New York, 1949, 226 pp.
62. MacIntyre, A., A Short History of Ethics, Macmillan, New York, 1966, 280 pp.
63. Mantell, M.I., Ethics and Professionalism in Engineering, The Macmillan Company, New York, 1964, 260 pp.
64. Martin, M., Everyday Morals, Wadsworth Publishing, Belmont, CA, 1989.
65. Martin, M. and Schinzinger, R., Engineering Ethics, 2nd Ed., McGraw-Hill, New York, 1988.
66. Mill, J.S., Utilitarianism, Hackett Publishing Company, Indianapolis, IN, 1979 Ed. edited by G. Sher, 63 pp.
67. Morton, R.J., Engineering Law, Design Liability, and Professional Ethics, 1983.
68. Mount, E., Professional Ethics in Context: Institutions, Images and Empathy, Westminster/John Knox Press, Louisville, KY, 1990, 176 pp.
69. Murdough Center for Engineering Professionalism, Independent Study and Research Program in Engineering Ethics and Professionalism, College of Engineering, Texas Tech University, Lubbock, Texas, October, 1998, 150 pp.
70. Nader, R., Petkas, P., and Blackwell, K., Whistleblowing, Grossman, New York, 1972.
71. National Academy of Science, Committee on the Conduct of Science, On Being a Scientist, National Academy Press, Washington, D.C., 1989, 22 pp.
72. National Research Council, Committee on the Education and the Utilization of the Engineer, Engineering Employment Characteristics, National Academy Press, Washington, DC, 1985, 94 pp.
73. Peterson, J.C. and Farrell, D., Whistleblowing: Ethical and Legal Issues in Expressing Dissent, Center for the Study of Ethics in the Professions, Kendall/ Hunt Publishing, Dubuque, IO, 1986, 31 pp.
74. Petroski, H., Beyond Engineering: Essays and other Attempts to Figure without Equations, St. Martins Press, New York, 1985, 221 pp.

75. Petroski, H., To Engineer is Human: the Role of Failure in Successful Design, St. Martins Press, New York, 1982, 256 pp.
76. Pletta, D.H., The Engineering Profession: Its Heritage and Its Emerging Public Purpose, University Press, Washington, D.C., 1984.
77. Porreco, R., Ed., The Georgetown Symposium on Ethics: Essays in Honor of Henry Babcock Veach, University Press of America, Lanham, MD, 1984, 293 pp.
78. Rachels, J., The Elements of Moral Philosophy, Random House, New York, 1986, 168 pp.
79. Schaub, J.H. and Pavlovic, K., Engineering Professionalism and Ethics, Wiley, 1983, 559 pp.
80. Schoumacher, B., Engineers and the Law, 1986.]
81. Singer, M.G., ed., Morals and Values, Charles Scribner's Sons, 1977.
82. Solomon, R.C. and Hanson, K.R., Above the Bottom Line: An Introduction to Business Ethics, Harcourt Brace Jonanovich, Inc., New York, 1983, 434 pp.
83. Streeter, H., Professional Liability of Architects and Engineers, 1988.
84. Strobel, L.P., Reckless Homicide? Ford's Pinto Trial, and Books, South Bend, IN, 1980.
85. Taylor, P.W., Principles of Ethics, An Introduction, Dickenson, Encino, CA, 1975.
86. Thomas, J. M., Ethics and Technoculture, University Press of America, Lanham, MD, 1987, 264 pp.
87. Unger, S.H., Controlling Technology: Ethics and the Responsible Engineer, Holt, Rinehart and Winston, New York, 1982, 192 pp.
88. Vaughn, R.C., Legal Aspects of Engineering, Kendall/Hunt Publishers, 1977.
89. Vogel, D.A., "A Survey of Engineering Ethics and Legal Issues in Engineering Curricula in the United States," Stanford Law School, Winter, 1991, 71 pp.
90. Weil, V., Report of the Workshops on Ethical Issues in Engineering, July 16-27, 1979, CSEP, IIT, 1980, 65 pp.
91. Weil, V., ed., Beyond Whistleblowing: Defining Engineers' Responsibilities, Proc. of the 2nd National Conf. on Ethics in Engineering, March, 1982, 334 pp.
92. Wells, P., Jones, H., and Davis, M., Conflicts of Interest in Engineering, Center for the Study of Ethics in the Professions, Kendall/Hunt Publishing, Dubuque, IO, 1986, 68 pp.
93. Westin, A.F., Individual Rights in the Corporation: A Reader on Employee Rights, Random House, New York, 1980.
94. Westin, A.F., Whistle Blowing: Loyalty and Dissent in the Corporation, McGraw-Hill, New York, 1981.
95. Wester, J.J., An Engineer Looks at the Law, 1975.
96. Westphal, D., Westphal, F., Planet in Peril: Essays in Environmental Ethics, Harcourt Brace College Publishers, Orlando, FL 32887, 1994, 265 pp.
97. Williams, O.F., Houck, J.W., Ed., A Virtuous Life in Business, Rowman & Littlefield Publishers, Inc., Lanham, MD, 1992, 185 pp.

Suggested Supplemental Readings

Note: Many of these books are referenced by the authors of the basic text for this course. They also are the source of many of the ideas that are discussed in the workbook. They are available in most research libraries. Reading them is an excellent way to enhance your awareness of ethical dilemmas that are common in the practice of engineering as well as to expose you to some of the best thinking as to how such dilemmas may be resolved.

Alcorn, Paul A.: Social Issues in Technology; Prentice Hall, Englewood Cliffs, NJ, 1986

Alger, Philip L., Christensen, N.A., and Olmsted, Sterling P.: Ethical Problems in Engineering; Wiley, New York, NY, 1965

Bayles, Michael D.: Professional Ethics; Wadsworth; Belmont, CA, 1981

Beauchamp, Tom L.: Case Studies in Business, Society and Ethics; Prentice Hall; Englewood Cliffs, NJ, 1989

Beauchamp, Tom L., and Bowie, Norman E.: Ethical Theory and Business; Prentice Hall; Englewood Cliffs, NJ, 1983

Bowie, Norman: Business Ethics; Prentice Hall; Englewood Cliffs, NJ, 1982

Callahan, Joan C.: Ethical Issues in Professional Life; Oxford University Press; New York, NY, 1988

DeGeorge, Richard T.: Business Ethics; Macmillan, New York, NY 1986

Flores, Albert: Ethics and Risk Management in Engineering; Westview Press, Boulder, CO, 1988

Fishkin, James S.: The Limits of Obligation; Yale University Press; New Haven, CT, 1982

Florman, Samuel C.: Blaming Technology; St. Martin's Press; New York, NY, 1981

Florman, Samuel C.: The Civilized Engineer; St. Martin's Press; New York, NY, 1987

Florman, Samuel C.: The Existential Pleasures of Engineering; St. Martin's, New York, NY, 1976

Hill, Stephen, and Johnson, Ron: Future Tense; University of Queensland Press; St. Lucia, Queensland, Australia, 1983

Intellectual Property Primer; Brooks & Kushman; Southfield, MI, 1990

Jackall, Robert: Moral Mazes; Oxford University Press, New York, NY, 1988

Johnson, Deborah G.: Computer Ethics; Prentice Hall; Englewood Cliffs, N.J., 1985

Johnson, Deborah G., and Snapper, John W.: Ethical Issues in the Use of Computers; Wadsworth; Belmont, CA, 1985

Jones, Donald G.: Doing Ethics in Business; Oelgeschlager, Gunn & Hain, Publishers, Inc.; Cambridge, MA, 1982

Layton, Edwin T., Jr.: The Revolt of the Engineers; Case-Western Reserve University Press; Cleveland, OH, 1971

Mantell, Murray I.: Ethics & Professionalism in Engineering; The Macmillan Co.; New York, NY, 1964

Martin, Mike W.; Self-Deception and Morality; University of Kansas Press; Lawrence, KS 1986

Rosenthal, David M., and Shehadi, Fadlou: Applied Ethics and Ethical Theory; University of Utah Press; Salt Lake **City,** UT, 1988

Schaub, James H., and Pavlovic, Karl: Engineering Professionalism and Ethics; Wiley; New York, NY 1983

Scherer, Donald, and Attig, Thomas: Ethics and the Environment; Prentice Hall, Englewood Cliffs, NJ, 1983

Thiroux, Jacques P.: Ethics, Theory and Practice; Macmillan Publishing Co.; New York, NY, 1986

Unger, Stephen H.: Controlling Technology—Ethics and the Responsible Engineer; Holt, Rinehart & Winston; New York, NY 1982

Vaughn, Richard C.: Legal Aspects of Engineering; Kendall/Hunt Publishing Co.; Dubuque, Iowa, 1974

Winner, Langdon: Autonomous Technology; The MIT Press; Cambridge, MA, 1977

APPENDIX C

CODES OF ETHICS

National Society of Professional Engineers Code of Ethics

American Society of Civil Engineers Code of Ethics & Guidelines

American Society of Mechanical Engineers Code of Ethics

American Institute of Chemical Engineers Code of Ethics

Institute of Electrical and Electronics Engineers Code of Ethics

NCEES Model Code of Ethics

National Institute for Engineering Ethics, Statement on Principles of Ethics

Principles of Conduct and Ethics in Engineering Practice under NAFTA

National Society of Professional Engineers
Codes of Ethics for Engineers

Revised January 2003

Preamble

Engineering is an important and learned profession. As members of this profession, engineers are expected to exhibit the highest standards of honesty and integrity. Engineering has a direct and vital impact on the quality of life for all people. Accordingly, the services provided by engineers require honesty, impartiality, fairness, and equity, and must be dedicated to the protection of the public health, safety, and welfare. Engineers must perform under a standard of professional behavior that requires adherence to the highest principles of ethical conduct.

I. Fundamental Canons

Engineers, in the fulfillment of their professional duties, shall:

1. Hold paramount the safety, health and welfare of the public.
2. Perform services only in areas of their competence.
3. Issue public statements only in an objective and truthful manner.
4. Act for each employer or client as faithful agents or trustees.
5. Avoid deceptive acts.
6. Conduct themselves honorably, responsibly, ethically, and lawfully so as to enhance the honor, reputation, and usefulness of the profession.

II. Rules of Practice

1. Engineers shall hold paramount the safety, health, and welfare of the public.
    a. If engineers' judgment is overruled under circumstances that endanger life or property, they shall notify their employer or client and such other authority as may be appropriate.
    b. Engineers shall approve only those engineering documents that are in conformity with applicable standards.
    c. Engineers shall not reveal facts, data, or information without the prior consent of the client or employer except as authorized or required by law or this Code.
    d. Engineers shall not permit the use of their name or associate in business ventures with any person or firm that they believe are engaged in fraudulent or dishonest enterprise.

e. Engineers shall not aid or abet the unlawful practice of engineering by a person or firm.

f. Engineers having knowledge of any alleged violation of this Code shall report thereon to appropriate professional bodies and, when relevant, also to public authorities, and cooperate with the proper authorities in furnishing such information or assistance as may be required.

2. Engineers shall perform services only in the areas of their competence.

   a. Engineers shall undertake assignments only when qualified by education or experience in the specific technical fields involved.

   b. Engineers shall not affix their signatures to any plans or documents dealing with subject matter in which they lack competence, nor to any plan or document not prepared under their direction and control.

   c. Engineers may accept assignments and assume responsibility for coordination of an entire project and sign and seal the engineering documents for the entire project, provided that each technical segment is signed and sealed only by the qualified engineers who prepared the segment.

3. Engineers shall issue public statements only in an objective and truthful manner.

   a. Engineers shall be objective and truthful in professional reports, statements, or testimony. They shall include all relevant and pertinent information in such reports, statements, or testimony, which should bear the date indicating when it was current.

   b. Engineers may express publicly technical opinions that are founded upon knowledge of the facts and competence in the subject matter.

   c. Engineers shall issue no statements, criticisms, or arguments on technical matters that are inspired or paid for by interested parties, unless they have prefaced their comments by explicitly identifying the interested parties on whose behalf they are speaking, and by revealing the existence of any interest the engineers may have in the matters.

4. Engineers shall act for each employer or client as faithful agents or trustees.

   a. Engineers shall disclose all known or potential conflicts of interest that could influence or appear to influence their judgment or the quality of their services.

   b. Engineers shall not accept compensation, financial or otherwise, from more than one party for services on the same project, or for services pertaining to the same project, unless the circumstances are fully disclosed and agreed to by all interested parties.

c. Engineers shall not solicit or accept financial or other valuable consideration, directly or indirectly, from outside agents in connection with the work for which they are responsible.

d. Engineers in public service as members, advisors, or employees of a governmental or quasi-governmental body or department shall not participate in decisions with respect to services solicited or provided by them or their organizations in private or public engineering practice.

e. Engineers shall not solicit or accept a contract from a governmental body on which a principal or officer of their organization serves as a member.

5. Engineers shall avoid deceptive acts.

a. Engineers shall not falsify their qualifications or permit misrepresentation of their or their associates' qualifications. They shall not misrepresent or exaggerate their responsibility in or for the subject matter of prior assignments. Brochures or other presentations incident to the solicitation of employment shall not misrepresent pertinent facts concerning employers, employees, associates, joint venturers, or past accomplishments.

b. Engineers shall not offer, give, solicit or receive, either directly or indirectly, any contribution to influence the award of a contract by public authority, or which may be reasonably construed by the public as having the effect of intent to influencing the awarding of a contract. They shall not offer any gift or other valuable consideration in order to secure work. They shall not pay a commission, percentage, or brokerage fee in order to secure work, except to a bona fide employee or bona fide established commercial or marketing agencies retained by them.

III. Professional Obligations

1. Engineers shall be guided in all their relations by the highest standards of honesty and integrity.

a. Engineers shall acknowledge their errors and shall not distort or alter the facts.

b. Engineers shall advise their clients or employers when they believe a project will not be successful.

c. Engineers shall not accept outside employment to the detriment of their regular work or interest. Before accepting any outside engineering employment they will notify their employers.

d. Engineers shall not attempt to attract an engineer from another employer by false or misleading pretenses.

e. Engineers shall not promote their own interest at the expense of the dignity and integrity of the profession.

2. Engineers shall at all times strive to serve the public interest.

   a. Engineers shall seek opportunities to participate in civic affairs; career guidance for youths; and work for the advancement of the safety, health, and well-being of their community.

   b. Engineers shall not complete, sign, or seal plans and/or specifications that are not in conformity with applicable engineering standards. If the client or employer insists on such unprofessional conduct, they shall notify the proper authorities and withdraw from further service on the project.

   c. Engineers shall endeavor to extend public knowledge and appreciation of engineering and its achievements.

3. Engineers shall avoid all conduct or practice that deceives the public.

   a. Engineers shall avoid the use of statements containing a material misrepresentation of fact or omitting a material fact.

   b. Consistent with the foregoing, engineers may advertise for recruitment of personnel.

   c. Consistent with the foregoing, engineers may prepare articles for the lay or technical press, but such articles shall not imply credit to the author for work performed by others.

4. Engineers shall not disclose, without consent, confidential information concerning the business affairs or technical processes of any present or former client or employer, or public body on which they serve.

   a. Engineers shall not, without the consent of all interested parties, promote or arrange for new employment or practice in connection with a specific project for which the engineer has gained particular and specialized knowledge.

   b. Engineers shall not, without the consent of all interested parties, participate in or represent an adversary interest in connection with a specific project or proceeding in which the engineer has gained particular specialized knowledge on behalf of a former client or employer.

5. Engineers shall not be influenced in their professional duties by conflicting interests.

a. Engineers shall not accept financial or other considerations, including free engineering designs, from material or equipment suppliers for specifying their product.

b. Engineers shall not accept commissions or allowances, directly or indirectly, from contractors or other parties dealing with clients or employers of the engineer in connection with work for which the engineer is responsible.

6. Engineers shall not attempt to obtain employment or advancement or professional engagements by untruthfully criticizing other engineers, or by other improper or questionable methods.

a. Engineers shall not request, propose, or accept a commission on a contingent basis under circumstances in which their judgment may be compromised.

b. Engineers in salaried positions shall accept part-time engineering work only to the extent consistent with policies of the employer and in accordance with ethical considerations.

c. Engineers shall not, without consent, use equipment, supplies, laboratory, or office facilities of an employer to carry on outside private practice.

7. Engineers shall not attempt to injure, maliciously or falsely, directly or indirectly, the professional reputation, prospects, practice, or employment of other engineers. Engineers who believe others are guilty of unethical or illegal practice shall present such information to the proper authority for action.

a. Engineers in private practice shall not review the work of another engineer for the same client, except with the knowledge of such engineer, or unless the connection of such engineer with the work has been terminated.

b. Engineers in governmental, industrial, or educational employ are entitled to review and evaluate the work of other engineers when so required by their employment duties.

c. Engineers in sales or industrial employ are entitled to make engineering comparisons of represented products with products of other suppliers.

8. Engineers shall accept personal responsibility for their professional activities, provided, however, that engineers may seek indemnification for services arising out of their practice for other than gross negligence, where the engineer's interests cannot otherwise be protected.

a. Engineers shall conform with state registration laws in the practice of engineering.

b. Engineers shall not use association with a nonengineer, a corporation, or partnership as a "cloak" for unethical acts.

9. Engineers shall give credit for engineering work to those to whom credit is due, and will recognize the proprietary interests of others.

a. Engineers shall, whenever possible, name the person or persons who may be individually responsible for designs, inventions, writings, or other accomplishments.

b. Engineers using designs supplied by a client recognize that the designs remain the property of the client and may not be duplicated by the engineer for others without express permission.

c. Engineers, before undertaking work for others in connection with which the engineer may make improvements, plans, designs, inventions, or other records that may justify copyrights or patents, should enter into a positive agreement regarding ownership.

d. Engineers' designs, data, records, and notes referring exclusively to an employer's work are the employer's property. The employer should indemnify the engineer for use of the information for any purpose other than the original purpose.

e. Engineers shall continue their professional development throughout their careers and should keep current in their specialty fields by engaging in professional practice, participating in continuing education courses, reading in the technical literature, and attending professional meetings and seminars.

—As Revised January 2003

"By order of the United States District Court for the District of Columbia, former Section 11(c) of the NSPE Code of Ethics prohibiting competitive bidding, and all policy statements, opinions, rulings or other guidelines interpreting its scope, have been rescinded as unlawfully interfering with the legal right of engineers, protected under the antitrust laws, to provide price information to prospective clients; accordingly, nothing contained in the NSPE Code of Ethics, policy statements, opinions, rulings or other guidelines prohibits the submission of price quotations or competitive bids for engineering services at any time or in any amount."

Statement by NSPE Executive Committee

In order to correct misunderstandings which have been indicated in some instances since the issuance of the Supreme Court decision and the entry of the Final Judgment, it is noted that in its decision of April 25, 1978, the Supreme Court of the United States declared: "The Sherman Act does not require competitive bidding."

It is further noted that as made clear in the Supreme Court decision:

1. Engineers and firms may individually refuse to bid for engineering services.

2. Clients are not required to seek bids for engineering services.

3. Federal, state, and local laws governing procedures to procure engineering services are not affected, and remain in full force and effect.

4. State societies and local chapters are free to actively and aggressively seek legislation for professional selection and negotiation procedures by public agencies.

5. State registration board rules of professional conduct, including rules prohibiting competitive bidding for engineering services, are not affected and remain in full force and effect. State registration boards with authority to adopt rules of professional conduct may adopt rules governing procedures to obtain engineering services.

6. As noted by the Supreme Court, "nothing in the judgment prevents NSPE and its members from attempting to influence governmental action . . ."

NOTE: In regard to the question of application of the Code to corporations vis-à-vis real persons, business form or type should not negate nor influence conformance of individuals to the Code. The Code deals with professional services, which services must be performed by real persons. Real persons in turn establish and implement policies within business structures. The Code is clearly written to apply to the Engineer and items incumbent on members of NSPE to endeavor to live up to its provisions. This applies to all pertinent sections of the Code.

American Society of Civil Engineers
Code of Ethics

As adopted September 25, 1976 and amended October 25, 1980 and April 17,1993

Fundamental Principles

Engineers uphold and advance the integrity, honor and dignity of the engineering profession by:

1. using their knowledge and skill for the enhancement of human welfare;

2. being honest and impartial and serving with fidelity the public, their employers and clients;

3. striving to increase the competence and prestige of the engineering profession; and

4. supporting the professional and technical societies of their disciplines.

Fundamental Canons

1. Engineers shall hold paramount the safety, health and welfare of the public in the performance of their professional duties.

2. Engineers shall perform services only in areas of their competence.

3. Engineers shall issue public statements only in an objective and truthful manner.

4. Engineers shall act in professional matters for each employer or client as faithful agents or trustees, and shall avoid conflicts of interest.

5. Engineers shall build their professional reputation on the merit of their services and shall not compete unfairly with others.

6. Engineers shall act in such a manner as to uphold and enhance the honor, integrity, and dignity of the engineering profession.

7. Engineers shall continue their professional development throughout their careers, and shall provide opportunities for the professional development of those engineers under their supervision.

ASCE Guidelines to Practice Under the Fundamental Canons of Ethics

CANON 1. Engineers shall hold paramount the safety, health and welfare of the public in the performance of their professional duties.

a. Engineers shall recognize that the lives, safety, health and welfare of the general public are dependent upon engineering judgments, decisions and practices incorporated into structures, machines, products, processes and devices.

b. Engineers shall approve or seal only those design documents, reviewed or prepared by them, which are determined to be safe for public health and welfare in conformity with accepted engineering standards.

c. Engineers whose professional judgment is overruled under circumstances where the safety, health and welfare of the public are endangered, shall inform their clients or employers of the possible consequences.

d. Engineers who have knowledge or reason to believe that another person or firm may be in violation of any of the provisions of Canon I shall present such information to the proper authority in writing and shall cooperate with the proper authority in furnishing such further information or assistance as may be required.

e. Engineers should seek opportunities to be of constructive service in civic affairs and work for the advancement of the safety, health and wellbeing of their communities.

f. Engineers should be committed to improving the environment to enhance the quality of life.

CANON 2. Engineers shall perform services only in areas of their competence.

a. Engineers shall undertake to perform engineering assignments only when qualified by education or experience in the technical field of engineering involved.

b. Engineers may accept an assignment requiring education or experience outside of their own fields of competence, provided their services are restricted to those phases of the project in which they are qualified. All other phases of such project shall be performed by qualified associates, consultants, or employees.

c. Engineers shall not affix their signatures or seals to any engineering plan or document dealing with subject matter in which they lack competence by virtue of education or experience or to any such plan or document not reviewed or prepared under their supervisory control.

CANON 3. Engineers shall issue public statements only in an objective and truthful manner.

a. Engineers should endeavor to extend the public knowledge of engineering, and shall not participate in the dissemination of untrue, unfair or exaggerated statements regarding engineering.

b. Engineers shall be objective and truthful in professional reports, statements, or testimony. They shall include all relevant and pertinent information in such reports, statements, or testimony.

c. Engineers, when serving as expert witnesses, shall express an engineering opinion only when it is founded upon adequate knowledge of the facts, upon a background of technical competence, and upon honest conviction.

d. Engineers shall issue no statements, criticisms, or arguments on engineering matters which are inspired or paid for by interested parties, unless they indicate on whose behalf the statements are made.

e. Engineers shall be dignified and modest in explaining their work and merit, and will avoid any act tending to promote their own interests at the expense of the integrity, honor and dignity of the profession.

CANON 4. Engineers shall act in professional matters for each employer or client as faithful agents or trustees, and shall avoid conflicts of interest.

a. Engineers shall avoid all known or potential conflicts of interest with their employers or clients and shall promptly inform their employers or clients of any business association, interests, or circumstances which could influence their judgment or the quality of their services.

b. Engineers shall not accept compensation from more than one party for services on the same project, or for services pertaining to the same project, unless the circumstances are fully disclosed to and agreed to, by all interested parties.

c. Engineers shall not solicit or accept gratuities, directly or indirectly, from contractors, their agents, or other parties dealing with their clients or employers in connection with work for which they are responsible.

d. Engineers in public service as members, advisors, or employees of a governmental body or department shall not participate in considerations or actions with respect to services solicited or provided by them or their organization in private or public engineering practice.

e. Engineers shall advise their employers or clients when, as a result of their studies, they believe a project will not be successful.

f. Engineers shall not use confidential information coming to them in the course of their assignments as a means of making personal profit if such action is adverse to the interests of their clients, employers or the public.

g. Engineers shall not accept professional employment outside of their regular work or interest without the knowledge of their employers.

CANON 5. Engineers shall build their professional reputation on the merit of their services and shall not compete unfairly with others.

a. Engineers shall not give, solicit or receive either directly or indirectly, any political contribution, gratuity, or unlawful consideration in order to secure work, exclusive of securing salaried positions through employment agencies.

b. Engineers should negotiate contracts for professional services fairly and on the basis of demonstrated competence and qualifications for the type of professional service required.

c. Engineers may request, propose or accept professional commissions on a contingent basis only under circumstances in which their professional judgments would not be compromised.

d. Engineers shall not falsify or permit misrepresentation of their academic or professional qualifications or experience.

e. Engineers shall give proper credit for engineering work to those to whom credit is due, and shall recognize the proprietary interests of others. Whenever possible, they shall name the person or persons who may be responsible for designs, inventions, writings or other accomplishments.

f. Engineers may advertise professional services in a way that does not contain misleading language or is in any other manner derogatory to the dignity of the profession. Examples of permissible advertising are as follows:

> Professional cards in recognized, dignified publications, and listings in roster directories published by responsible organizations, provided that the cards or listings are consistent in size and content and are in a section of the publication regularly devoted to such professional cards.
>
> Brochures which factually describe experience, facilities, personnel and capacity to render service, providing they are not misleading with respect to the engineer's participation in projects described.
>
> Display advertising in recognized dignified business and professional publications, providing it is factual and is not misleading with respect to the engineer's experience and participation in projects described.
>
> A statement of the engineers' names or the name of the firm and statement type of service posted on projects for which they render services.

Preparation or authorization of descriptive articles for the lay or technical which are factual and dignified. Such articles shall not imply anything more than participation in the project described.

Permission by engineers for their names to be used in commercial advertisements such as may be published by contractors, material suppliers, etc., only by means of a modest, dignified notation acknowledging the engineers' participation in the described. Such permission shall not include public endorsement of proprietary products.

g. Engineers shall not maliciously or falsely, directly or indirectly, injure the professional reputation, prospects, practice or employment of another engineer or indiscriminately criticize another's work.

h. Engineers shall not use equipment, supplies, laboratory or office facilities of their employers to carry on outside private practice without the consent of their employers.

CANON 6. Engineers shall act in such a manner as to uphold and enhance the honor, integrity and dignity of the engineering profession.

a. Engineers shall not knowingly act in a manner which will be derogatory to the honor, integrity, or dignity of the engineering profession or knowingly engage in business or professional practices of a fraudulent, dishonest or unethical nature.

CANON 7. Engineers shall continue their professional development throughout their careers, and shall provide opportunities for the professional development of those engineers under supervision.

a. Engineers should keep current in their specialty fields by engaging in professional practice, participating in continuing education courses, reading in the technical literature, and attending professional meetings and seminars.
b. Engineers should encourage their engineering employees to become registered at the earliest possible date.

c. Engineers should encourage engineering employees to attend and present papers at professional and technical society meetings.

d. Engineers shall uphold the principle of mutually satisfying relationships employers and employees with respect to terms of employment including professional grade descriptions, salary ranges, and fringe benefits.

American Society of Mechanical Engineers
Code of Ethics of Engineers

The Fundamental Principles

Engineers uphold and advance the integrity, honor, and dignity of the Engineering profession by:

I. using their knowledge and skill for the enhancement of human welfare;

II. being honest and impartial, and serving with fidelity the public, their employers and clients; and

III. striving to increase the competence and prestige of the engineering profession.

The Fundamental Canons

1. Engineers shall hold paramount the safety, health and welfare of the public in the performance of their professional duties.

2. Engineers shall perform services only in areas of their competence.

3. Engineers shall continue their professional development throughout their careers and shall provide opportunities for the professional development of those engineers under their supervision.

4. Engineers shall act in professional matters for each employer or client as faithful agents or trustees, and shall avoid conflicts of interest.

5. Engineers shall build their professional reputation on the merit of their services and shall not compete unfairly with others.

6. Engineers shall associate only with reputable persons or organizations.

7. Engineers shall issue public statements only in an objective and truthful manner.

8. Engineers shall consider environmental impact in the performance of their professional duties.

9. Engineer shsall consider sustainable development in the performance of their professional duties.

American Institute of Chemical Engineers Code of Ethics

(Revised January 17, 2003)

Members of the American Institute of Chemical Engineers shall uphold and advance the integrity, honor and dignity of the engineering profession by: being honest and impartial and serving with fidelity their employers, their clients, and the public; striving to increase the competence and prestige of the engineering profession; and using their knowledge and skill for the enhancement of human welfare. To achieve these goals, members shall

- Hold paramount the safety, health and welfare of the public and protect the environment in performance of their professional duties.
- Formally advise their employers or clients (and consider further disclosure, if warranted) if they perceive that a consequence of their duties will adversely affect the present or future health or safety of their colleagues or the public.
- Accept responsibility for their actions, seek and heed critical review of their work and offer objective criticism of the work of others.
- Issue statements or present information only in an objective and truthful manner.
- Act in professional matters for each employer or client as faithful agents or trustees, avoiding conflicts of interest and never breaching confidentiality.
- Treat fairly and respectfully all colleagues and co-workers, recognizing their unique contributions and capabilities.
- Perform professional services only in areas of their competence.
- Build their professional reputations on the merits of their services.
- Continue their professional development throughout their careers, and provide opportunities for the professional development of those under their supervision.
- Never tolerate harassment.
- Conduct themselves in a fair, honorable and respectful manner.

Institute of Electrical and Electronics Engineers
Code of Ethics

We, the members of the IEEE, in recognition of the importance of our technologies in affecting the quality of life throughout the world, and in accepting a personal obligation to our profession, its members and the communities we serve, do hereby commit ourselves to the highest ethical and professional conduct and agree:

1. to accept responsibility in making engineering decisions consistent with the safety, health and welfare of the public, and to disclose promptly factors that might endanger the public or the environment;

2. to avoid real or perceived conflicts of interest whenever possible, and to disclose them to affected parties when they do exist;

3. to be honest and realistic in stating claims or estimates based on available data;

4. to reject bribery in all its forms;

5. to improve the understanding of technology, its appropriate application, and potential consequences;

6. to maintain and improve our technical competence and to undertake technological tasks for others only if qualified by training or experience, or after full disclosure of pertinent limitations;

7. to seek, accept, and offer honest criticism of technical work, to acknowl edge and correct errors, and to credit properly the contributions of others;

8. to treat fairly all persons regardless of such factors as race, religion, gender, disability, age, or national origin;

9. to avoid injuring others, their property, reputation, or employment by false or malicious action;

10. to assist colleagues and co-workers in their professional development and to support them in following this code of ethics.

The National Council of Examiners for Engineering and Surveying

Rules of Professional Conduct

240.15 Rules of Professional Conduct

A. Licensee's Obligation to Society

1. Licensees, in the performance of their services for clients, employers, and customers, shall be cognizant that their first and foremost responsibility is to the public welfare.

2. Licensees shall approve and seal only those design documents and surveys that conform to accepted engineering and surveying standards and safeguard the life, health, property, and welfare of the public.

3. Licensees shall notify their employer or client and such other authority as may be appropriate when their professional judgment is overruled under circumstances where the life, health, property, or welfare of the public is endangered.

4. Licensees shall be objective and truthful in professional reports, statements, or testimony. They shall include all relevant and pertinent information in such reports, statements, or testimony.

5. Licensees shall express a professional opinion publicly only when it is founded upon an adequate knowledge of the facts and a competent evaluation of the subject matter.

6. Licensees shall issue no statements, criticisms, or arguments on technical matters which are inspired or paid for by interested parties, unless they explicitly identify the interested parties on whose behalf they are speaking and reveal any interest they have in the matters.

7. Licensees shall not permit the use of their name or firm name by, nor associate in the business ventures with, any person or firm which is engaging in fraudulent or dishonest business or professional practices.

8. Licensees having knowledge of possible violations of any of these Rules of Professional Conduct shall provide the board with the information and assistance necessary to make the final determination of such violation.

(Section 150, Disciplinary Action, NCEES Model Law)

B. Licensee's Obligation to Employer and Clients

1. Licensees shall undertake assignments only when qualified by education or experience in the specific technical fields of engineering or surveying involved.

2. Licensees shall not affix their signatures or seals to any plans or documents dealing with subject matter in which they lack competence, nor to any such plan or document not prepared under their direct control and personal supervision.

3. Licensees may accept assignments for coordination of an entire project, provided that each design segment is signed and sealed by the licensee responsible for preparation of that design segment.

4. Licensees shall not reveal facts, data, or information obtained in a professional capacity without the prior consent of the client or employer except as authorized or required by law. Licensees shall not solicit or accept gratuities, directly or indirectly, from contractors, their agents, or other parties in connection with work for employers or clients.

5. Licensees shall make full prior disclosures to their employers or clients of potential conflicts of interest or other circumstances which could influence or appear to influence their judgment or the quality of their service.

6. Licensees shall not accept compensation, financial or otherwise, from more than one party for services pertaining to the same project, unless the circumstances are fully disclosed and agreed to by all interested parties.

7. Licensees shall not solicit or accept a professional contract from a governmental body on which a principal or officer of their organization serves as a member. Conversely, licensees serving as members, advisors, or employees of a government body or department, who are the principals or employees of a private concern, shall not participate in decisions with respect to professional services offered or provided by said concern to the governmental body which they serve.

(Section 150, Disciplinary Action, NCEES Model Law)

C. Licensee's Obligation to Other Licensees

1. Licensees shall not falsify or permit misrepresentation of their, or their associates', academic or professional qualifications. They shall not misrepresent or exaggerate their degree of responsibility in prior assignments nor the complexity of said assignments. Presentations incident to the solicitation of employment or business shall not misrepresent pertinent facts concerning employers, employees, associates, joint ventures, or past accomplishments.

2. Licensees shall not offer, give, solicit, or receive, either directly or indirectly, any commission, or gift, or other valuable consideration in order to secure work, and shall not make any political contribution with the intent to influence the award of a contract by public authority.

3. Licensees shall not attempt to injure, maliciously or falsely, directly or indirectly, the professional reputation, prospects, practice, or employment of other licensees, nor indiscriminately criticize other licensees' work.

(Section 150, Disciplinary Action, NCEES Model Law)

NIEE Statement of Ethics Principles

The privilege of practicing engineering is entrusted to those qualified and who have the responsibility for applying engineering skills, scientific knowledge and ingenuity for the advancement of human welfare and quality of life. Fundamental principles of conduct of engineers include truth, honesty and trustworthiness in their service to society, and honorable and ethical practice showing fairness, courtesy and good faith toward clients, colleagues and others. Engineers take societal, cultural, economic, environmental and safety aspects into consideration, and strive for the efficient use of the world's resources to meet long term human needs.

In the practice of engineering:

1. Engineers shall hold paramount the health, safety and welfare of the public in the practice of their profession.

2. Engineers shall practice only in their areas of competence, in a careful and diligent manner and in conformance with standards, laws, codes, and rules and regulations applicable to engineering practice.

3. Engineers shall examine the societal and environmental impact of their actions and projects, including the wise use and conservation of resources and energy, in order to make informed recommendations and decisions.

4. Engineers shall issue public statements only in an objective and truthful manner. If representing a particular interest, the engineer shall clearly identify that interest.

5. Engineers shall sign and take responsibility for all engineering work which they prepare or directly supervise.

6. Engineers shall act as faithful agents for their employers or clients and maintain confidentiality; they shall avoid conflicts of interest whenever possible, disclosing unavoidable conflicts.

7. Engineers shall ensure that a client is aware of the engineer's professional concerns regarding particular actions or projects, and of the consequences of engineering decisions or judgments that are overruled or disregarded. An employee engineer shall initially express those concerns to the employer.

8. Engineers shall appropriately report any public works, engineering decisions, or practice that endanger the health, safety and welfare of the public. When, in an engineer's judgment, a significant risk to the public remains unresolved, that engineer may ethically make the concerns known publicly.

9. Engineers shall commit to life-long learning, strive to advance the body of engineering knowledge and should encourage other engineers to do likewise.

10. Engineers shall promote responsibility, commitment, and ethics both in the education and practice phases of engineering; they should enhance society's awareness of engineers' responsibilities to the public and encourage the communication of these principles of ethical conduct among engineers.

APPROVED BY THE NIEE BOARD OF DIRECTORS; OCTOBER 28, 2000

Principles of Ethical Conduct in Engineering Practice Under the North American Free Trade Agreement (NAFTA)

The privilege of practicing engineering is entrusted to those qualified and who have the responsibility for applying engineering skills, scientific knowledge and ingenuity for the advancement of human welfare and quality of life. Fundamental principles of conduct of engineers include truth, honesty and trustworthiness in their service to society, and honorable and ethical practice showing fairness, courtesy and good faith toward clients, colleagues and others. Engineers take societal, cultural, economic, environmental and safety aspects into consideration, and strive for the efficient use of the world's resources to meet long-term human needs. In the practice of engineering:

1. Engineers shall hold paramount the health, safety and welfare of the public in the practice of their profession.
2. Engineers shall practice only in their areas of competence, in a careful and diligent manner and in conformance with standards, laws, codes, and rules and regulations applicable to engineering practice.
3. Engineers shall examine the societal and environmental impact of their actions and projects, including the wise use and conservation of resources and energy, in order to make informed recommendations and decisions.
4. Engineers shall issue public statements only in an objective and truthful manner. If representing a particular interest, the engineer shall clearly identify that interest.
5. Engineers shall sign and take responsibility for all engineering work which they prepared or directly supervised. An engineer may sign work prepared by others, but only with their knowledge and after sufficient review and verification to justify taking responsibility for that work.
6. Engineers shall act as faithful agents for their employers or clients and maintain confidentiality; they shall avoid conflicts of interest whenever possible, disclosing unavoidable conflicts.
7. Engineers shall ensure that a client is aware of the engineer's professional concerns regarding particular actions or projects, and of the consequences of engineering decisions or judgments that are overruled or disregarded. An employee engineer shall initially express those concerns to the employer.

8. Engineers shall appropriately report any public works, engineering decisions or practice that endanger the health, safety and welfare of the public. When, in an engineer's judgment, a significant risk to the public remains unresolved, that engineer may ethically make the concerns known publicly.

9. Engineers shall commit to life-long learning, strive to advance the body of engineering knowledge and should encourage other engineers to do likewise.

10. Engineers shall promote responsibility, commitment, and ethics both in the education and practice phases of engineering; they should enhance society's awareness of engineers' responsibilities to the public and encourage the communication of these principles of ethical conduct among engineers.

Developed by the Murdough Center for Engineering Professionalism,
College of Engineering Texas Tech University,
Lubbock, Texas

Under a grant from the United States National Science Foundation
Jimmy H. Smith, P.E. Project Director & Trish Barrington, Project Coordinator
Murdough Center for Engineering Professionalism

Team Members from Canada: Garry Wacker, P.Eng. and Jack Bordan, ing.
Team Members from Mexico: Ing. Humberto Peniche Cuevas and Dr. Felipe Ochoa Rosso
Team Members from the United States: E. D. "Dave" Dorchester, P.E. and
John Steadman, Ph.D., P.E.

1994 - 1996

Appendix D
Information about the Murdough Center for Engineering Professionalism

Founded in 1988, the Murdough Center is named in honor of Professor James H. Murdough, the first head of the Texas Tech Civil Engineering Department.

Mission and Goal

The Mission of the Murdough Center is to provide engineering ethics and professionalism education, research, and communication to students, faculty, staff, and engineers in industry, government and private practice, other professionals, and citizens in the community, state, and nation. The goal of the Center is to increase the awareness of the professional and ethical obligations and responsibilities entrusted to individuals who practice engineering.

To help accomplish our mission and goal, the Center operates two other major programs promoting engineering ethics — the National Institute for Engineering Ethics and the Applied Ethics in Professional Practice (see www.niee.org Welcome Page). In addition, the Center has a strong base of support and encouragement from Texas Tech University, private donors, state agencies, national professional societies, industries, and foundations.

We are committed to the values of:

- Honesty and trustworthiness;
- Ethical autonomy;
- Cooperation and communication;
- Creativity and innovation; and
- Ethical leadership.

These values help create and maintain an environment where engineers and engineering students feel safe in bringing up ethical issues without fear of retribution.

Symposia, workshops, and seminars have been conducted nationally and internationally for government agencies, industry, professional societies, and academic institutions. The Murdough Center and its affiliated programs are designed to:

- Prepare students to be ethical leaders and decision makers, articulate and principled, innovative and confident, and able to think critically with sound reasoning ability;
- Be recognized as the top public educational and research center in the United States promoting the study and communication of engineering professionalism and ethics;
- Be widely recognized internationally; and
- Be engaged in local, regional, state, national, and international activities promoting ethics within the engineering profession.

Appendix E

Information about the National Institute for Engineering Ethics

The Founding of NIEE

The National Institute for Engineering Ethics (NIEE) was created by the National Society of Professional Engineers (NSPE) in 1988 and became an independent organization in 1995. In October 2001, with the approval of the NIEE Board of Directors and the Texas Tech University Board of Regents, NIEE became an official component of the Murdough Center for Engineering Professionalism in the College of Engineering at Texas Tech University. Texas Tech provides a permanent home for the Institute as well as financial, administrative, management, and leadership responsibilities.

NIEE Mission

The Mission of the Institute is to provide opportunities for education in the field of engineering ethics and to promote the understanding and application of ethical processes within the engineering profession and the public. In implementing this mission, the Institute will:

- Provide a recognized forum for technical/professional societies, corporations, firms, and individuals to exchange educational and other information on ethics activities;
- Serve as a coordinator for ethics conferences, workshops, etc.;
- Serve as a resource for educational materials on ethics for those societies and employers of engineers requesting such assistance;
- Provide a forum for participation in international ethics issues and activities; and
- Serve as the coordinator of joint and/or special ethics projects which several engineering societies and organizations develop together.

NIEE Programs

The principal aims of NIEE are education, communication, program and project development, and practice applications in the area of engineering ethics. A primary role of the Institute is to encourage cooperation among individuals, universities, professional and technical societies, and business organizations with regard to engineering ethics and professionalism issues.

NIEE National Activities

NIEE is a co-sponsor of the "On-Line Ethics Center Help Line" developed and operated by Dr. Caroline Whitbeck at Case Western Reserve University. NIEE also sponsors, develops, and implements many engineering ethics workshops, seminars and symposia.

NSPE-Board of Ethical Review Cases

NIEE has placed over 200 NSPE-BER Cases on its web site (www.niee.org). These are free when used for educational purposes as long as NSPE is acknowledged for developing them. A consolidated Table of Contents allows the visitor to search for cases by topic.

NIEE Ethics Videos

NIEE owns and markets two engineering ethics videos: the 1989 video "Gilbane Gold" and the 2003 "Incident at Morales." Brief clips from these videos as well as a significant amount of free downloadable ethics material is available from www.niee.org.

NIEE's Applied Ethics in Professional Practice Program

Created in 1997 by Ronald E. Bucknam, P.E., Ph.D., the Applied Ethics in Professional Practice Program is used by more than 100 universities and the website has had more than 50,000 hits. The program presents ethics cases taken from professional practice in order to stimulate greater emphasis on ethical issues in a real-world context.

The internet site can most easily be accessed at www.niee.org. The site contains an ethics case with several alternate approaches or solutions. The learner can vote for the alternate approach which best fits his or her own opinion, or can supply another approach as a comment.

The National Institute for Engineering Ethics gratefully acknowledges the cooperation of the University of Washington's Professional Engineering Practice Liaison Program in transitioning the program to the NIEE, as documented in the Memorandum of Understanding between NIEE and the University of Washington executed in November 2001.

Engineering Ethics Live!

NIEE co-sponsors this high-energy forum each year at the American Society of Civil Engineers National conference. Based on NIEE's Applied Ethics in Professional Practice program, Engineering Ethics Live! features live performances of ethics skits by ASCE student members, as well as audience participation and commentary from a distinguished panel of ethics experts.

Appendix F

About the NIEE Staff

Jimmy H. Smith, Ph.D., P. E., F.ASCE, F.NSPE

Jimmy H. Smith is a Professor of Civil Engineering and Director of the National Institute for Engineering Ethics and the Murdough Center for Engineering Professionalism at Texas Tech University, Lubbock, Texas. He received his BS and MS in Civil Engineering from Texas Tech University and his Ph.D. from the University of Arizona. He is an active member of TSPE, NSPE, ASEE, and ASCE. He has served the profession in several capacities, including President, Texas Society of Professional Engineers, President, National Institute for Engineering Ethics, Member, NSPE Board of Ethical Review, and Member and Team Chair, ABET/Engineering Accreditation Commission. His professional ethics activities have spanned 15 years and have included over 300 presentations, seminars, and workshops throughout the United States, Russia, Mexico, and Canada. Jimmy has received several awards, including:TSPE Engineer of the Year, Distinguished Engineer of the Foundation - Texas Engineering Foundation, Engineer of the Year – ASCE High Plains Branch, NSPE Distinguished Service Award, 2001 TSPE Engineering Dream Team, and Gonzaga University Engineering Ethics Award.

William D. Lawson, P.E., M.ASCE, M.NSPE

William D. Lawson earned his BS and MS degrees in Civil Engineering from Texas A&M University with a focus on structures and foundations and later studied theology at Dallas Theological Seminary. Prior to becoming Deputy Director of the National Institute for Engineering Ethics and the Murdough Center for Engineering Professionalism at Texas Tech University, Bill worked as a consulting engineer. With over 20 years consulting, research and teaching experience, Bill has provided project management and technical oversight for geotechnical, construction materials, transportation and facilities projects nationwide. He began his professional career with Trinity Engineering Testing Corporation, Austin, Texas, and later held positions as principal engineer, project manager and national client manager for the global engineering firm, Law Engineering and Environmental Services, Inc. Bill is active in various professional and technical societies including the NIEE, ASCE, NSPE, TSPE and ASFE. He serves as a corresponding editor for the ASCE Journal of Professional Issues in Engineering Education and Practice; he is the faculty advisor for the Texas Tech Student Chapter of the Texas Society of Professional Engineers; and member and chair of the NSPE Board of Ethical Review.

Patricia M. Harper

Patricia M. Harper, Program Coordinator and Assistant to the Director for the NIEE and the Murdough Center for Engineering Professionalism, has been assisting in developing and presenting professional ethics for the Institute and Center for over four years. Patti is responsible for coordinating the NIEE/MC Correspondence Courses in Engineering Ethics which have enrolled over 1,800 engineers and students from 49 states, Puerto Rico, Guam, and Mexico. Patti is also the Coordinator of the NIEE/MC Professional Development Programs including the TxDOT Professional Ethics courses. She recently served as the Assistant to the Executive Producers of the video "Incident at Morales."

NIEE Products & Services

(Details available at www.niee.org)

Incident at Morales

Recently Released Video (2003)

An Engineering Ethics Story

VHS and Interactive DVD Available

36 minutes + Interactive Material

Includes a 24-page Study Guide

Gilbane Gold

Classic Video (1989)

Engineering Ethics - Focus on the Environment

VHS and DVD Available - 24 minutes

Includes Study Guide

Engineering Ethics by Correspondence

Distance Learning Opportunities

Academic Credit (3 Credit Hours)

Professional Development Hours (Several Options)

Professional Ethics Workshops, Seminars and Presentations

Resource Guide and Reference Lists

NSPE BER Cases in Book Form